CIVILIZING IRELAND
Ordnance Survey 1824–1842:
Ethnography, Cartography, Translation

STIOFÁN Ó CADHLA
University College, Cork

Foreword by
Éamon Ó Cuív T.D.

IRISH ACADEMIC PRESS
DUBLIN · PORTLAND, OR

First published in 2007 by
IRISH ACADEMIC PRESS
44, Northumberland Road, Dublin 4, Ireland

and in the United States of America by
IRISH ACADEMIC PRESS
c/o ISBS, Suite 300, 920 NE 58th Avenue
Portland, Oregon 97213-3644

© 2007 Stiofan O Cadhla

WEBSITE: www.iap.ie

British Library Cataloguing in Publication Data
An entry can be found on request

        ISBN 0 7165 3372 3 (cloth)
        ISBN 978 0 7165 3372 4
        ISBN 0 7165 2881 9 (paper)
        ISBN 978 0 7165 2881 2

Library of Congress Cataloging-in-Publication Data
An entry can be found on request

This book has received funding from College of Arts, Celtic Studies and Social Sciences' Publication Fund, National University of Ireland, Cork

All rights reserved. Without limiting the rights under copyright reserved alone, no part of this publication may be reproduced, stored in or introduced into a retrieval system, or transmitted, in any form or by any means (electronic, mechanical, photocopying, recording or otherwise), without the prior written permission of both the copyright owner and the above publisher of this book.

Typeset by ForDesign
Printed by Creative Print and Design, (Wales), Ebbw Vale

I gcuimhne ar
Phádraigín

O! Boylagh, sweet country of O'Boyle, wert thou ever civilized? Thy mountains nod declaring no! Not since fair Kasser with her band of antediluvians saw this land, until the red-coated sapper dragged the chain along thy dreary glens and o'er thy azure peaks. Thy lowly cabins tell the tale of old, here we stand, fair models of the time when Connell preached, and built his little Church, and Croan refused to multiply her kind, deeming it sinful to contribute to the increasing of a race that time had proved a wicked race ... Boylagh now is the very same as it was two thousand years ago; the same Celtic voice that shouted on its hills in the time of Conall Gulban, is still the only sound that is heard. It is the language of nature without culture, forcible and rough, but not as rough as the Caffer's lashing talk! And why, because the bard took up his pen to smooth it; it is true he failed to plane it for modern ears.

John O'Donovan, Dunglow, County Donegal,
12 October 1835

Contents

Acknowledgements	vi
Foreword	vii
Archival Sources	viii
Order and Survey: An Introduction	1

Chapter One
A Full Face Portrait of the Land: Topographic
and Antiquarian Vision — 13

Chapter Two
The Habits of the People: The Emergence of Folklore
and Ethnology — 40

Chapter Three
Colonialism's Culture: Ethnography, Cartography
and Translation — 73

Chapter Four
The Bonnet and the Brogue: Race, Place and Language — 102

Chapter Five
Old Rhymes and Rags of Legends: Tradition
and Civilization — 133

Chapter Six
Traversing the Country: Fieldwork, Mimicry
and Method — 170

Chapter Seven
Derry with Derrida: Translation, Anglicization
and Culture — 218

Select Bibliography — 253

Index — 265

Acknowledgements

My thanks to all who showed interest in this research since I began work on it around 1995. Although family are often last to be thanked in acknowledgements such as this I would like to begin by thanking Sally, Seán Óg, Liadain and Éabha for their patience and the heart-ening magic of spirit and mind. I can only apologize for the amount of time it consumed. I thank my colleagues in *Roinn an Bhéaloidis*, the Department of Folklore and Ethnology in University College Cork, Veronica Fraser, Maeve McDevitt, Dr Marie Annick Deplanques, Diarmuid Ó Giolláin and Dr Clíodhna O'Carroll for encouraging and facilitating my work at different times. Thanks also to my former colleague and teacher Associate Professor Gearóid Ó Crualaoich. I am grateful to everyone who recommended readings, criticized or gave advice along the way. Mícheál Briody of the University of Helsinki, Paul Ferguson, Map Librarian in Trinity College Dublin, Jason Martin, Manuscripts Editor in the History of Cartography Project, Louise Shabat Bethlehem at the Porter Institute for Poetics and Semiotics in Tel Aviv, Associate Professor Kay Schaffer, Department of Social Inquiry-Women's Studies, University of Adelaide, Maura O'Connor, Map Curator in the National Library of Australia and Mathew Edney, Associate Professor of Geography–Anthropology in the University of Southern Maine in America. Thanks to Dr Michael Cronin, Director of the Centre for Translation and Textual Studies in Dublin City University. My thanks to Dr Margaret A. Mackay, Morag MacLeod and all staff and students at the former School of Scottish Studies in Edinburgh where there was much encouragement of, and interest in, my early ideas during enjoyable visits there. My appreciation also to the Department of Anthropology at National University of Ireland, Maynooth, whose lively conference on the theme of Culture, Space and Representation sharpened my interest at a crucial stage. I would like to extend my appreciation to the College of Arts, Celtic Studies and Social Sciences' Publications Fund, National University of Ireland, Cork, for their generous support of this book. Thanks to the staff of the Boole library in University College Cork. Thanks to my parents Liam and Úna, and finally this book is ultimately dedicated to the memory of my sister Phádraigín, always gentle and kind, who left us behind so suddenly while I was preparing it for publication.

Foreword

> To depend on documentary evidence alone is to see Ireland
> through the eyes of her conquerors
>
> Eslyn Evans

In this book Ó Cadhla uncovers the interesting history of Irish folklore. To find that what we know to be Irish folklore, is not simply an inheritance of our past selves, but a version of our past colonisers, is fascinating and somewhat disturbing. The English Ordnance Survey of Ireland attempted to access an understanding of Irish culture through the transcription and translation of Irish folklore. Although the stance of science was utilised, the English account of Irish folklore remained subjective, a culturally influenced report of 'true' Irish folklore. Early Ordnance Survey in Ireland pickled Irish folklore; it might look like the real thing, but the flavour has certainly been tampered with. Our outstanding legends were scientifically preserved and classified and as a result we are left with the English colonial perspective of Ireland. Although one is left slightly bewildered at the magnitude of Ó Cadhla's assertions, it allows for a greater understanding of the English colonial mind.

Ó Cadhla's achievement not only challenges the imposition of cultural categorisation, but challenges us all to rethink our set notions of folklore, history and memory.

Minister Éamon Ó Cuív T.D.
June 2006

Archival Sources

Rev. Michael O'Flanagan (ed.) *Ordnance Survey Letters 1834–1841,* OSL in the foornotes, forty-three volumes typeset and bound in Boole Library, University College Cork, (Bray, 1927–35). These were written by the civilian assistants to the survey, notably John O'Donovan, Eugene O'Curry, George Petrie, Patrick O'Keefe and Thomas O'Connor. They became known collectively as the Topographical or Antiquarian Department. There are 127 bound volumes of these letters in the Royal Irish Academy in Dublin. Forty-one volumes of Ordnance Survey Letters for twenty-nine counties (not including Antrim, Cork or Tyrone) were originally deposited in the Royal Irish Academy. In the years between 1927 and 1935 these were reproduced in typescript in the same sequence by the Rev. Michael O'Flanagan. It is thought that O'Flanagan was continuing the work of the Irish Topographical Society. Fifty copies were put on sale in twenty-seven parts, each containing one or two counties with some counties represented by three and more by two volumes. Five of the twenty-nine sets of letters have been published by Enda Cunningham of Cathach Books in Dublin. They are being edited by Michael Herrity and published by Fourmasters Press. The published sets to date are Dublin, Donegal, Meath, Kildare and Kilkenny. Seventy-six volumes of material from the Ordnance Survey Name Books were also reproduced and put on sale by O'Flanagan comprising twenty counties. The memoir material has been edited by Angélique Day and Patrick McWilliams and published by the Institute of Irish Studies in Belfast entitled the *Ordnance Survey Memoirs of Ireland* or OSM in the footnotes. There is a set of these, as there is of the letters and Name Books, in the Boole Library, University College Cork.

Order and Survey:
An Introduction

What is the Ordnance Survey, what does it mean and where did it come from? It was established in 1791 with its headquarters in the Tower of London under the Master General of the British Board of Ordnance.[1] *Ordnance Survey* is a historic amalgam of two key terms in the history of ideas and the governance of knowledge. Used since the eighteenth century in combination with the term survey, it remains the name for the three official map-making bodies of Britain and Ireland.[2] The first term, ordnance, or ordinance, is derived from the French *ordenance* and the Latin *ordināre* meaning to order, ordain, arrange, regulate or rule. From the fourteenth century it referred to militaristic warlike provisions or the decrees of a sovereign. Ordnance came to refer specifically to the artillery corps of the army and the British military remained in the survey's Dublin headquarters until independence in 1922.

Survey means to oversee, the French *surveoir* combines the stem *sur*, or over, and *veoir*, to see, from the Latin *vidēre* or *vĭdĕo*. From the seventeenth and eighteenth centuries it had the sense of a view from a commanding position, an inspection or an examination. A survey was also a comprehensive mental view, a written description or the measurement of a tract of ground. It is a coordination of hand, eye and mind. The contemporary sense refers to any systematic collection or analysis of data, attitudes or opinions.[3] In the first half of the nineteenth century the British Ordnance Survey of Ireland combined all of these senses from ordering to authoring, from inspection to measurement and from voyaging to voyeurism. Not unlike a nineteenth-century *Guinness Book of Records*, it includes information such as the fact that Peggy Frizell was a public character and an idiot and that a woman in Aghagallon had twenty-four children. It notes that Fanny Marlin of the village of Curran was a great fighter and a hermaphrodite. At present the Ordnance Survey of Ireland continues to map and pursue raw information in Cork under the aegis of the Central Statistics Office, unaware of their long lost cousins the folklorists and ethnologists. One of the aims of this work is to retrace this genealogy.

The popular mock-up of the survey conjures up a providential information juggernaut rolling through nineteenth-century Ireland with its divisions of red-coated soldiers busily measuring boundaries, mapping townlands, sketching ruins and fossils, writing notes and letters and bottling insects. This image of a nomadic, information-hungry mechanism of empire is only slightly far-fetched. In Co Derry people suspected that the soldiers massing on the hilltops were preparing for an end game which would see the surrounding countryside pounded by artillery. The fact that this was one of the roles of the ordnance corps of the British military was not lost on them. It was two hundred years, not too distant to survive in memory perhaps, since Essex's massacre of six hundred men, women and children on Rathlin Island. The bemusement of the surveyors must have been matched by the awe of the onlookers as they fixed the sights of their theodolites. The survey carried an impressive array of technology. It was a self-conscious harbinger of a brave new world with the steamships, spinning machines, railways, the telegraph, washing machines, mangles, power looms, pumps and locomotives of the nineteenth-century, all contributing to the novelty.

The technological advances were matched by increasingly sophisticated methods of cultural description. The nineteenth century also saw the birth of 'the science of savages' and, as progress admired itself in the looking glass, it imagined Red Indians, West Africans, Bushmen, Kaffirs, Malays, Dyaks, Papuans, Australians, Maoris and the Irish queuing up behind it. The orderly and crystalline vision of enlightened science was obscured only by Luddite natives who also had to be accounted for. This book is not concerned with the technical or technocratic aspects of the survey but rather with the cultural and ethnographic aspects. The popular opinion that the survey is a salvage yard of Irish folklore is what first drew my attention to it. Before too long it became apparent that the genesis of this opinion might lie beyond the survey in a discourse that survived on the silence, rather than the eloquence, of the other. The vagaries of living with two unequal languages interested me in the understudied relationship between English and Irish. The bilingual practice of the discipline of *béaloideas*, folklore and ethnology at University College Cork constantly accentuates the vexed questions that representing others can broach.[4] The sometimes haphazard translative to and fro out of which the broad theory here took shape is a workaday problem rather than a purely philosophical conundrum.

There are some obvious aspects of the survey that make it especially relevant to folklore and ethnology. Apart from sharing similar intellectual origins the contemporary discipline of folklore and ethnology is arguably defined by the creativity and duplicity of interpretation and translation. As well as mirroring the contemporary life of the discipline, the method, theory and practice of the survey reflects the social and cultural life of nineteenth-century Ireland. Describing the culture of any group in any society is not a simple matter. In this case, although it has been downplayed, it is not just a question of translation within a single language but translation between two different languages. A common solution has been to simply evade one of the languages, to pronounce the Irish language dead or moribund. Having done this much, the coast is clear to proceed with a self-regarding anglocentric view. The opposite also occurs when the Irish language is studied as an arcane subcultural phenomenon somehow sequestered from the real world. These are more than mere oversights, they are rehearsals of older colonialist propensities.

Monolingualism, bilingualism or any divergence from Standard English is stigmatized like poet Patrick Kavanagh's 'thick tongued mumble' or the begorraghs of the Stage Irishman. The effect of this, as Kennedy noted in the work of Edgeworth, 'is to downplay the difference between Ireland and England by more or less completely effacing the differences of language'.[5] The identity of the colonized simply disappears. In an almost evangelical model of translation it is not simply useful to translate, but rather it is amoral not to translate. That which is perceived as oppositional is brushed aside, rendered still and unmoving as though it was unreal. As if it never was or is not there at all. When this amounts to a few centuries of the social, cultural and linguistic life of a people 'large swathes of the cognitive, aesthetic and affective experience of the people who have lived on the island become invisible ... by a foreshortened sense of historical time which only admits the experience of English monoglossia'.[6]

This point is important in this work, as much writing on the survey has originated from such a perception. This, however, also explores the survey's vision or conceptualization of itself vis-à-vis Ireland and the Irish. Contemporary opinion has borrowed much from the original ideological manifesto. Is the survey really the pre-Armageddon atlas of star-crossed Gaeldom? Kevin Myers refers to it in sacerdotal language as 'the Dead Sea Scrolls of the scriptures of Irish identity' that are the only trace of 'a world about to vanish'.[7]

John B. Keane considered the survey's letters to be 'priceless beyond estimate'.[8] Leersen says it was 'an enormous salvage operation to collect and preserve the broken remains of Ireland's native culture', it 'lifted the heritage of Gaelic antiquity from the ruins of the colonised countryside and its pauperised peasantry'.[9] The folklorist or ethnologist could be forgiven for having great expectations of such a survey.

This research was initially envisaged as a contextualized presentation of just such a prophesied panorama of Irish folklore. Apart from a handful of cameos, inserted often as mere padding or embellishment, there was simply too little to work with. Humble Irish folklore was obscured at almost every turn by bombastic and grandiose imperial pronouncements on man. This is only partly explained by the confusion of nineteenth-century ideas about antiquities, the measurement of old churches, castles, stone cells and so on, with more recent or contemporary ideas about folklore. The certainty of the survey's scientific accomplishment is not carried across to the more open-ended field of cultural description. It is also argued here that 'the concepts of ethnography, fieldwork, participant observation, and even culture and history themselves have to be put in historical context'.[10] These techniques and methods cannot produce anything at all without hand, eye or mind to guide them.

In any case pretensions to humility do not originate within the survey itself where ideas of grandeur, both real and imagined, predominated. These ideas combined the art of science, including technology, triangulation and theodolites, with the science of art, including antiquarianism, ethnology and the picturesque. What if folklore did not disappear? What if the Irish language did not disappear? What if contemporary fieldwork produces a greater profusion of place names than the survey? What if the imaginary world about to vanish forever did not vanish at all? What if some of the place names rescued from the ruins were wrong or assigned to the wrong rivers or wrong lakes? In mapping an Irish-speaking south Connemara in contemporary times Tim Robinson found that the survey ignored the living language and he was able to solve many supposed riddles by simply asking local people.[11] Was it a rerun of the never-ending story, a trick pulled like a dazzled rabbit from the imperial magician's hat? If the survey was not the saviour of Irish folklore then what was it? The reflection of the anonymous eighteenth-century Irish poet, that Alexander and Caesar, Tara and Troy had all disappeared, and that perhaps the Empire would also disappear, seems appropriate.[12] Perhaps it is better understood as an exemplar of nineteenth-century English language discourse on Ireland.

When you look at the survey, instead of through it, there is more to be learned about the surveyor than the surveyed, more about the mammoth metropolis of London than the humble hovels of Galbally. The presence of the ostensible subject of the survey, Ireland or the Irish, is shadowy at best. It is undeniable that it was an engagement with vernacular culture but, and this is the important point, it was a particular kind of engagement that was articulated in a particular dialect of an imperious and almost impervious evolutionary science. Once, or if, this is accepted then the survey becomes an interesting focal point for mapping official discursive configurations of Irishness taking just one point in an ongoing process from Giraldus Cambrensis to Father Ted and on to Terry Eagleton. It offers insights into the ever-changing official responses to popular culture in general. If surveyors noted Halloween as a primitive survival for example, a hundred years the later the Irish Folklore Commission noted it as a proud national tradition, almost another century later a native Irish government would all but ban it. After two centuries of official interest and interference it has not gone away.

The survey produced more than maps: it also produced prose. The primary evidence in this book is drawn from the memoirs and the letters that make up the bulk of the ethnographical content of the survey's archive. The letters consist of the observations of surveyors working on the orthography of Irish place names. The memoirs are a more comprehensive survey of statistical, historical, economic, religious, botanical, geological and ethnographic information. They were written mainly by the officers of the British Royal Artillery and Engineers until the project was abandoned in 1840 with only the province of Ulster covered. This book is neither a history of the Ordnance Survey nor a history of cartography, it aims to place the survey in an ethnographic context.

The modern lionization of the survey's technological achievements is understandable, that of its cultural achievements less so. It did engage Irish scholars in its own variant of contemporary and emergent ideas of social research in ways that had an enduring impact. There is no doubting the sound scientific accomplishments of the survey but there is an abundance of evidence, both historical and contemporary, that neither technological nor methodological excellence is a guarantee of knowledge or understanding. The novelties of the survey were quickly and cataclysmically overshadowed by disease, death and starvation both during its work and after its completion, most gravely and starkly in the famines of the 1840s. As Kiely says 'it was a sharp test for the nineteenth century, the challenge to ships and mines and machinery and talk of progress and Macaulay prophesying a new Manchester in the wilds of Connemara'.[13] The opposite may well have been the actual outcome, a new Connemara in the wilds of Manchester.

This book blends, of necessity as much as of desirability, some of the theoretical implications of ethnography, cartography and translation. Each of these focuses on the cultural, spatial and symbolic dimensions of life. The survey embodied all three theories simultaneously. This combination is used to locate and contextualize what Spivak calls the 'palimpsestic narrative of imperialism' and its subtext or referent, the subjugated knowledge. It is not about authenticity unless that is understood as elite efforts to posit science, truth or fact as a counter-vernacular authenticity. The idea of so called peasants' belief in fairies is matched by the surveyors' own belief in the eloquence of the skull. The survey represents the folklore of science as much as it represents the science of folklore. It is not always obvious who is the naive one and who is gullible one. Informants who, in harmony with survey technique, 'interview' the fairies contrast with those described as being experts 'in the art of earning a few pence'. The hustle and bustle of a busy nineteenth-century Ireland contrasts with the empty landscape of the maps.

Quite simply this book views certain key aspects of the survey as intellectual or imaginative work. If the maps are artefacts then the discourse surrounding them is the art of facts. In the world of the nineteenth-century surveyor the map is a fact and the facts are mapped. Such glittering confidence is summarized in Leopold von Ranke's nineteenth-century aphorism *wie es eigentlich gewesen* or 'simply to show how it really was'.[15] This work asks how did it show it and what did it show? An account of the survey from an ethnographic perspective must focus on the obvious question of representation: who or what is represented and who or what is representing them.

The more discursive aspects of the survey offer primary evidence not about the Irish but about the rationalization of the project itself. It was, quite literally, a defining moment of Irish modernity and should not be underestimated. Neither should it be overestimated; the concept of modernity is as problematic as the concept of traditionality. Interestingly the achievement of the survey is often expressed in both terms. It is credited with the advancement of progress and also with the rescue of its antecedent, tradition. The common perception is that the Empire brought the English language, synonymous with trailblazing modernity, to Ireland overtaking the Irish language, synonymous with lumpen traditionality.

It is true that both modernity and tradition suggest a similitude to imagined states of existence in the past and the present, but neither is reliable. Although proudly regaled in the emperor's new clothes, the survey itself could also be described as traditional. Although innovative in its method and theory, it also drew upon the traditional techniques and theories of imperial science. That such ossified oppositions continue to inform cultural theory in the present highlights the continuing relevance of the survey to contemporary discussion. Just as elements of the culture described in the survey continue to be a part of the cultural life of the country, so too elements of the theory informing it continue to circulate in academic and cultural discourse.

The survey was not as confident or as assured as it appears in hindsight. The simple, factual, realistic or black and white outlook that underpinned its structure was fuzzy and blurred at times. What Sahlins aptly calls 'the risk of categories in action' or the expression of cultural synthesis often unsettled the surveyors in their encounter with the vernacular culture.[16] The apparently messy, displaced or arbitrary character of everyday life sometimes confused the fixated modernizers. The quick fix replacement of the traditional by the modern replicates the tendency mentioned above to eliminate one language or another. This was intended as an empowered inscription, an act of accession and a significant step towards political, social and cultural hierarchization in Ireland. In order to achieve this, in order to outline what would appear as clear and transparent lines, the rougher and more resistant qualities of culture had to be smoothed over.

People are described, interpreted and translated as the land itself is outlined, interpreted and translated. The local, the vernacular, the subject, the other language is rendered pale and translucent. The discourse of the survey manoeuvres it strategically further and further from view as it announces, the converse of the mottoes of contemporary Irish language revival, that English is spoken here. What the survey yields, with the aid of an elastic theory and method, is not Irish folklore but a vague cultural residue it classifies as folklore. While the technology, methodology and execution were new in ethnographic, cartographic or translative terms, administrative and cultural transparency was the primary goal. In the commanding practice of colonial ethnography exposition coincided with ellipsis.

Whether or not Ireland can be considered a true or proper colony of Britain has been questioned by historians and litterateurs. Some of these engage with elite literary expressions, while others proceed from privileged positions within the hegemonic historiography of contemporary progress. Almost all work exclusively in the English language. Apart from the desirability of more bilingual approaches to an overwhelmingly anglophone Irish studies, the discipline of folklore and ethnology, focused on culture as a lived-in experience, can contribute much to this analysis.[17] This is particularly so in the case of a large scale administrative, colonial or state ethnography such as the survey. It could be said that literary or historical discourse represents collectives while ethnographic discourse collects representations. The conclusions may or may not be similar in the end but the critical assumption that the space in question (the field, Ireland) is empty prior to any overriding literary or learned signature is short-sighted. Any discourse that is falsely and wrongly premised upon the end of one culture and the beginning of another, discourses that begin after the fact, simply take up where the colonizer left off.

As the core discipline informing this book, folklore and ethnology has at least one benefit: it can re-enliven the processes within the survey that created the texts and observe the ethnographic inskilment of their authors. My approach to the survey is based broadly on the three phases of ethnographic process outlined by Pels. The first is the *préterrain* or power relations within which the ethnographer works institutionally, professionally or in the field. This necessitates a preparatory overview of the conception and formation of the survey in its Irish context. The second is the social organization of the ethnographic occasion itself and the encounter between the ethnographer and those represented. This book presents the first exploration of the conduct, role, reception and experiences of the surveyors. The third is the authoring of the ethnography or the ways in which knowledge of the cultural life of others is translated into specific ethnographic traditions directed at specific audiences.[18] In presenting and examining the actual translation, transaction or use of the knowledge attained, this forms the core element the book.

Along with the commissions, reports, registers, evaluations, decrees, treaties, rent-rolls, blue books, leases, tithes, letters, estate maps and diaries, the survey was enshrined in what Carr calls the temple of facts bequeathed by the bureaucracy of empire.[19] The provenance of such data has not been the subject of critical inquiry itself and thus many would argue that it enjoys an underserved degree of impunity. The question of just whose history it is has not been asked too often, as Estyn Evans wrote 'to depend on documentary evidence alone is to see Ireland through the eyes of her conquerors'.[20] That the imperial imprimatur, the plans, commissions, orders, ordinances and conditions that led to such archival hoards are primarily exogenous is veiled in English language discourse as needs must. By an anachronistic process of cultural and linguistic osmosis such evidence could be described as continuing to fulfil its original goal. Often used without any disclaimer it continues to speak about uncivilized natives as if they were there in the first place or are still here today. It objectifies antique antipathies and transubstantiates racialist otherness. The Irish, lacking as they are in true or proper symbolic resources of their own, simply are how others described them. The skewed difference which the survey sought to demonstrate and domesticate, in a magical retrospective advent, is ushered in. The past is re-presented and 'the Irish' are other to themselves.

What could be called the ongoing Ordnance Survey retrospective turns on three main issues: colonialism, the provenance of the survey and the Irish language.[21] Whether academic, literary or journalistic, most debates centre on one or more of these issues and develop their critical standpoints from them. It has an enduring topicality and afterlife that integrates it into the contemporary life of Ireland as a symbol or symptom of something in the cultural process. This found expression in the debate between John H. Andrews and the playwright Brian Friel. Friel's play *Translations* is perhaps an example of a view that might only have come from outside the Procrustean quarters of a conservative Irish academy. It could be said that one advocated the arts of science while the other argued the science of art.

The long deep shadow cast by the survey may be disproportionate to the reality and, in a project-saturated world, there have been many analogous large-scale local and global plans since that passed more or less unnoticed. It could even be argued that the nineteenth-century surveyor attracted no more notice that the surveyor of the present day, just another delay caused by the trafficking of the plans and schemes of government or developer. It is as an Old World social and cultural laboratory, a site of experimentation, a theatre of theory, a workshop of ethnography, that it has become a New World nexus of ideas critical to Irish identity.

Since all interpretation is inseparable from ideology it is not a radical but a moderate and modest proposal to suggest the relevance of theory to Ireland; it is rather the denial of ideas that is worrying.[22] While some have felt that the survey's poor reputation is undeserved or even due to 'unguarded', 'misguided' or 'misleading' interpretations, this book, the first sustained critique of the survey, while rejecting the heroic image of it as an imperial saviour and culture giver, also attempts to re-place it within the wider history of ideas such as British evolutionary thought, emergent folkloristic, ethnological traditions and the interpretation of culture in general. The survey is clearly inseparable from this complex of ideas and any treatment of it, no matter how specific it is, can only be enriched by an account of them.

Finally a few words on the organization of the text. I felt it appropriate, given the interconnectness of the central concepts, to arrange the chapters thematically rather than in strict chronological order. Since this is not intended as a history, and in order to draw the different strands together and to introduce the reader to the main ideas and protagonists, each chapter has a specific focus. The first chapter outlines briefly the origin of the survey and introduces the main protagonists to the reader. It traces the emergence of images and visions of Ireland through various genres showing how a map could be considered as a perspective on the world. The second chapter outlines some of the key developments in the emergence of the disciplines of ethnology and anthropology. These informed and shaped the ethnographic survey, the main focus of this work, summarized as the 'habits of the people'. The idea of civilization is central to this and the criteria that defined it. The third chapter takes a closer look at the close relationship between colonialism, cartography and translation. It discusses MacLeod's telling statement that science has no nation but nations have science. In the fourth chapter the key question of representation is raised with evidence drawn from the survey's memoirs. One of the themes that emerge from this, the uses of the concept of tradition to define Irishness, is developed further in the fifth chapter. The sixth chapter looks at the fieldwork of the survey and the construction of the ethnographic occasion itself. This highlights the enactment of nineteenth-century ethnographic theory and methodology and the crucial role of the Irish fieldworkers. It is argued that the role of these fieldworkers, the first professional Irish ethnologists, and that of John O'Donovan in particular, contributed to the eventual emergence of Irish folklore and ethnology. The last chapter realigns the ethnographic aspects of translation with the translative aspects of ethnography. It offers a final reflection on translation as both a technique of, and a metaphor for, the survey.

NOTES

1. Ordnance Survey of Ireland, *An Illustrated Record of Ordnance Survey in Ireland* (Dublin, 1991), p.12.
2. *Oxford English Dictionary:* Second Edition prepared by J.A. Simpson and E.S.C. Weiner (Oxford, 1989), pp.909–10.
3. Ibid., p.309.
4. The first known written form appearing in the seventeenth century, *béaloideas* is the established modern Irish-language equivalent for the nineteenth-century English word, folklore. A compound of *béal* the word for mouth, and *oideas*, a word for pedagogy, it has been interpreted as traditional or hereditary knowledge.
5. Valerie Kennedy, 'Ireland in 1812: Colony or Part of the Imperial Main? The "imagined community" in Maria Edgeworth's *The Absentee*', in Terence McDonough (ed.) *Was Ireland a Colony? Economics, Politics and Culture in Nineteenth-century Ireland* (Dublin, 2005), p.266.
6. Michael Cronin, *Irish in the New Century: an Ghaeilge san Aois Nua* (Dublin, 2005), p.28.
7. Kevin Myers, 'An Irishman's Diary', *The Irish Times* (17 Dec. 2003), p.17.
8. John B. Keane, 'Introduction', in John O'Donovan, *The Antiquities of the County of Kerry* (Cork, 1983), p.10.
9. Joep Leersen, 'Petrie: polymath and innovator', Introduction to Peter Murray, *George Petrie (1790–1866): The Rediscovery of Ireland's Past* (Cork, 2004), p.8.
10. Peter Pels, 'The Anthropology of Colonialism: Culture, History and the Emergence of Western Governmentality', *The Annual Review of Anthropology* 26 (1997), p.167.
11. Tim Robinson and Liam Mac Con Iomaire, *A Twisty Journey Mapping South Connemara: Camchuairt Chonamara Theas* (Dublin, 2002).
12. 'A Wild Hope', translated in Frank O'Connor, *A Book of Ireland* (London, 1971), p.91.
13. Benedict Kiely, *Poor Scholar: A Study of William Carleton* (Dublin, 1972), p.122.
14. Gayatri C. Spivak, 'Can the Subaltern Speak?', in Bill Ashcroft, *et al.* (eds), *Post-Colonial Studies Reader* (London, 1995), p.25.
15. Edward H. Carr, *What is History?* (London, 1961), p.8.
16. Marshall Sahlins, *Islands of History* (Chicago, 1985), p.145.
17. See Colin Graham, 'Post-colonial Theory and Kiberd's "Ireland"', *The Irish Review* 19 (Spring/Summer 1996), pp.62–7; Steven G. Ellis, 'Writing Irish History: Revisionism, Colonialism and the British Isles', *The Irish Review* 19 (Spring/Summer 1996), pp.1–22. See also David Cairns and Shaun Richards, *Writing Ireland: Colonialism, Nationalism and Culture* (Manchester, 1988) and David Lloyd, *Anomalous States: Irish Writing and the Post-colonial Moment* (Dublin, 1993).

18. Peter Pels, 'The Construction of Ethnographic Occasions in Late Colonial Uluguru' in Peter Pels and Oscar Salemink (eds), *Colonial Ethnographies*, a special issue of the journal History and Anthropology 8, 1–4 (1994), p.322.
19. Carr, *What is History?* p.16.
20. Emyr Estyn Evans, *Ireland and the Atlantic Heritage* (Dublin, 1996), p.42.
21. Stiofán Ó Cadhla, 'Mapping a Discourse: Irish Gnosis and the Ordnance Survey 1824–1841', in Jamie Saris and Steve Coleman (eds), *Culture, Space and Representation*, a special issue of the *Irish Journal of Anthropology*, 4 (1999), pp.84–109.
22. John H. Andrews, 'Preface', in *A Paper Landscape: The Ordnance Survey in Nineteenth-century Ireland* (Dublin, 2001, 2nd Edn), p.vi (i).

CHAPTER ONE

A Full Face Portrait of the Land: Topographic and Antiquarian Vision

The earliest known map of Ireland was made by Claudius Ptolemy of Alexandria in 150 AD. Although there were many maps subsequent to Ptolemy there are not, until recent times, too many Irish maps. There has never been a comprehensive official cartography of the island of Ireland, or any other island, in the Irish language. There is no early Irish language term for a map, the modern word *mapa* is a transliteration of the English map while *léirscáil* is a recent neologism. The semantic range of the older Irish *cairt* includes parchment, manuscript, book, charter, right, claim, contract or conquest. It cannot be presumed, however, that the Irish lacked some geographic or topographic interest.

Virgil of Salzburg, who had been compared with Copernicus himself, spoke of *ilthuatha faoi mhuir*, or an inhabited southern hemisphere. The Irish intelligentsia accepted the Antipodes, inhabitants of the opposite hemisphere, combining indigenous ideas of the Otherworld or simply other worlds with new Graeco-Roman science.[1] In the eighth century Dicuil, a geographer of Irish origin, drew upon the personal observations of a man called Suibhne in describing an island to the north of Britain.[2] Saints such as Brendan or Colum Cille journeyed to Scotland, England and the Continent in search of 'deserts in the ocean'. The Faroe Islands are described in the *Nauigatio* of St Brendan. Curiosity, and an interest in what anthropology might call the other, is implied in '*oilithreacht*', the Irish word for pilgrimage. It combines the earlier sense of *eile*, literally 'other', with *ithir*, meaning soil or land and actually meant to be in another country. The early learned verse or narrative genre of *iomramh*, *eachtra* or *loingeas*, all translate as voyage, expedition or exile. The *Voyage* of Bran mentions islands *ins an oceon uainn thiar*, or to the west using the loan term from Latin *oceanus*.[3] This may have been used to separate the known world from the unknown.

In the early modern period in particular the history of cartography in Ireland is part of the history of British cartography and is written most noticeably in English. From the moment of colonial incursion into Ireland the island was mapped as an island in an English-dominated British archipelago. Much as the early mapmakers looked at Ireland from their own particular perspective, culture and language, the contemporary historian now looks back on Ireland from the same perspective, culture and language. What happened in-between seldom merits more than a footnote in an otherwise universalistic and univocal description.

The idea that the Irish language or what was painted as a lack of civility kept the English from venturing too far into the interior of Ireland until the sixteenth century is a little fay. No language is in itself an obstacle to invasion or conquest. Conquerors seldom delay to allow the local population to brush up on their civility. It is more likely that they would be delayed by fears for their own lives. The unfolding of cartographic history in Ireland charts the advances and adventures of British colonialism and the country appears more often than not as a barbarous outpost on the homeliest horizon of its expanding Empire. Giraldus Cambrensis or Gerald of Wales' *Topographia Hiberniae*, better known as the *Topography of Ireland*, initiated 'a discourse that extends and criticizes the portrayal of the Irish as barbarians'. Ireland was placed outside European culture 'in an exotic place of the bizarre and unknown'.[4] It was still in use as a reference work by the nineteenth-century surveyors.

Modelling their plans for conquest and plantation upon the Spanish example in the Americas, the colonial government had an unequivocal policy of Anglicization in those areas where it wielded power.[5] There are many examples that illustrate this:, in 1366 an ordinance was enacted that 'every Englishman do use the English language, and be named by an English name'.[6] In 1536 Henry the Eighth wrote of Galway, 'that every inhabitant within the saide towne indevor theym selfe after the Engliyshe Facion; and specyally that you, and every of you, do put forth your childe to scole, to lerne to speke Englyshe'.[7] A recommendation was made in 1665 'that the barbarous and uncouth names . . . be replaced by others "more suitable to the English tongue"'.[8] For Geoffrey Keating, writing in the Irish language in the seventeenth century, the difference between the Norman conquest of Ireland and the Elizabethan conquest was that the Normans did 'not suppress the language of the people and expel them from their land.'.[9] The mid-sixteenth century saw an attempt at a 'thoroughgoing and systematic policy of Anglicization . . . to enforce inheritance of land through primogeniture, to abolish Gaelic law and language, and to transplant Irish inhabitants'.[10]

The production of maps was not an innocent or passive activity, it was enmeshed in the political and military manoeuvres of governments. In Ireland many mapmakers contorted their maps in an effort to extend, if only pictorially, the area of English influence. They highlighted areas of strategic importance such as the residences of kings, chiefs and monastic settlements. Mapping has always flourished during periods of resistance, war and colonization. A map was 'part of the "knowing" that was essential to political and military conquest'.[11] Schama writes of the 'peculiar alliance between drawing and subjugation'. In Scotland Thomas Sandby was attached to the camp of 'the Butcher' of the Jacobites, the Duke of Cumberland 'and through the duke's influence won an appointment in the Ordnance Office in London, drafting maps and surveys of the conquered territory'. This survey's aim was to provide information in support of an extension of a system of strategic forts, roads, and bridges.[12]

In the centuries preceding the survey mapping was not the result of a casual interest in insular topography. One of the goals of surveying was to obtain knowledge about other peoples and to incorporate them into the system of governance of those who did the depicting and classifying.[13] In Ireland mapping 'provided the cartographical underpinnings for a whole system of English rule, and English nation building, in seventeenth and eighteenth-century Ireland. They implied that all the latent characteristics of the Irish people were unchanging because they were rooted in the geography of the country as a "barbarous place"'.[14] Mapping allowed for better knowledge of the realm and military advantage in terms of conquest, it 'staked out the claims of Empire'.[15]

The mapmaker was often unwelcome and the fact that many of them risked life and limb in Ireland is highlighted by the decapitation of Richard Bartlett in Donegal in 1604. Bartlett accompanied Charles Blount or Baron Mountjoy in his campaign against Hugh O'Neill.[16] Several surveyors were beheaded during the seventeenth-century Down Survey that followed Oliver Cromwell's military campaign. Colonization went hand in hand with the collection of maps, a term used by Robert Devereaux's mapmaker Babtista Boazio.[17] Maps were often accompanied by cultural or ethnographic descriptions of people. The mapping of territory by imperial powers such as Portugal, France, Britain, Russia, Siam or Qing China, claimed the physical extent and demarcated the boundaries of emergent nation-states or incorporated them into the world of Empire, 'as early modern maps helped to determine the territorial boundaries of the modern nation-state, ethnographies would help to define its citizens'.[18] In the early seventeenth century John Speed depicted the 'wild' and 'civil' Irishmen and Irishwomen on the margins of his maps. In his depictions the obedient or civil Irishman is extending a hand of friendship to Britain.

Maps and other representations had their part to play in the battle for hearts and minds. The maps were fashionable in London at the time and enjoyed much the same popularity as a newspaper. In the seventeenth century the appearance of forts on maps provided an assurance of safety and assuaged the fears of the consumer. New maps of new colonies were lucrative novelties, the stock and trade of Empire. Early metropolitan English language consumerism was influential in the construction of sanitized and saleable images of Ireland. From the beginning of the nineteenth century print-capitalism played a significant role in the imagining of both the Empire and indigenous peoples within it. Teachers, writers, parsons, philologists, antiquaries, lexicographers and surveyors worked through the medium of print both to produce novel portraits of other peoples and to popularize these portraits.[19] The map was one of many genres that represented other countries and populations.

Maps could show where the other was geographically but current ethnological methods and emergent ideas about folklore classified what the other was like. Contemporaneous with the survey, from the 1820s to the 1850s, dozens of books of folklore were issued by London publishers. Field-based tale collections from outside England were transcribed, published, translated and consumed as contemporary examples of primitive culture.[20] Alongside these on the shelves stood 'books on the experience of travellers, explorers, naturalists, missionaries, and colonial officials'.[21] Between 1826 and 1855 over 300 annuals and mass-produced albums of paintings and drawings appeared 'establishing a canon of sentimental and picturesque taste amongst the middle classes which has hardly altered to this day. This included historical tales and easy versions of national history'.[22]

A smorgasbord of the embryonic genre of evolutionary science blended language and race to create a colonialist and commoditized English image of Ireland. In the early nineteenth century tourism and literacy created a new demand for travel guides, 'during the Napoleonic Wars, with the Continent off-limits to tourists from Britain or Ireland, the affluent middle-class and aristocracy began to explore scenic areas such as the Lake District, the Scottish Highlands, Wales and the west of Ireland'.[23] In 1824, foreshadowing the Ordnance Survey to a certain extent, Thomas Crofton Croker published *Researches in the South of Ireland, Illustrative of the Scenery, Architectural Remains, and the Manners and Superstitions of the Peasantry with an Appendix, Containing a Private Narrative of the Rebellion of 1798*. He married Marianne Nicholson, daughter of the landscape artist Francis Nicholson. The different genres of representation mingled and, in the mind of the mapmaker, the landscape was inhabited by people of archaic manners and ways. In 1825 *Fairy Legends and Traditions of the South of Ireland* appeared invoking established colonialist rhetoric in placing 'the reader . . . squarely in a position of critical authority and cultural superiority'.[24] During these same years T. K. Cromwell published his *Excursion Through Ireland*, James Brewer his *Beauties of Ireland* and G.N. Wright his *Guide to the Giant's Causeway*, *Guide to Wicklow* and *Guide to Killarney*.[25] The delineation of Ireland in watercolour as well as on the map coincided with its incorporation within the British Empire. The emphasis on maps shifted to postal stations, roads and military barracks. These signified improvement and civilization to the furtive eye of the colonizer.[26]

In the nineteenth century there were renewed fears of rebellion, agrarian revolt, and the mass movement for Catholic Emancipation gained momentum. Questions of civilization, knowledge and taste were politically resonant.[27] Croker, whose father Major Thomas Croker suppressed many such revolts, was aware of the charged nature of political and cultural relations between England and Ireland. Two decades before the English word folklore was coined, antiquities, travel, survey, painting, sketch, translation and 'traditionary tales' blended fantasy with design and imagination with politics in a visionary imperial science. Since the time of Gerald of Wales topography was an umbrella term for combined ethnographic and territorial description. The topographical tradition of drawing landscape had its origin in mapmaking and military draughtsmanship.[28]

The antiquarian, folkloristic or ethnographic branch of this science viewed the culture of the other primarily as an index of race. The nineteenth-century survey of Ireland utilized this branch in creating its own images of Ireland.

A Full-face Portrait of the Land

The survey was not solely an ethnographic project; its origins lay in more pragmatic issues of governance. In 1824 a select committee chaired by Thomas Spring Rice, a landowner and member of the British parliament, recommended that a townland survey of Ireland be carried out to enable a valuation for taxation purposes. Land surveying was a prerequisite for the administration of a sound system of land tenure.[29] Local taxes based on townland units, called the county cess, were thought to be inequitable. In popular speech 'bad cess to you' remains a serious insult. Successive committees of the British House of Commons debated the problem and concluded that although the names and outlines of these townlands were documented in previous surveys, the acreages and rateable valuations were uncertain.[30]

The detailed valuation of land and buildings was a preliminary to altering the taxation system. Parliament passed the legislation and delegated the matter to the British military's Board of Ordnance. Richard Griffith, an Irish civil engineer, headed the Boundary Commission established in 1825 that demarcated the boundaries while the Ordnance Survey mapped them in a comprehensive topographical survey.[31] Precise measurements of the country's 62,000 townlands and 2,500 civil parishes were taken. Meresmen marked out the meerings or boundaries; these were often the collectors of the county cess, unpopular figures compelled to work at 2s. per day to conduct the surveyors along the boundaries.[32] The survey, authorized ultimately by the Duke of Wellington, Prime Minister of Britain, produced maps at the scale of 6".

The surveyors had considerable powers to set up marks, enter land or force co-operation 'of all subjects of his Majesty's realm'.[33] The staff of the survey comprised of divisions of British soldiers and subordinates led by a lieutenant of the Royal Engineers or Royal Artillery. Colonel Thomas Colby, who directed the survey from 1820 to 1827, first sent them to England on a preparatory course of instruction in surveying and mathematics. Colby, and later Major General Sir Thomas Larcom, commanded the officers, miners, country labourers and sappers. In 1826 the numbers of sappers and labourers increased from 87 and 53 to 132 and 133 respectively. Mountjoy House, later Mountjoy Barracks, in the Phoenix Park was the survey's headquarters.[34] This is close to the Viceregal Lodge, today called *Áras an Uachtaráin*, the residence of the President of Ireland in the Phoenix Park, Dublin.

Early teething pains beset the survey and, following an investigation, a select committee on public income and expenditure almost dissolved it in 1828. Colby was subject to the authority of a polygon of forces controlling the survey. Andrews relates how a complaint about him made by Major Reid 'passed from the chief secretary of Ireland to the lord-lieutenant, Lord Anglesey, and from Anglesey . . . to another and even more distinguished ex-master-general, the Duke of Wellington, who had lately become prime minister'.[35] Colby was against the recruitment of civilians to the survey as he did not consider them capable of the rigour and discipline required for the job. He advocated the military system, 'forms filled in, signatures appended, responsibilities assigned and accepted; instruments disposed of by the hundred, men by the thousand; the corporate professionalism of the well-drilled multitude'.[36]

The survey did not enjoy a free reign and can be considered as a part of the wider configuration of scientific practice in the nineteenth-century. A distinctive feature of this was 'its association with a particular cultural identity or social interest'.[37] London was considered the hub and home of scientific endeavour and Colby may have felt unhappy in Dublin where he was 'bottled up' and 'buried'.[38] Governmental control was exerted by the approval or disapproval of grants. In this regard Bennett points to the establishment of the Queen's Colleges, the Museum of Irish Industry, the Department of Science and Arts, the Royal College of Science and the Dublin Society.[39] Although the tendency over time has been to call it the Irish Survey or the Irish Ordnance Survey, no such organization existed at the time. The British Ordnance Survey of Ireland was controlled by select committees and commissions reporting ultimately to the British parliament in London.

Ireland in the Limelight

Triangulation, the precise calculation of a network of points or stations on which the mapping could be based, was the first task.

Triangulation-based topographic surveys were undertaken across Europe in the late-eighteenth and early-nineteenth centuries. In 1825 the surveyors began mapping from north to south transfixing Ireland in the crisscrossing webs of a futuristic science from Lough Foyle to Hungry Hill and from Mount Brandon to the Hill of Howth. Sightings were taken to distant mountaintops using a theodolite. The survey used cutting-edge technology. Lieutenant Thomas Drummond's invention of the limelight enabled measurement over greater distance and later became synonymous with popular culture. The maps themselves, showing 'a full face portrait of the land', were available in 1846 and were 'briskly canvassed among potential purchasers by the survey's Dublin agents, Messrs Hodges and Smith of Grafton Street'.[40]

These are not the primary concern of the present work, however, but the body of prose, consisting of memoirs and letters, that emerged in the course of the survey. The memoirs are a multifarious body of information of a social, cultural and economic nature. This information reflects the interests of many, if not all, major branches of emergent nineteenth-century evolutionary science. It includes geological, botanical, historical, geographical, zoological, archaeological, economical, ecological, ethnographic and statistical data. The memoirs, and to a lesser extent, the letters, sought to catalogue countryside and town, buildings and antiquities, land-holdings and population, employment and livelihood as well as the religions, languages and lifestyles of the people.

The officers of the Royal Engineers authored many of the memoirs while conducting fieldwork in the province of Ulster. The first of them were written in the 1830s and continued throughout that decade. The first volume, a memoir for the parish of Templemore, Co Derry, was published in 1837. This met with a hostile reception due to its size, 350 pages for one parish, its disjointed compilation and the fact that it cost more than three times the original budget for one county.[41] The scheme fell out of favour between 1839 and 1840 with three of the four provinces uncovered. Under Robert Peel's leadership the British Government considered the expense and effort involved to be excessive. The extant memoirs, however, provide a unique example of both the ethnographic aspects of the survey and the evolutionary colonialist discourse informing its conceptualization and practice in the field (see Chapter 4).

The Topographical Department

Ethnography, translation or cartography might all be characterized as negotiations of the tensions between the strange and the familiar in cultural, linguistic or physical terms. In this sense what follows is a reverse triangulation of both tensions and intentions within the survey. These had both constructive and destructive outcomes and were the result of various public and private, amateur and professional, local and global relations. They reveal both predictable and unpredictable relations between countries, cultures and languages, as well as between individuals. At times the protagonists do not say or do what they could well be expected to say or do, the native almost goes foreign and the foreigner almost goes native. The survey neither glorified nor ignored the Irish language or culture, rather it drew upon contemporary ethnographic theory and practice to categorize and catalogue them.

The department given the responsibility to negotiate the logistical problem posed by the Irish language for the survey has been variously, and perhaps romantically, styled the Orthographical, Historical, Antiquarian or Topographical Department. The relationship between that department, represented here mainly by the prominent Irish scholars John O'Donovan and Eugene O'Curry, and the superintendents of the survey was not a simple one. Although O'Donovan's circumstances were comfortable Major General Sir Thomas Larcom recalled his 'peasant garb' and while O'Donovan initially met Larcom to teach him Irish, Larcom spoke of teaching English to O'Donovan.[42] Larcom rebuked him during the survey for his criticism of it and pressed upon him not to 'disfigure official documents' pointing out that this 'lowered him in everyone's estimation'.[43] From this initial meeting to its conclusion there were tensions within the organizational structure of the survey that belie its monolithic image.

The ambiguous position of this somewhat *ad hoc* department resulted from its falling between many uneven stools. It nestled between empowerment and subordination, peasant and upper class, pandit and pundit, Irish and British. The negotiation of the tensions between imperial government and it's subjects, who happened to be compatriots, working with the signs, practices, axioms and aesthetics of Empire, was riven with anomaly. This expressed itself in many ways in the course of the survey but is best exemplified in the mimicry of O'Donovan himself. The names of both O'Donovan and O'Curry later became synonymous with the origins of professional

Irish learning. They bridged the gap between the invisible and subjugated indigenous universe of discourse under colonialism and the emergent postcolonial disciplines within the new universities in the middle of the nineteenth century.

The precarious nature of their employment is highlighted by various incidents and instances in the 1860s, twenty odd years after the completion of the survey. While native servants of the Empire could expect reward or recognition, the Irish were all but redundant. O'Donovan and O'Curry had their joint editorialship of the Brehon laws taken from them and given to William Neilson Hancock and Thomas Busteed, neither of whom were competent in the Irish language. It has been suggested that they were unacceptable due to their race and religion and may even have been viewed much like 'criminal suspects'.[44] O'Donovan took insult at an offer of a £50 a year pension, 'Her Majesty has been graciously pleased to grant me a pension of £50 a year! In all my life I never felt so marked out for degradation.' When William Carleton, author of *Traits and Stories of the Irish Peasantry*, was given £200 a year, O'Donovan commented that it was 'well earned from them by his production of Bostoon MacFlail'.[45] In this he was not only conscious of the precariousness of imperial patronage and privilege but critically aware of the nature of the British market for representations of Ireland. The letters sent to the survey's headquarters during the survey, particularly those of O'Donovan, give an insight into the fieldwork of the surveyors as well as the tensions that simmered beneath its otherwise smooth surface (see Chapter 5 and 6).

Eugene O'Curry

One of the better known workers in the orthographical department of the survey, O'Curry was born in Doonaha near Loop Head in County Clare in 1796 and was educated locally in a hedge-school. An Irish speaker from the cradle it is thought that he acquired English as his second language by conversing with tourists in Kilkee.[46] He joined the staff of the survey in 1834 and from 1837 was employed to catalogue manuscripts in the Royal Irish Academy in Trinity College Dublin and the British Museum. He was later appointed Professor of Irish Archaeology and History in the Catholic University in 1854. Along with his friend and neighbour Tadhg Ó Mathúna, who was still alive in Kildysart Workhouse in 1886, he contributed many airs to George Petrie's 1855 *Ancient*

Music of Ireland.⁴⁷ O'Curry was vocal in his opposition to the Empire and he was a strong supporter of Catholic Emancipation that was passed in 1829. Although he described the survey as a 'great national undertaking' he was also unhappy at what he called the sudden and wrongful termination of his seven-year connection with it.⁴⁸

In 1825 O'Curry wrote a political ballad in Irish in which he stated that Robert Peel, Secretary of State in Ireland from 1812 to 1818, and Henry Goulburn, Secretary of State from 1821 to 1827, would be brought under 'the yoke of misery' and 'yellow' Gouldburn brought to hell.⁴⁹ Goulburn was the Minister for Finance in the Duke of Wellington's government that lasted until 1830. Towards the end of 1834 Peel was made Prime Minister of Britain and Goulburn was appointed as his Secretary. Peel, nicknamed 'orange peel' by the Irish, was convinced of the moral depravity of Ireland. Both he and the Duke of Wellington, along with most of the Tory leadership, viewed Daniel O'Connell, who had a survey station in Donegal named after him, as 'a minister of sedition'. Peel actually went so far as to challenge O'Connell to a duel in the 1830s.⁵⁰

The survey was not exempt from the political, racial and sectarian tensions of the nineteenth century. After the fall of the British government in August 1835, the year that O'Curry was employed by the survey, a meeting of the British Association was held in Dublin. The Templemore memoir was published two years later and the survey was ordered to return to its 'original duties' in July 1840. Peel was Prime Minister again that year and Goulburn was Minister for Finance. By September 1841 the Antiquarian or Topographical Department was undergoing an internal investigation. Larcom was ordered to account for the expenditure of the department. More tellingly Goulburn received an anonymous complaint that further underlines the cultural, political, religious and racial tensions of the time:

> Of all the abuses and misapplications of public funds, which have taken place for a considerable time past, I know not a more palpable instance than in the Topographical Department of the Ordnance Survey of Ireland. About the year 1829 an Irish Scholar was employed by Captain Larcom ([sic]) to settle the names of the Townlands &c. for the Ordnance Survey maps who received 3/- (three shillings) a day. This employment though carried out with such trifling expense, was upon examination, considered worse than useless by the Right Honourable Lord Stanley when Chief Secretary of Ireland. A few years ago however a Mr. Petrie discerned that this subject was of the most vital importance to the Survey, and suggested

to the late Government the necessity of establishing a district office for this purpose, and to be entitled The Topographical Department of the Ordnance Survey. He got himself of course appointed to superintend it with an allowance of 13/- a day together with the patronage of appointing persons in this office, which I have known at one time to have been unlimited. It is remarkable that all were Papists with 2 or 3 exceptions. I happened to be one of the exceptions and you may therefore rely on the following to be an unvarnished statement of facts. Mr. Petrie is a Roman Catholic in disguise, his wife and family profess that faith publicly. His politics you may infer from the fact that on the eve of great decisions taking place, during the late Government in the House of Commons, he was often heard to say in a vituperative tone 'I very much fear that those rascally Tories will come into power'. He generally spent his day out attending his tuitions in drawing, which profession he follows and the office was a regular school for theatrical declamation and acting, but more generally for religious and political controversy, in which the fewer in number always came off second best in as much as they seldom got a hearing, and were by degrees removed from the office by the cunning schemes of a low Irish person of the name of Curry, whom I would take to be as determined a rebel as ever shot a Protestant in this Kingdom, of which acts he generally approved by saying they must have occurred thru' their own insufferable deeds. In the course of time this fellow took the whole management of Mr. Petrie's office into his own hands and at the late reduction of that office, he got every person turned off, while he and his brother in law O'Donovan, the essence of a bigoted Papist, are the only persons remaining with Petrie, each receiving 13/- a day.

The letter alleged that Scotsman Dr George Petrie, in whose home the fieldworkers were based, spent his days giving art lessons, O'Donovan educating his relatives and O'Curry 'walking about the streets' in search of information. It described O'Curry as a politician and 'one of the most dangerous men of his caste'. The memoir was described as 'all trash and nonsense' and 'the composition of bigoted Papists' in which there was a 'tissue of Popery and Monkery'. The balance between mathematical, scientific or statistical accuracy and ethnographic or antiquarian research was crucial. The complainant stated that they were 'engaged in taking down the pedigree of some Irish chief whose ancient estate they most carefully mark by boundaries, and they have actually in several instances, as I have seen by their letters, nominated some desperate characters as the rightful heirs of those territories'.[51] The possibility that an overly explicit or enthusiastic description of manners and customs could lead to the imagining of another, a different or even a rival civilization was problemmatic.

The complaint was shuffled from Captain W. G. Boldero to Colby who in his turn ordered Larcom to look into it. Goulburn, writing from Downing Street, replied that he knew Ireland well enough to know that 'all is not gold that glitters'. In June 1843 Peel set up a commission of three members to investigate the issue and make a report. The report was published by HMSO in London in 1844 under the title *Report of the Commissioners appointed to inquire into the facts relating to the Ordnance Memoir of Ireland: together with the Minutes of Evidence, Appendix and Index.* It recommended that the memoirs be revived on a more modest scale, but the Antiquarian or Topographical Department, inactive since 1842, disappeared.[52] The ethnographic methods of the survey were perhaps too successful in leading to the description of an equal civilization and this could not be countenanced. By 1838 there were only three collectors kept on for the purpose of producing memoirs. It was perceived to fall between the increasingly opposed stools of historical–antiquarian or quantitative–statistical work. The senior superintendents of the survey wanted inquiries limited to those 'of obvious concern to agriculturalists, engineers, industrialists, and other practical men; and those purely scientific studies, if any, that could be pursued without distracting the surveyors' attention from their principal task'.[53]

In 1840 Sir Hussey Vivian announced that the survey was to return to its original purpose and the emphasis shifted to a stricter form of statistical inquiry, which by then was detaching itself from the ethnographic. Larcom was appointed as commissioner of the 1841 census, and a statistical department was established after the survey in 1844. Andrews notes that, 'it was felt in some quarters that the historical and social sections of the memoir might have an exacerbating effect on Irish patriotic feeling and so accentuate the divisions between planter and native, protestant and catholic, government and governed'.[54]

It was Larcom, administrator at Mountjoy since 1828, who decided what was engraved on the final maps. It had been his wish from an early stage to broaden the survey's scope to include commercial, historical and geological details. After a brief attempt to learn the Irish language himself he recognized the need for a person competent enough in it to work on the place names. It was this recognition that heralded the employment of lexicographer Edward O' Reilly, the first Irish field worker. O'Reilly, however, died after six months' service and O'Donovan replaced him in October 1830. His role was to go into the countryside to consult people with a view to investigating the etymology and orthography of place names prior to Anglicizing them. The Anglicized spellings were institutionalized as official versions for the valuation office and other government departments. O'Donovan's letters, 'the single most entertaining, instructive and comprehensive correspondence written in this island,', resulted from seven years of fieldwork throughout mid-nineteenth-century Ireland.[55]

He was one of a team of Irish scholars employed by the survey that, at various times, had a staff of up to twenty-four under the tutelage of George Petrie (1790–1866). Petrie was a writer, painter, musicologist, archaeologist and historian.[56] He is regarded as 'belonging to a new generation of antiquaries, whose methods were relatively more scientific'. He had progressed from the fanciful ideas of Miss Beaufort and Charles Vallancey. His writings and engravings influenced the Hiberno-Romanesque revival that led to buildings like the Honan Chapel in University College Cork and St Enda's in *An Spidéal*, Conamara.[57] His accounts of the Aran Islands influenced the playwright John Millington Synge who in turn influenced many other artists, folklorists and anthropologists, not necessarily positively. O'Donovan, O'Curry, Patrick O' Keefe and Thomas O'Connor were the most prominent of the Irish fieldworkers. The poet James Clarence Mangan gave his services as a scrivener from August 1838 to Christmas 1841.[58] Petrie also employed William Frederick Wakeman and geologist George du Noyer as draughtsmen to make sketches. Positioned at the interface of vernacular culture and international cultural crosscurrents such researchers played a significant part in the emergence of an independent Irish scholarship not the least of which was the appearance of professional Irish folklore and ethnology at the start of the twentieth century (see Chapter 6).

John O'Donovan

O'Donovan was born on a small farm in Atateemore, County Kilkenny in 1809. It is thought that he was educated initially at a hedge school and later in Waterford city. In 1822 he opened his own hedge school in his native place but moved to Dublin the following year.[59] He studied Latin for a time with the intention of becoming a priest but on finding that he had no vocation for this he turned his attention instead to the study of the Irish language. In 1826 he was employed in the Irish Record Office transcribing old Irish manuscripts for the antiquary James Hardiman. He did similar work for Myles John O'Reilly and it also appears that he spent some of his time employed in a school in Dublin.[60] As a result of this work he was introduced to the Royal Irish Academy, Petrie and Sir Samuel Ferguson. He edited and translated the Charter of Newry in the *Dublin Penny Journal* of the 22 September 1832 and acquired a good reputation for his work on Irish manuscripts and antiquities.

After a period at Mountjoy House the Topographical or Antiquarian department made its offices in 'TeePetrie', a rendering in English orthography of the Irish language for 'Petrie's house', this was Petrie's home at 21 Great Charles Street in Dublin. O'Donovan was employed from 1830 to 1842 as etymologist and orthographer. He worked for the survey from the age of 24 to the age of 36, a total of twelve years.[61] His death of rheumatic fever in 1861 was attributed to his years of fieldwork for the survey. While in the countryside he wrote prolifically to headquarters with various queries, requests, descriptions of antiquities and personal commentaries that amount to 103 volumes in the Royal Irish Academy.

The letters were typed and bound by Father Michael O'Flanagan in the late 1920s and have been in more or less constant use by researchers since.[62] O'Flanagan brought to fruition the work of the Irish Topographical Society who set about copying the survey's name-books and letters in 1911. Members of this society included Arthur Griffith, the founder of Sinn Féin and Michael Joseph O'Rahilly who was shot at the General Post Office during the Easter Rising in 1916.[63] The popular perception that O'Donovan and his colleagues were employed to rescue Irish place names from oblivion or that they represented a purely indigenous survey in their own right is misleading. The letters themselves reveal the often contested and pregnant nature of the images and representations that emerged from the survey. It is through this tensioned relationship or engagement itself that the foundations of later schools of Irish learning begin to configure themselves (see Chapter 5).

Letters from the Field

Begun in County Down the letters continued apace between 1834 and 1843 as O'Donovan 'kept up with the Survey' and 'laid down' the Anglicized orthography for the names of places. They include occasional ethnographic narratives, humorous and lugubrious reflections on the process of fieldwork as well as copious descriptions of an archaeological nature. They were often addressed to Larcom, and many were written in preparation for future work, requesting support or sourcing material from Dublin libraries.

The work of eliciting toponyms and arriving at standardized Anglicizations involved mediating the signal culturally specific element of all cartographic projects, the indigenous language. The survey could not have proceeded without confronting the Irish names of townlands. Many names were written initially by the officers of the Royal Artillery or Engineers into name books and, in their Anglicized form, they are amongst the most prominent features on the maps. Andrews captures the sense of the place names being a vast and comprehensive cultural inventory in their own right:

> The maps make reference to ball-courts, the birthplace of Oliver Goldsmith, bleach-greens, the burial place of a ship's crew, a candlewick factory, corn kilns, a 'deep hole in the rock', fair greens, flax-kilns, freestone-quarries, graveyards (with children's graveyards separately distinguished), intended roads, bridges, railways and canals, limestone-quarries, market houses, several kinds of mill, a natural bridge, ponds, houses, threshing machines, underground rivers, and the wreck of the frigate Saldanha in 1811.[64]

As an ethnographic index of cultural process, place names are always a quagmire of cultural decay and growth. The interpretative and translative processes at the heart of the survey sought to fulfil one of its key imperatives, an Anglicization of the field. The field was Ireland and its residual and emergent personal, local and national identity-forming expressions. It domesticated, sanitized and stratified Irish culture in order to streamline British administration and governance.

The survey was conceived of from the outset as a great work, it celebrated itself as history in the making and science in the doing. Colby conceived of his own role in terms of nationalistic fervour and wished to work 'in a manner that should be worthy of my country'.[65] At best it was a bureaucratic gift and at worst a discursive deceit.

Since English was to be the new Irish, those in the past who did not have it were going to be given it (see Chapter 7).

Proceeding to the Ground

The survey developed a systematic questionnaire based on contemporary models. Larcom's heads of inquiry had three major sections, Natural Features and History, Modern and Ancient Topography and Social and Productive Economy. The statistical remarks or accounts, reminiscent of Sinclair's survey of Scotland, were similar to the memoirs and copies or early drafts were made in the office for administrative purposes. Captain Robert Kearsley Dawson corrected, emended and commented on the Derry, Antrim and Donegal memoirs; he also 'questioned matters of fact and style'.[66] Larcom himself later edited these reports. Many of the civilian assistants wrote memoirs to the stage of a final draft and supervised juniors in their collection of information.

The survey neither looked for nor found Irish folklore ready made in the field. It was mainly through the construction of ethnographic occasions that it encountered vernacular culture and oral tradition. It was selective both in what it chose to see and what it chose to ignore. In terms of standard Western logic and theory, science met tradition, modernity met its antecedent, the colonizer met the colonized and the English language met the Irish language. The accumulation of a considerable amount of knowledge under various headings lent itself readily to interpretation. It was not simply a matter of quantifying or reporting this or that fact, this or that statistic. Along the way data was translated, dictated, transcribed and refined in fair sheets, statistical reports, memoirs, letters, working papers and drafts.

Translation was pivotal in this processing of knowledge. MacLeod has noted how 'the ordered regularities of the library and the museum, the inventory and the balance sheet, became the hallmarks of the civilising mission . . . these institutions . . . create a new . . . English speaking culture'.[67] The subjects of colonial science were generally thought to be unable to represent themselves, 'indigenous traditions and belief systems were reduced to artefacts, culturally inaudible except through their colonial interlocutors'.[68] The culture and language that lay behind the inventory and the balance sheet was silenced or voiced over.

Apart from the explicit power relations underpinning the survey some implicit methodological and theoretical questions are raised by it and foremost among these is 'the problematic phenomenon of the observer'.[69] It cannot simply be taken for granted that the surveyor, as a contemporary newscast might put it, was our man on the spot. An observer does not simply look at something passively; he or she complies with, or conforms to, some code and is 'embedded in a system of conventions and limitations . . . if it can be said that there is an observer specific to the nineteenth century, or to any period, it is only as an effect of an irreducibly heterogeneous system of discursive, social, technological, and institutional relations. There is no observing prior to this continually shifting field . . . there never was or will be a self-present beholder to whom a world is transparently evident.'[70]

The subjectivity of vision in the nineteenth century has been linked with the reorganization of knowledge, languages, networks of spaces and communications. The nineteenth-century observer was a modernizer but was also modernized in the process. Signs and codes were loosened from their hierarchical or socially structured bonds and new social classes became capable of producing imitations, copies, counter texts and equivalents. Fashions began to emerge in the media and in advertising through 'flows of typographic and visual information'.[71]

In the nineteenth century vision was removed from the *camera obscura* and relocated in the human body as concern grew about the eye itself and its defects. New technologies imposed a corrective or 'normative vision' on the observer. Industrialization created a new observer, a new individual who was free from old allegiances. Crary states that:

> The management of subjects depended above all on the accumulation of knowledge about them . . . crucial to the development of these new disciplinary techniques of the subject was the fixing of quantitative and statistical *norms* of behaviour. The assessment of "normality" in medicine, psychology, and other fields became an essential part of the shaping of the individual to the requirements of institutional power in the nineteenth century, and it was through these disciplines that the subject in a sense became visible.[72]

The structured and organized accumulation of information in nineteenth-century Ireland led to an assessment of the normality or abnormality of the population and their state of progress or civilization relative to British imperial norms and standards. It provided something more than an expansive view or prospect, it provided and projected a particular vision, a prospectus of progress.

Ethnology

Contemporary ethnology has an almost painfully acute sensitivity to questions surrounding inquiry and investigation and the social or psychological implications of the enactment of knowledge. Sometimes carried to the extreme of a Shakespearean dilemma, to do ethnography or not to do ethnography, it questions the social and historical origins of discourses. To put it more obviously, it asks how we know what we know. The factuality and certitude of a singular universal reality was questioned by what has variously been called the post-positivist, post-modern or post-structuralist re-assessment of the grand narratives and intellectual architecture of modernity. Kuper says, 'the classical ethnographer had represented himself as an authoritative scientific observer, who crossed cultural barriers while retaining a heroic detachment, and who reported facts in objective language. This image could now be exposed as an illusion.'[73] This view concentrates on the ethnographer as a creator:

> Caught up as they were in the colonial projects of the great powers, the classical ethnographers were all concerned to impose an order on the actual chaos of voices, perspectives, and situations that they confronted in the field – to inscribe one point of view on history. In this way, they served the interests of a political class that wished to impose an alien order on colonial subjects abroad, or on minorities at home.[74]

Ethnography is a written account that 'focuses on a particular population, place and time with the deliberate goal of describing it to others. So, often, did the writings of nineteenth-century explorers, missionaries, military agents, journalists, travellers, and reformers.'[75] The preoccupation with observation and archiving was illustrative of the modern European or Enlightenment's conception of knowledge creation in which the cosmos was held to be observable, measurable and understandable. As we will see in the following chapter, its origins lie in natural history and the study of the peoples encountered along the frontiers of Empire. Understood as being located at earlier stages in the development or evolution of man they were defined by notions of the exotic or primitive.[76]

The survey is an example of the various branches of the evolutionary science of man that included elements of physical anthropology, prehistoric archaeology or pre-history and the colonialist British sense of ethnology as 'the science which classifies peoples in terms of their racial and cultural characteristics, and attempts to explain these by reference to their history or their pre-history'.[77] Most members of the early statistical societies were doctors, scientists, government officials and members of the industrial and commercial elite who 'believed that statistics, whether quantitative or non-quantitative, would provide scientific, i.e. objective, answers to the great social questions of the day, and would either provide evidence of the need for reform and the direction that it should take, or alternatively . . . that statistics would justify the status quo'.[78] The information enabled the construction of databases for the purposes of taxation, military preparedness and the enforcement of religious orthodoxy.[79] The gradual increase in direct observation and independent verification along with the privileging of scientific methodology and inquiry led to the eventual emergence of ethnography as a separate or distinct scientific discipline.[80]

The museum or archive is physical testimony to the appropriative nature of this knowledge accumulation. Along with the social surveys collections of ethnographic, mineralogical, botanical and zoological artefacts were collected. Captain J. E. Portlock gave instructions that 'every person keep a small dry bottle in his pocket to receive any beetles or other insects'.[81] O'Donovan came across a museum of exotic ethnographica in Castle Caldwell in 1834 that boasted a collection of stuffed exotic birds and animals, American–Indian weapons and dress, a Negro-beheading axe and, of most interest to O'Donovan, the skull of the well-known harpist O'Carolan.[82] The museum founded by the Danish king, Frederick III in 1650 is one of the earliest known examples of such museums that became more prominent in the 1800s.

Large national museums were established in London (1753), Paris (1801) and Washington DC (1843). The first specialized ethnographic museums were established in German-speaking areas such as Vienna (1806), Munich (1859) and Berlin (1868). German academics had begun to collect 'folktales and legends, dress and dance, craft and skills'.[83] Ó Giolláin points to the establishment of the Museum voor Volkenkunde in Leiden in 1837 as the first ethnographic collection 'that resulted from the travels of explorers or the expansion of colonial Empires'.[84]

Petrie's collection of antiquities, some gathered while working with the survey, became the basis of the National Museum of Ireland. The survey was an intertextual and interdisciplinary exercise producing sketches, letters, memoirs and maps as representations of what was observed. Like the *Description de l'Égypte* (1802–22) it provided both the text itself as well as the context for interpretation. More specifically it appropriated knowledge and reproduced it 'for the understanding and edification' of the English.[85]

Cartography

As a canonization of landscape the map is an eminent example of these processes, as Monmonier says, 'the map is the perfect symbol of the state'.[86] The term map derives from terms relating to the material on which drawings were made. *Mappa* was the Latin word for cloth while khartes was the Greek word for papyrus but it also referred to pictures or illustrations in general.[87] Stevenson points out that 'just as a "description" might be a map, so a '"carte"' might be a written description or exposition and not a map'.[88] It appears that there may be no such thing as pure cartography, maps and charts are images or expressions of territorial and cultural desires. For eighteenth-century philosophers mapmaking was the epitome of the ordered and structured creation of a coherent archive of knowledge.[89]

Surveyors in India are remembered in historical literature as heroes-of-science braving the elements, tolerating official interference and hostile natives in order to advance science or human knowledge. Cartography, the term was coined by the Viscount de Santarém in 1839, was understood as unproblematic, neutral, truthful, rational and scientific. The use of instruments like the camera, the telescope and the microscope to extract meaning from the world served to reinforce this. Triangulation led to the possibility of a scale of 1:1 where the map could be the same size as the land itself. The map was understood as mimetic, an imitation of an external objective reality.[90] This was questioned in the 1980s when maps were described as 'ethnocentric images, and part of the apparatus of cultural colonialism'.[91] In this sense a map is an empowered ethnograph that gauges past and present cultural and political hegemony.

A map's accuracy, fidelity, objectivity or neutrality, is not a given. It is not a non-indexical informative device but a social construction that projects worlds in a favourable or unfavourable light. It externalizes meanings that are formed in constellations, patterns and contours of discourse and power. The map is at the same time historical, social, political, cultural, intellectual, and scientific. Cartography did not shed its artistic or creative elements in the 1700s. By the 1800s it was regarded as the epitome of encyclopaedic knowledge and one of the most successful forms of knowledge creation. Seeing maps as ethnocentric images and part of the apparatus of colonialism Harley says they are 'perspectives on the world at the time of their making'.[92] The map, in what Turnbull calls a realist illusion, has been a metaphor for science itself as a progressive, cumulative and especially accurate representation.[93]

Silences in maps are equally as significant as the exclusion of elements in discourse, they become statements in themselves and part of the reproduction, reinforcement and legitimization of values. Andrews suggests that the selection of the six-inch scale was 'a cartographic expression of the union of two Kingdoms . . . for extending statute measure to Ireland'.[94] The use of triangulation could be seen as a dramatic extension of the state's epistemological power reaching deep into different landscapes across the territory of a state or Empire.[95] It represents the Enlightenment's encyclopaedism and the new ideology behind systematic and disciplined observation.

The demesnes of the gentry are prominent on the maps of the survey and landowners were often the sole authorities for their names. The survey in this case privileges the position of the gentry and authorizes the private naming of land. Land valuation and taxation lies at the heart of the survey. Boundaries are configured and reconfigured to reflect political, administrative or cultural imperatives. The townlands of the survey became units of estate management and local taxation as well as postal addresses. Hamer points out that the maps lack colour and seem magnified due to the scale. She notes their similarity to the cell under the microscope, the cell having become visible in the 1820s.[96] They have an empty quality in the sense that 'they deny presence – the sharing of time and space by the observer and the observed – so that through its representation, the observed is removed, appropriated from its original context, and recast within the archive's discourse'.[97]

The survey has some of the characteristics of a panopticon. Foucault spoke of the panopticon, the spy hole in the door of a prison cell, as the exemplar for all technologies of state surveillance, control, and discipline in society. The conviction that representations can be actual or objective copies of a rigid reality is problematic. They can also be viewed as inseparable from the conditions of observation and inscription and the subjective condition of the observer. There is a cumulative effect whereby all representations 'melt together into a vast discursive web which defines *in toto* not what the world is but what it ought to be'.[98]

Did the survey show or describe Ireland as it really was or did it create it in its own image, as it thought it should have been, according to its own view of civilized English society? Did it rescue elements of an ancient Ireland or reproduce a new England? Was it, as Hamer says, 'a new Ireland, and one . . . subtly regulated by the discourse of ethnology?'.[99] When it is understood as a form of writing, as a literary genre concerning the cultures of others, ethnography could be described as dominative and, as Fabian says, 'to become a victim the Other must be written at (as in "shot at") with literacy serving as a weapon of subjugation and discipline'.[100]

The survey constitutes a field of knowledge-representations that were ethnographic, statistical, historical, archaeological, antiquarian, linguistic and literary. In its desire to produce images, information or knowledge of both land and people it was an extension of colonial power. It rationalized administrative control and took stock of potentially exploitable resources. Mixing and matching patterns from the global pastiche of evolutionary and colonialist discourse on subject populations it privileged certain registrations of specific cultural elements. The appropriative nature of this discourse is veiled and abstracted in the twin processes of interpretation and translation. In terms of surveying, antiquarianism, ethnography, agriculture or industry this discourse was transformative, reformative and even revolutionary. This is qualified, however, by the underlying imperative to gauge levels of compliance, docility or resistance to the hegemonic English language culture.

NOTES

1. John Carey, 'Ireland and the Antipodes: The Heterodoxy of Virgil of Salzburg', in Jonathan M. Wooding (ed.), *The Otherworld Voyage in Early Irish Literature* (Dublin, 2000), pp.133–42.
2. Jonathan M. Wooding, 'Monastic Voyaging and the Nauigatio', in Jonathan M. Wooding (ed.), *The Otherworld Voyage in Early Irish Literature* (Dublin, 2000), pp.226–45.
3. Kuno Meyer (ed), *The Voyage of Bran* (Felinfach, 1994). *Iomramh* suggests an expedition by sea.

4. Clare Carroll, 'Barbarous Slaves and Civil Cannibals: Translating Civility in Early Modern Ireland', in Clare Carroll and Patricia King (eds), *Ireland and Postcolonial Theory* (Cork, 2003), p.67.
5. Seán Ó Tuama, *Cúirt, Tuath agus Bruachbhaile* (Baile Átha Cliath, 1991). See also Clare Carroll, 'Barbarous Slaves and Civil Cannibals', p.66. Carroll is quoting from the work of D. B. Quinn, 'Sir Thomas Smith and the Beginnings of English Colonial Theory', *Proceedings of the American Philosophical Society*, 89:, 4 (1945), pp.543–6.
6. Tony Crowley, '"The Struggle Between the Languages": The Politics of English in Ireland', *Bullán: An Irish Studies Journal*, 5, :2 (Winter/Spring 2001), p.5.
7. J. E. Caerwyn Williams and Máirín Ní Mhuiríosa, *Traidisiún Liteartha na nGael* (Baile Átha Cliath, 1979), p.311.
8. John H. Andrews, *Shapes of Ireland: Maps and their Makers 1564–1839* (Dublin, 1997), p.22.
9. Carroll, p.79.
10. Carroll, 'Barbarous Slaves and Civil Cannibals', p.69.
11. Jacinta Prunty, *Maps and Map-making in Local History* (Dublin, 2004), p.42.
12. Simon Schama, *Landscape and Memory* (London, 1995), p.466.
13. Laura Hostetler, *Qing Colonial Enterprise: Ethnography and Cartography in Early Modern China* (Chicago, 2001), p.208.
14. Jim MacLaughlin, *Reimagining the Nation-state* (London, 2001), p.234.
15. Hostetler, *Qing Colonial Enterprise*, p.79.
16. Andrews, *Shapes of Ireland*, p.23.
17. Ibid., p.58.
18. Hostetler, *Qing Colonial Enterprise*, pp.208–9.
19. Benedict Anderson, *Imagined Communities: Reflections on the Origin and Spread of Nationalism* (New York, 1983).
20. Jennifer Schacker, *National Dreams: The Remaking of Fairy Tales in Nineteenth-century England* (Philadelphia, 2003), pp.6–7.
21. George W. Stocking Jr., *Victorian Aanthropology* (New York, 1987), p.79.
22. David Brett, *The Construction of Heritage* (Cork, 1996), p.66.
23. Peter Murray, *George Petrie 1790–1866: The Rediscovery of Ireland's Past* (Cork, 2004), p.53.
24. Schacker, *National Dreams*, p.48.
25. Murray, *George Petrie*, p.55.
26. Andrews, *Shapes of Ireland* (1997).
27. Schacker, *National Dreams*, p.63.
28. Murray, *George Petrie*, p.49.
29. Elri Liebenberg, 'Mapping British South Africa: The Case of G.S.G.S. 2230', *Imago Mundi*, 49 (1997), p.133.
30. Ordnance Survey of Ireland, *An Illustrated Record of Ordnance Survey in Ireland* (Dublin, 1991), p.13.
31. John H. Andrews, *History in the Ordnance Survey Map: An Introduction for Irish Readers* (Dublin, 1974), p.1.
32. Michael Herrity (ed.), *Ordnance Survey Letters Meath* (Dublin, 2001), p.vii.
33. Prunty, *Maps and Map-making*, p.131.
34. John H. Andrews, *A Paper Landscape; The Ordnance Survey in Nineteenth Century Ireland* (Oxford, 1975).
35. Andrews, *Paper landscape* Ibid., p.68.
36. Andrews, *Paper landscape* Ibid., p.88.
37. James Bennett, 'Science and Social Policy in Ireland in the Mid-nineteenth Century', in Peter J. Bowler and Nicholas Whyte (eds), *Science and Society in Ireland* (Belfast, 1997), p.40.

38. Andrews, *Paper Landscape*, p.89.
39. Bennett, 'Science and Social Policy', p.39.
40. Andrews, *History in the Ordnance Survey Map*, (Dublin, 1974), p.2.
41. Ordnance Survey of Ireland, *An Illustrated Record*, p.24.
42. Patricia Boyne, *John O'Donovan (1806–1861): A Biography* (Kilkenny, 1987), p.10.
43. Ibid., p.15.
44. Ibid., p.102.
45. Ibid., p111.
46. Michael Herrity (ed.), *Ordnance Survey Letters Dublin* (Dublin, 2001), p.xi.
47. Séamus Ó Duilearga, 'Notes on the Oral Tradition of Thomond', *The Journal of the Royal Society of Antiquarians*, 95 (1965), p.136.
48. Eugene O'Curry, *Lectures on the Manuscript Materials of Ancient Irish History* (New York, 1861), p.370–1.
49. Art Ó Maolfabhail, 'Eoghan Ó Comhraí agus an Suirbhéireacht Ordanáis', in Pádraig Ó Fiannachta (eaged.), *Ómós do Eoghan Ó Comhraí* (Maynooth, 1995), p.148.
50. Richard N. Lebow, *White Britain and Black Ireland: The Influences of Stereotypes on Colonial Policy* (Philadelphia, 1976), p.55.
51. Ó Maolfabhail, 'Eoghan Ó Comhraí', pp.172–74.
52. Boyne, *John O'Donovan*, p.22.
53. Andrews, *Paper Landscape*, p.163.
54. Ibid., p.173.
55. Kevin Whelan, 'Beyond a Paper Landscape: John Andrews and Irish Historical Ggeography', in Kevin Whelan and F. H. A. Aalen (eds), *Dublin City and County: From Prehistory to Present* (Dublin, 1992), p.395.
56. Damien Murray, *Romanticism, Nationalism and Irish Antiquarian Societies, 1840–80* (Maynooth, 2000), p.16.
57. Murray, *George Petrie*, p.39.
58. Ruaidhrí de Valéra, 'Seán Ó Donnabháin agus a Lucht Cúnta', in *The Journal of the Royal Society of Antiquarians of Ireland*, 79 (1949), pp.146–159.
59. Robert Welch, *The Oxford Companion to Irish Literature* (Oxford, 1996), p.424.
60. Art Ó Maolfabhail, 'Éadbhard Ó Raghallaigh, Seán Ó Donnabháin agus an Tsuirbhéireacht Ordanáis 1830–4', in *Proceedings of the Royal Irish Academy*, 91, :C, :4 (1991), p.78.
61. Boyne, *John O'Donovan*, p.xiii.
62. Five of the twenty-nine sets of letters have been published by Enda Cunningham of Cathach Books in Dublin. They are being edited by Michael Herrity and published by Fourmasters Press. The published sets to date are Dublin, Donegal, Meath, Kildare and Kilkenny.
63. William Nolan, 'Introduction', in John O'Donovan and Eugene O'Curry, *The Antiquities of County Clare* (Clare, 2003, 2nd Edn), p.i.
64. Andrews, *Paper Landscape*, p.126.
65. Ibid., p.89.
66. Angélique Day, '"Habits of the people": Traditional Life in Ireland 1830–1840 as Recorded in the Ordnance Survey Memoirs', *Ulster Folklife*, 30 (1984), p.24.
67. Roy MacLeod, 'On Science and Colonialism', in Peter Bowler and Nicholas Whyte (eds), *Science and Society in Ireland: The Social Context of Science and Technology in Ireland 1800–1950* (Belfast, 1997), p.13.
68. Ibid., p.13.

69. Jonathan Crary, *Techniques of the Observer; on Vision and Modernity in the Nineteenth Century* (London, 1996), p.5.
70. Ibid., p.6.
71. bid., p.11.
72. bid., p.15
73. Adam Kuper, *Culture: The Anthropologists' Account* (Harvard, 1999), p.207.
74. Ibid., p.207.
75. Alan Barnard and Jonathan Spencer (eds), *Encyclopedia of Social and Cultural Anthropology* (London, 1996), p.193.
76. Bill Ashcroft *et al.* (eds), *Key Concepts in Post-Colonial Studies* (London, 1998), p.85.
77. John Beattie, *Other Cultures: Aims, Methods and Achievements in Social Anthropology* (London, 1966), p.19.
78. Mary Daly, *The Spirit of Earnest Inquiry: The Statistical and Social Inquiry Society of Ireland 1847-1997* (Dublin, 1997), p.11.
79. David Englander and Rosemary O'Day, *Retrieved Riches: Social Investigation in Britain 1840-1914* (Aldershot, 1998), p.3.
80. Hostetler, *Qing Colonial Enterprise*, pp.80–3.
81. Angélique Day and Patrick McWilliams (eds), *Ordnance Survey Memoirs of Ireland* (Belfast, 1990–98), forty volumes, Vol. 36, p.118. Hereafter abbreviated OSM.
82. J. B. Cunningham, 'The Letters of John O'Donovan in County Fermanagh: dogs, turkeycocks and ganders', *Ulster Local Studies*, 14:, 2 (1992), p.29.
83. Thomas H. Eriksen and Finn S. Nielsen, *A History of Anthropology* (London, 2001), p.15.
84. Diarmuid Ó Giolláin, *Locating Irish Folklore: Tradition, Modernity, Identity* (Cork, 2000), p.55.
85. Mathew Edney, 'Reconsidering Eenlightenment Geography and Map-making: Reconnaissance, Mapping, Archive', in Charles W. J. Withers and David N. Livingstone (eds), *Geography and Enlightenment* (Chicago, 1999), p.6. Edney is referring to the European interpretation of Egypt.
86. Mark S. Monmonier, *How to Lie with Maps* (Chicago, 1991), p.88.
87. Hostetler, *Qing Colonial Enterprise*, p.3.
88. David Stevenson, 'Cartography and the Kirk: Aspects of the Making of the First Atlas of Scotland', in *Scottish Studies*, 26 (1982), p.3.
89. Mathew Edney, *Mapping an Empire: The Geographical Construction of British India 1765–1843* (Chicago, 1997), p.18.
90. Christian Jacob, 'Toward a Cultural History of Cartography', *Imago Mundi*, 48 (1996), p.192.
91. John B. Harley, 'Silences and Secrecy: The Hidden Agenda of Cartography in Early Modern Europe', *Imago Mundi*, 40 (1988), p.70.
92. Ibid., p.70.
93. Edney, *Mapping an Empire*; also Edney, 'Mathematical Cosmography and the Social Ideology of British Ccartography, 1780–1820', *Imago Mundi*, 46 (1994), pp.101–16; Edney, 'Reconsiderieography and map-making: reconnaissance, mapping, archive', in Charles W. J. Withers and David N. Livingstone (eds), *Geography and enlightenment* (1999); J. B. Harley, 'Silences and Secrecy: the hidden agenda of cartography in early modern Europe', *Imago Mundi 40* (1988)', pp.57–76; David Turnbull, 'Cartography and Science in Early Modern Europe: Mapping the Construction of Knowledge Spaces', *Imago Mundi*, 48 (1996), pp.5–24; Christian Jacob, 'Toward a Cultural History of Cartography', *Imago Mundi* 48 (1996), pp.191–97; Denis Wood, *The Power of Maps* (London, 1992).

94. Andrews, *Paper Landscape*, p.24.
95. Edney, 'Reconsidering Enlightenment Geography and Map-making'.
96. Mary Hamer, 'Putting Ireland on the Map', *Textual Practice,* 3 (1989), p.198.
97. Edney, *Mapping an Empire*, p.41.
98. Ibid., p.41.
99. Hamer, 'Putting Ireland on the Map, p.188.
100. Johannes Fabian, 'Presence and Representation: The Other and Anthropological Writing', *Critical Inquiry,* 16 (1990), p.760.

CHAPTER TWO

The Habits of the People: The Emergence of Folklore and Ethnology

While the improved methods and techniques of mapping brought the topography of Ireland into ever-sharper focus, the refinement of social research produced a plethora of ethnographic impressions of its inhabitants. Stagl says that 'between ca. 1550 and 1800, social research (I also include under this term political and cultural research) became more systematic than it had ever been before, and its field of application was extended throughout the world. This prepared the emergence of the new socio-cultural sciences (ethnology, folklore, sociology, political science) in about the year 1800.'[1] This period in Ireland, from the time of Henry VIII to the Act of Union, from Hugh O'Neill to Robert Emmet, is marked by colonization and a contested colonialist discourse based to a considerable extent on ethnographic detail. The Ordnance Survey was conceived at a time when Britain had mastered the technologies and techniques of social, cultural and political surveillance, observation and description.

The *Edinburgh Review* stated in 1836 that there was general ignorance about the state of Ireland in official British circles, 'the advantages of remaining *terra incognita*, at least to English statesmen, Ireland has, till lately, possessed almost as fully as the interior of Africa'.[2] Although Crofton Croker reiterated this view in *Researches in the South of Ireland* there was an 'abundance of moralistic and politically charged portraits' of Ireland in the English press.[3] Cullen says that the growth of interest in the exotica of Irish rural life can be explained in terms of 'metropolitan interest in anthropological detail'.[4] The image of Ireland portrayed by political appointees, adventurers, authors, artists and travel writers 'coincided with contemporary anthropological ideas of savagery'. Africans on the Gambian River and American–Indians provided 'an index of comparison'.⁵ The Irish were viewed in terms of barbaric and exotic natives. Stocking says that 'for Englishmen at home and abroad, domestic class and overseas colonial society were linked by the "internal colonialism" of the Celtic Fringe. Thus Ireland, especially, had since Elizabethan times provided a mediating exemplar for both attitude and policy in relations with "savages" overseas'.⁶

Curiosity and inquisitiveness are innate human characteristics and as a result social research may well be as old as humanity itself. As Eriksen and Nielsen point out 'people have always been curious about their neighbours and more remote people. They have gossiped about them, fought them, married them and told stories about them. Some of these stories or myths have been written down . . . some stories were compared with others, about other peoples, leading to more general assumptions about "people elsewhere".'⁷ The Irish word for the English language, *béarla*, could also refer to the sounds of animals or birds. Prior to their modern equivalents anthropological and ethnological description can be found diffused throughout many different genres. In the survey literature of the English Elizabethans Ireland was 'transformed into a wandering discursive entity between Britain and the Americas'.⁸ The works of Gerald of Wales, the *Topography of Ireland* in particular, can be counted 'among the first "ethnographies" of the post-classical world'.⁹ Ireland, like the East, was described as an uncultivated place of plenty that needed colonization.¹⁰

Since the putative discovery of America and the peoples whom Columbus called Indians, believing that he had found the route to India, ideas of progress, evolution and civilization preoccupied philosophers across Europe. These ideas shaped the early ethnographic imagination and served to allegorize colonization itself. As Nigerian writer Achebe says: 'Let us imagine that someone has come along to take my land from me. We would not expect him to say he is doing it because of his greed, or because he is stronger than I . . . so he hires a storyteller with a lot of imagination to make up a more appropriate story which might say, for example, that the land in question could not be mine because I had shown no aptitude to cultivate it properly for maximum productivity and profitability.'¹¹

Over these three centuries natural scientists, cartographers, explorers and missionaries 'shipped out, Darwin-like . . . to classify and compare societies like plants and animals and to note how a single culture evolved from the savage to the civilized'.¹² In 1775 the English naturalist Gilbert White proposed a Pacific-type expedition to Ireland envisaging it as a new field where the 'manners of the wild natives, their superstitions, their prejudices, their sordid way of life' could be usefully observed.[13] Three years earlier Samuel Johnson had refused to admit the word civilization into his dictionary, a work consulted by O'Donovan in the field.[14] In 1779 Gabriel Beranger, a Huguenot artist born in Rotterdam who moved to Ireland in about 1750, travelled to the west of the country to make sketches of antiquities and monuments for William Burton Conyngham of the Hibernian Antiquarian Society. Beranger saw his own role as similar to that of the artists employed by Captain James Cook. He likened female islanders on Ennismurray to the women of Tahiti.[15] In the course of his work with the survey John O'Donovan wrote of what he styled the indolent lifestyle of the Ennismurray people. Driven by new economic imperatives and policies, indolence was an antonym of industry in the jargon of the new industrial bourgeois society.

The ongoing discussion of the idea of civilization was central. Still a neologism in the 1770s it often carried a nuance of cultural superiority. As the idea emerged more fully into English language discourse it led to the outlining of 'a universal history in which savagery led to barbarism, and barbarism to civilization'.[16] The meaning that emerged in the eighteenth century 'emphasized not so much a process as a state of social order and refinement, especially in conscious historical or cultural contrast with *barbarism*'.[17] For French and Scottish writers like Baron de Montesquieu or Adam Smith civilization was naturally determined by factors like heredity, climate, temperament or topography.[18] If a map could encompass a territory then ethnography could encompass its people. The word did not acquire any sense of plurality until the nineteenth century: there was only one civilization. The first half of this century produced a considerable body of British writing that was preoccupied with the progress of civilization. Stocking notes that 'the verb "civilize", the participle "civilized", and the noun "civility" had long been used to express a contrast between European and "savage" or "barbarous" manners and social life'.[19]

The relative degree of civilization also found physiological expression and could be mapped on the human body itself. Ethnologists and anthropologists mixed anthropology, ethnogeny, craniology, physiognomy and ancient history to produce typologies of the races of Britain. The phrenology of Franz Gall and Doctor J.C. Spurzheim was influential at the beginning of the nineteenth century. In his correspondence O'Donovan frequently interprets peoples' personalities on the basis of phrenology, the belief that different human faculties resided in different parts of the brain and were reflected in the contours of the skull. After examining the skull of renowned harpist Turlogh O'Carolan he attributed benevolence, reverence and gaiety to him. He also suggested that people might study their own skulls.[20] The authority of Englishness was underpinned by the assumption of the backwardness of others, as Hall says, 'it was the space between Ireland and England, between the West Indies and England, between the domestic and the colonial, which confirmed the right of Anglo-Saxons to rule'.[21]

Racialist discourse tended to be reserved for the colonies. Dr John Beddoe used the term Africanoid to refer to the people of Wales and Munster, contrasting them to the Anglo-Saxon. This placed the Irish on 'the lower limbs of the tree of human civilization' along with the Negroes, Chinese, Indians and other non-Caucasians.[22] The Irish were frequently depicted as bestial and simian, being called Calaban, Frankenstein, Yahoo or gorilla. Anglo-Saxon racialism contrasted the self-reliant self-controlled liberty-loving Saxon with the impulsive, childish Celt. James Cowley Prichard, the Bristol physician and pioneer of British ethnology, drew on a Celtic analogy to defend Africans from the charge of mental inferiority. He noted that the projecting mouths, advanced cheekbones and depressed noses of the Celt were evidence of their barbarism.[23] The exploitative nature of such quasi-scientific argument is evident in the racialist physical anthropology of the early nineteenth century. As Richards states:

> The sciences of Empire all become pseudo-sciences in their imperial fields of application. All knowledge becomes marginal at the margins. The methods, means, and procedures for constructing a positive knowledge of the world at its European centre become the material for constructing a mythology of the world at its colonial periphery. This process can often be pushed so far that knowledge lapses completely into myth, as when the means of a traditional physiognomy became the tools of a racist phrenology.[24]

Ethnographic or racial information provided the moral justification for the establishment of colonial superiority and 'cultural phenomena were readily translatable into "racial" tendencies'.[25] Contemporary Irish novelist Roddy Doyle plays with this recurrent trope in his novel *The Commitments* where one of his characters ecstatically exclaims 'say it loud I'm black and I'm proud'.[26]

The alliance of science and surveillance is exemplified by the early relationship between geography and ethnology. The Geographical and Philological Societies in Britain discussed topics that would be regarded as anthropological or ethnological today. The 1830s saw an increase in scientific and scholarly societies.[27] Many members of the Ethnological Society of London were also fellows of the Royal Geographical Society.[28] Prior to the Secret Service 'an unofficial network of Victorian learned societies acted as a central clearing-house for British imperial intelligence'.[29] The British Museum, the Royal Geographic Society and the Royal Asiatic Society were participants in the collection of classified knowledge. Technologies of surveillance were derived from 'the geographic, demographic, and ethnographic practices' of the learned societies engaged in producing and classifying a comprehensive knowledge of the Empire.[30] Geography was 'the queen of all imperial sciences in the nineteenth century' and was 'inseparable from the domain of official and unofficial state knowledge'.[31] It was not sufficient on its own and was 'accompanied by the imperatives of state ethnography, which territorialize a domain not only by mapping it but by producing all manner of "thick" description about it. The survey is only one form of ethnographic surveillance, and it tends to produce a homogeneous, geometric space rather than an accidental heterogeneous space.'[32] These disciplines and discourses have diverged and reconverged in interesting ways since the sixteenth century.

The study of man has been called *anthropologia* or anthropology since 1501. The name was coined in Germany and was restricted to that region until the eighteenth century. Interest in the physiology of humans combined with an interest in the universal progress of humankind in general introduced the new science to Western Europe. The eighteenth century saw the emergence of new disciplinal names. Some of these were based on the idea of man or *ánthropos* and others derived from people or *éthnos*. It appears that anthropology had a recognizable identity in the English-speaking world by 1805. The *Edinburgh Review* published the word that year to denote the occupant of a specialized role, 'its intellectual roots drew upon classics, biblical studies, and philosophy, but it is best appreciated as a type of natural history, one species of a class of knowledge – also including geology, botany, zoology, and geography'.[33]

The earliest recorded use of the word *ethnos* in Homer describes swarms of animals or warriors where their number, size, amorphous structure and unpredictable movement posed a threat. It described groups of like people whose location and conduct placed them beyond the threshold of social normality, 'aspects of naturality, of non-legitimate social organization, of disorganization, and of animality, are strong in *ethnos*'.[34] Its combination with the Latin *graphos*, from the Greek *graphein* to write, in the word ethnography suggests written description. Fabian says, however, that 'the element of description, writing about, had from the beginning a nominal slant, suggestive less of the activity of writing than of its products: descriptions, tableaux, in short, representations'.[35]

Ethnography, ethnology, *Vökerkunde* and *Völkskunde* were also coined in Germany between 1771 and 1783 and spread across Europe in the nineteenth century.[36] Suggesting knowledge of peoples, the terms appeared in geographic textbooks and travel reports. Johann Christoph Gatterer 'linked Geography closely with *Statistik*, the collection, classification and enumeration of empirical information on countries, peoples and constitutions, also called *Staatenkunde*'.[37] These theories and methods, the manuals, glossaries, handbooks, tableaux, the development of taxonomies, lists, questionnaires, descriptions and observations of diverse natural and cultural phenomena in the field, all shaped and informed the survey of Ireland.

It is useful to ask how these ideas manifest themselves in Irish scholarly discourse. Was there no indigenous concept of learning or folklore or antiquarianism? That O'Curry and O'Donovan 'were more familiar with modern methods of information gathering and analysis than other Gaelic Irish scholars' is at least partly due to the influence of the survey.[38] Douglas Hyde wrote in 1910 that Ireland had 'no folklorist who could compare with a man such as Iain Campbell of Islay' in Scotland. It could be said that Hyde was overlooking the work of O'Donovan, O'Curry and his colleagues, as did Dorson.[39] Campbell's position as secretary to several government commissions made his lot in life more secure and his varied interests in geology, meterology, art, sports, travel, invention, writing and languages enjoyed a free rein.[40] This was not the case for scholars in Ireland.

The marginal, if not suspect, status of the Irish language and Irish scholars within the Empire at least partly explains the lacking that Hyde perceived. Access to education, understood as instruction in the English language, was limited by race, class and religion. From the last quarter of the sixteenth century, and throughout the seventeenth and eighteenth centuries, 'Irish Catholics who wished to provide a higher education for their sons usually sent them to universities and colleges in mainland Europe. Attending a university at home was not an option, since the university founded in Dublin in 1592 was firmly Protestant, indeed Puritan, in ethos.'[41] It wasn't until the years immediately following the survey that formal university education was available to the majority of the Irish people. The Colleges (Ireland) Act, a conciliation measure introduced by Robert Peel to appease Catholics, did not become law until 1845.[42] Smith argues that 'the transplanting of research institutions, including universities, from the imperial centres of Europe enabled local scientific interests to be organized and embedded in the colonial system'.[43] Her argument about the lack of formal training and hobbyist aspect of local scholars applies to Irish researchers at the beginning of the nineteenth century. The answer is complicated insofar as this was a meeting or negotiation of two different cultures.

In eighteenth and nineteenth-century Ireland there was a proliferation of learned and antiquarian societies such as the Royal Dublin Society, the Hibernian Antiquarian Society, the Royal Irish Academy and the Physico-Historical Society. Some were related to the survey through their membership. Edward O'Reilly was a member the Gaelic Society of Ireland. The Iberno-Celtic Society published his work and the Celtic Society, later the Ossianic Society, published the work of O'Donovan. Colonel Portlock of the Ordnance Survey was a member of the Cork Cuvierian Society for the Cultivation of the Sciences that was a forerunner of Queen's College Cork, presently University College Cork.[44] Georges Cuvier was a French comparative anatomist whose racialist theory of phrenology was influential in the early nineteenth century. Many of these societies were patronized by aristocrats, nobles, politicians and dignitaries.[45] Their antiquarian work contributed to the establishment of the Irish universities and formed 'a major basis of the more fashionable Celtic and literary revival' at the end of the nineteenth century.[46]

The voices of Irish scholarship are often muted by the English language historiography of disciplines and ideas. If it were otherwise that historiography would appear somewhat contradictory and illogical. It becomes almost *de rigueur* to overlay Irish language terms with specialist English significance. An overtly colonial figure like Charles Vallancey is enlisted as a forerunner of Irish ethnology and folkoristics while Keating, O'Donovan or O'Curry are marginal. Even their role within the survey itself is diminished by purblind accounts of their ability, not at all remarkable it must be said, to understand the Irish language. The genteel title of antiquarian is often reserved for the English researcher while the Irish, to all intents and purposes divested of intellectual agency and efficacy, are mere scribes, interpreters or translators. They appear as footmen to colonial Celticists and Orientalists with bigger intellectual fish to fry. As many of these were simply unable to read Irish language manuscripts, translation was essential to their work.[47] In an epoch-making change of order the English language was institutionalized as the medium for the interpretation of indigenous culture.

Many of the Irish employed to facilitate the demand for translation appear as socially, racially, culturally and institutionally peripheral shadows. Maurice O'Gorman, just one of the ghettoized Gaelic minions, leaves the pages of English language historiography of antiquarianism as just another Irish drunk, his story untold. In 1791, just three decades before the commencement of the survey, politician Henry Flood bequeathed an endowment in his will for the establishment of an Irish language Chair in Trinity College Dublin. This was contested under 'the sixteenth-century "Act for English order, habite and language" (28 Henry VIII c.15), which equated loyalty with Anglicisation'.[48] Irish language and culture at the time was still 'linked with sedition'.[49] Indigenous discourses of knowledge existed, including systematized collection and classification, but perhaps without the modern powers or practices of political or governmental control.

The modern Irish language term for folklore *béaloideas*, old Irish *bélaiteas*, appears in a dictionary published in Paris in 1732.[50] Even if this year marked its coinage it preceded Gilbert White's proposed expedition in search of manners and superstitions by almost half a century. In this dictionary it is given as a close equivalent of the English word tradition. This term, like its popular synonym *seanchas*, evolves from an Irish discourse of cultural knowledge although both were subterrestrial to a largely anglophilic survey. The old Irish word *coimgne*, or *comheagna* as it would appear in the modern language, 'was used a thousand years ago to mean ... a shared universe of cultural discourse'.[51] This has been translated variously as great or historical knowledge or antiquarian lore.[52] In the survey, which all but denies the existence of alternative cognitive or affective achievement, such terms are displaced and denied correlation or equivalence. They are regarded as anterior to an ascendant English civilization and as such are the antonym or opposite of culture.

The Irish scholar Geoffrey Keating used *béaloideas* in the early seventeenth century in justifying his use of oral testimony to rebut some parts of the commentaries of English antiquaries and polemicists like Cambrensis and Stanhihurst.[53] He also wrote in Irish of *béaloideas cháich a coichinne* giving a strong sense of widespread general or common knowledge.[54] Ó Buachalla points out that Keating had, prior to O'Donovan, 'assimilated the various strata and components (mythology, hagiography, genealogy, folklore, chronology, topography) of traditional lore'.[55] The contribution of Keating as an early Irish ethnographer lies in his considered engagement with contemporary European discourse based to a significant extent on ethnographic detail, however contentious or contested that was. This engagement grew from a critical consciousness of the provenance and precedence of varied sources and the vernacularizing of hybrid knowledge traditions as well as identities in Ireland. In their own contestation of, and collaboration with, the imperial endeavours of the survey, O'Donovan and O'Curry blended such vernacular strata of knowledge accumulation in a modern idiom contributing ultimately to the recasting of such terms in contemporary Irish ethnographic tradition. They mixed them with elements of English antiquarianism and emergent European methods and theories of knowledge like statistics, folkloristics and ethnography.

Nation Description and Natural History

Statistics derives from the word statist, meaning politician or statesman. The word statistics entered the English language in 1770 and originally suggested the political arrangements of modern states. Sir John Sinclair, a student of Adam Smith, coined the word statistical in the English language and began a 'revolution in the gathering and aggregation of information'.[56] Like the other Scottish and French natural historians of progress Smith shared the assumption that 'in the absence of traditional historical evidence, the earlier phases of civilization could be reconstructed using data derived from the observation of peoples still living in earlier "stages" of development'. In *The Wealth of Nations* he distinguished hunters, shepherds and husbandmen from 'opulent and civilized nations'.[57]

Statistics was not confined solely to the enumeration of phenomena, 'in addition to topography, natural history and antiquities, taxation, local customs, diet and general living conditions were all regarded as important subjects for investigation. Reports which encompassed all these topics were described as "statistical" accounts.' By the 1820s it had come to be associated in Britain, France and Belgium with the numerical representation of social phenomena, a kind of political arithmetic. James Anthony Lawton, one of the founders of the Dublin Statistical Society defined statistics as 'the contents of all the blue books which are issued to both houses of Parliament' or 'the collecting of facts which relate to man's social condition'.[59]

Napoleon Bonaparte's 'Bureau de Statistique' in France, supervised by Joseph Marie de Gérando, carried out a census to estimate manpower needs during the Napoleonic Wars. In 1833 the British Association for the Advancement of Science established a statistical section. The same year saw the establishment of the Manchester Statistical Society. The London Statistical Society was founded in 1834 and by the 1840s similar societies emerged in many major urban centres across Britain.[60] Customs and beliefs were incorporated into official calendars and were elicited as part of cartographic projects, statistical or ethnographic surveys or censuses of populations. Renowned pioneer of folklore Jacob Grimm issued a circular in 1815 that contained an explanation of folklore and suggested ways to collect it.[61]

In England the county agricultural surveys were carried out in the eighteenth century while Sinclair's controversial *Statistical Account of Scotland* (1791–99), with reports on provincial customs, reported on 938 Scottish parishes. Between 1786 and 1787 he made a Grand Tour of Northern Europe, including France, Scandanavia and Russia.[62] Using a plan or questionnaire statistical 'missionaries' inquired after 'antiquities and records; characteristics of the people'.[63] In Ireland the Civil and Down Surveys of the eighteenth century resulted from the plantation and confiscation of land. William Petty, doctor, public servant, surveyor, geographer, statistician and political economist believed that the end of politics was to 'preserve the subject in peace and plenty'.[64] He was a founder member of the Royal Society. It has been suggested that the Down Survey was Colby's model in developing the British Ordnance Survey of Ireland.[65] The *New Statistical Account of Scotland* (1834–45) had also impressed Thomas Larcom, the director of the survey in Ireland from 1828.[66] In the forty-five years since the Union there was an explosion of ethnographic information and '114 commissions and 61 special committees were instructed to report on the state of Ireland'.[67]

In this formative period contemporary social, political and cultural elements are combined in the survey of Ireland. Folklore or antiquarianism formed one piece in a digest of regulation and governance. Customs and manners were but one expression of the populace, crime, sexuality, poverty and disorder were others. Day mentions Larcom's own library:

> They were books which indicate an interest in the contemporary mode of statistical enquiry then engaging interested individuals and groups like the statistical societies of Manchester and London (founded in 1833 and 1834 respectively), and include works by Sadler and Malthus, the Belgian statistician Lambert Quetelet's *Essai de Physique Sociale*, criminological and other surveys, as well as ethnographic and etymological works.[68]

These works influenced and shaped the interests of the survey. Thomas Robert Malthus, author of *Essay on Population* (First Essay 1798; Second Essay 1803), argued that mankind was endangered by the growth of population and that disease, war and famine would keep the population at near starvation level. Michael Thomas Sadler, author of *Ireland: its Evils and their Remedies* in 1828, was a politician and philanthropic businessman. He introduced poor law legislation for Ireland although it was not passed until 1838. The Belgian astronomer Adolphe L. Quetelet designed the Belgian census in the 1820s, he believed in what he called the laws of social physics and 'perceived correlationships between the physical attributes of a people and, for instance, rates of crime, marriage, suicide and disorder'.[69] Along with Malthus, Richard Jones, Charles Babbage, W.H. Sykes and John Elliot Drinkwater, he was a founder of the statistical section of the British Association for the Advancement of Science.[70] Babbage and Adam Sedwick reported to this section on the work of the survey in Ireland.[71] The application of such theory outside the heart of the empire could serve to classify culture as pathological behaviour. In the parlance of the time it becomes an array of habits, uses, conditions or customs symptomatic of a barbarous state of civilization.

The blue books, the royal commissions that employed question and answer techniques, were well established in the eighteenth century. The notion of a scientific collection of data based on personal testimony was already widespread. Colby's blue book directed that each officer was to enter in a journal 'all the facts he could obtain about communications, manufactures, geology, antiquities'.[72] Such a methodology was a routine part of the repertoire of official state and learned investigatory techniques. The county surveys of the Royal Dublin Society, numbering twenty-three in all, had suggestions of inquiry appended covering topics such as habitation, food, clothing, education, habits of industry, historical events, ancient buildings and the use of the English language, 'whether general or how far increasing'.[73] William Shaw Mason's *Parochial Surveys* can be traced back to the Physico Historical Society of 1714. Samuel Lewis's *Topographical Dictionary* of 1837 and Charles Smith's eighteenth-century county histories (Cork, Waterford and Kerry) are others. The censuses of 1813, 1821, 1831, 1841 and 1851 also reveal the increased use of questionnaires. Enumerators of the 1821 census 'required answers to a standard set of questions set out in prepared note books'.[74]

The *First Report of his Majesty's Commissioner for Inquiring into the Condition of the Poorer Classes in Ireland of 1835* is particularly interesting. Ó Cíosáin says that 'the nineteenth century was the first great age of social inquiry by the state, and Ireland was no exception. It was the subject of innumerable investigations throughout this period, due to a large extent to the growing economic and social crisis in rural Ireland after about 1820.'[75] This commission, chaired by the Anglican Archbishop of Dublin Richard Whately, 'took lengthy testimony from hundreds of witnesses' and reported it 'as nearly as possible in the words of each witness'.[76] Shaw Mason spoke of the transformative role of such surveys.[77]

George Everest considered the British survey of Ireland as an exemplar. Colin Mackenzie's early nineteenth-century survey of Mysore in India inquired after 'peculiar customs of the natives'. Francis Buchanan's surveys of Mysore and Bengal gathered information on the 'conditions of the inhabitants', 'history' and 'antiquities'.[78] The increasingly efficient statecraft of social research in the nineteenth century began to hone in on the minutiae of social, cultural and economic life. In the English-speaking world the terms civilization, statistics, folklore, ethnology, anthropology and cartography emerge in the eighteenth and nineteenth centuries. *Völkskunde*, a term that may have influenced William Thoms' 1846 coinage of the English term folklore, first appears in the journal *Der Reisende* in the year 1782. Thoms' coinage came one year after the Colleges (Ireland) Act was passed in 1845. The pioneering *Kilkenny Archaeological Society* would mention the '*folk-lore of Ireland*' only seven years later in 1853 when the *Transactions of the Kilkenny*

Archaeological Society was first published containing its own folklore questionnaire.[79] In the 1820s, while the survey plotted its course across Ireland, *ethnos*-terms appeared across Europe, *národpis* (ethnography) in Czech, *ethnográphiai* in Hungarian, *ethnologie* in French Switzerland and *volkenkunde* in the Netherlands. Ethnography emerged in the context of knowledge that was 'acquired "on the road" ... and provided a classification of knowledge that made it transferable and exchangeable from one context to the other'.[80]

Travel became associated with the discipline of statistics and the word *ethnographie* was first mentioned in this context as early as 1771. The term spread across Europe popping up in different languages at different times. By the 1780s it was commonly used by German scholars.[81] The Slovak historian Adam Frantisek Kollár first used the term ethnology (*ethnologia*) in 1783. It appeared four years later, apparently independently of each other, in the works of Johann Ernst Fabri and Alexandre César de Chevannes.[82] Pels and Salemink date its appearance in the English language to 1834 with the meaning of nation-description.

By the time the survey of Ireland was complete the disciplines of anthropology and folklore were establishing themselves. W. F. M. Edwards, an Englishman living in Paris, founded the *Société Ethnologique de Paris* in 1839 and the American Ethnological Society was established in 1842.[83] The *Ethnological Society of London* emerged in 1842 out of the *Aborigines Protection Society* of 1838 and later became the *Anthropological Society of London* in 1863. Both merged into the *Anthropological Institute of Great Britain and Ireland* in 1870. The British view of ethnology is particularly relevant to the Ordnance Survey. Pels says that this 'emerges from moral concerns and the belief that the 'truth' or 'facts' as provided by science would convert the ignorant to the moral necessity of change ... truth – scientific fact – was thought to be both politically neutral and morally compelling. Truth was itself moral: fact was value.'[85]

The term ethnology appeared in French around 1819 when the Napoleonic Wars had come to an end. Fabian gives the appearance of *ethnographie* in French as 1823 and *ethnologie* as 1834. In any case it was increasingly suggestive of descriptive details, expeditions, travel and 'the taxonomic organization of knowledge derived from Ramism and summarized by the term "natural history"'.[87] French philosopher Petrus Ramus or Pierre de la Ramée developed a method for the arrangement of useful knowledge and investigation by means of standard questions.⁸⁸ In the nineteenth century ethnography was understood more or less as:

> The collection of 'manners and customs', an activity for which the current questionnaires provided the models, even when the taxonomy of the questionnaire had now been transposed to an evolutionary taxonomy of 'stages' in the development of mankind. Still, ethnographic knowledge took the form of bits and pieces of knowledge that, by being classified in a questionnaire, could be transferred to another realm of thought.⁸⁹

Prichard was influential in the introduction of the terms ethnology and ethnography into the English language in the 1830s. At the start of the nineteenth century he published influential works such as the 1813 *Researches into the Physical History of Man*, the 1819 *An Analysis of the Egyptian Mythology* and the 1831 *The Eastern Origins of the Celtic Nations*. He was a student of moral philosopher Dugald Stewart in Edinburgh who had lectured on the origins of the native Americans.⁹⁰ In colony-free Germany the terms referred to all peoples but around 1800, the year of the Act of Union and some twenty years before the British survey of Ireland, ethnology in England came to mean the study of dark-skinned, non-European, uncivilized peoples. It became under conditions of colonialism the study of the 'races of men'.⁹¹

Thoms, a clerk in the British House of Lords, coined the English word folklore to replace popular antiquities. By the turn of the nineteenth century the English word supplanted the many other names that had been preferred at first in continental Europe. Many were based on the Greek term *demos* or people, which took the place of folk. In France *démologie*, *démopsychologie* or *anthropopsychologie* were used until the 1880s. Similarly in Spain *demología*, *demosofía* and *demotecnografía* were used. In Portugal this became *demotica* while in Italy *demologia* and *scienza demica* or popular science occurred.⁹² For Thoms and other English antiquaries folklore was akin to contemporary archaeology and was found in ancient derelictions on the land as well as in ancient beliefs and practices. It was imagined as a scattered collection of noble or sacred survivals. On the maps of the survey Ireland would appear as a museum, a repository of an ancient culture and language belonging in the past. André Varagnac considered folklore as *archéocivilisation* by which he meant the popular imitation of aristocratic modes representative of more or less ancient traits or *civilisation traditionnelle*.[93] This meaning persists in both popular and academic discourses in the present.

The survey's treatment of oral tradition and its cataloguing of antiquities had a long pedigree in Britain. Antiquities included oral tradition and William Camden included in his *Britannia* a section on the 'manners and customs of the ancient Irish'.[94] Camden went on walking tours observing topography and interviewing residents. He developed an interest in the ongoing debate about the usefulness of oral testimony to historians. Furthering the work of George Buchanan, Edward Lhuyd published *Archaeologia Britannica* in 1707 and, through the use of literary sources, he established that Irish, Welsh and Gaulish belonged to the Celtic family of languages.[95] Throughout the seventeenth and eighteenth centuries antiquarian research grew in popularity and recognition in England where 'the Roman fort, the Saxon Church, the Celtic burial mound proved alluring to a new genus of historical antiquary, whose researches took the form of walking tours during which he surveyed the local trophies – urn, coin, barrow, monument – and added his notes and sketches on these artefacts to his collection of manuscripts and monkish chronicles'.[96]

By the eighteenth century the antiquary had attained a degree of social and national status in Britain. Antiquarian research was seen as important nationally, Abrahams says, 'folklore . . . was pursued in the service of the British Union, and by extension, of the imperial crown'.[97] Specialized expertise was a hallmark of respectability and the words expert and technician themselves enter the English language in 1825 and 1833 respectively.[98] Within an English-dominated British tradition of antiquarianism Irish scholars, although experts in their own field, remained déclassé poetasters. In Ireland the Anglo-Irish were in the ascendancy, Sylvester O'Halloran published his *Ierne Defended* and *General History of Ireland* in 1774. Charlotte Brooke's *Reliques of Ancient Irish Poetry* appeared in 1789 and Vallancey edited the *Collectanea de rebus Hibernicis* between the years of 1770 and 1804. In England Reverend Joseph Brand published *Popular Antiquities* in 1774 revamping the 1725 work of Reverend Henry Bourne titled *Antiquitates Vulgares*. This was a compilation of calendar customs, beliefs, ceremonies, superstitions, witchcraft, charms, ghosts and life cycle rituals.[99]

After the Act of Union the interest of the Protestant elite in antiquities declined and the Royal Irish Academy experienced a slack period until the election to its membership of Sir William Betham in

1827 and Dr George Petrie in 1828.[100] In the opening decades of the nineteenth century a new era of Irish learning began as scholars began to turn away from the distortions of Betham and Edward Ledwich and aspired to a more objective methodology.[101] The emerging scholarship attempted to redress the inaccuracies, imbalances and racialist opinions of many of the English antiquarians. The varying interpretations of folklore itself could be expressed as a constant and shifting recasting of social, cultural and racial relations. Abrahams says:

> The European nationalist agenda called for the maintenance of the social messages contained within a class structure that gave country people a strange social placement as exotic and marginal *indigenes*. As with ruins or manuscripts, their lives served as a palimpsest through which the past still might be dimly observed . . . because of their status as native figures representing the earlier lifeways of indigenous peoples, these country dwellers filled the role of native-exotics.[102]

Natural history, exploration, mapping, statistics, antiquarianism, surveying and travel developed shared theories, techniques and technologies that were utilized by the empire as well as within burgeoning nation-states.

The Art of Travel

Travel stimulates curiosity just as curiosity stimulates questions, and the *ars apodemica* or art of travel led to the creation of the encyclopaedic manual of travel. This in turn gave rise to other strictly anthropological manuals such as De Gérando's *Considérations sur les méthodes à suivre dans l'observation des peuples sauvages* of 1800. This is regarded as 'the first methodology for anthropological field work employing an encyclopaedic schema of description'.[103] This was followed by the *Académie Celtique*, of which Constantin-François Volney was a pillar. It distributed many questionnaires published in its *Mémoires*. They referred to the manners, customs, dialects, antiquities and social conditions of the regions of France. The 1820s and 1830s in Ireland were marked by a curiosity about selective vestigial cultural elements by antiquaries of different religious and political persuasions, 'eighteenth century Gothicism and MacPherson's popular, though fake, Ossianic poems created a new interest in Britain's Celtic fringe'.[104] In a sense reviving the earlier eastern comparisons of Gerald of Wales the curiosity of Celticism coloured the debate about civilization at the time. Irish Orientalism 'stems from ideas about the nature of civilization and conceptions of Oriental and Celtic cultures which emerged during the English enlightenment and developed during Britain's period of imperial growth'.[105] On the map Ireland is west of Britain but in the nineteenth-century English imagination it was east.

Celtomania was spreading across Europe enthralling Bonaparte himself, as well as surveyors like Thomas Mitchell. Surveying in Australia Mitchell 'found "much sentimental satisfaction" from reading "the melancholy bard", Ossian, while exploring the banks of the River Murray'.[106] Likewise Sinclair collected at least one Ossianic lay while conducting his *Statistical Account* of Scotland.[107] Celticism served to define Ireland and the Irish in partly exotic and partly romantic terms. Although they inspired creative and oppositional expressions of identity the Siamese twin enthusiasms of Indomania and Celtomania were not always innocent avocations, they were profoundly colonial at times. O'Halloran points out that we see in both cases 'the manipulation of the subject culture and heritage by officers of the imperial power for purely domestic purposes'.[108]

The survey was not *sui generis*, it coincides with an intensified production of English language representations of Ireland. One year after Larcom issued his heads of inquiry, James Hardiman published *Irish Minstrelsy*. Hardiman's work received a hostile reaction in the *Dublin University Magazine* for what was perceived as its nationalistic agenda. Hardiman, a friend of O'Donovan, wrote significantly in his introduction of the mouldering membranes of Irish culture. Referring to Sir William Jones' work on Persia he made telling comparisons with the Orient. Jones was president of the Asiatic Society, a philologist and jurisprudence expert employed by the British East India Company. He was responsible for the introduction of textualized India to Europe. In the nineteenth century his translations were widely read by people such as poet Johann Wilhelm Goethe and influential German philosopher and critic Johann Gottfried von Herder. He was the first to posit the idea of an Indo-European language family and did so 'as part of an ethnological rather than philological project'.[109]

Something of the origins of ethnography and folkloristics lie in the encyclopaedic stocktaking of cultures viewed as exotic or the cultures of marginal groups at home. In the first half of the nineteenth century 'many private persons influenced by the romantic movement collected popular customs, superstitions, songs, tales, dialects, dances and other "antiquities" ... they did this by methodically questioning local experts, such as country squires, parsons, teachers or doctors, or by observing the people themselves on the spot'.[110] Since the eighteenth century traveller-explorers like William Rufus Chetwood (*A Tour Through Ireland* 1746), Richard Pococke (*Tour in Ireland* 1752) or John Bush (*Hibernia Curiosa* 1768) had been exploring Ireland.[111]

The various strands and branches of the emerging social research culminate in the *Notes and Queries in Anthropology, for the Use of Travellers and Residents in Uncivilized Lands* published by Thoms in 1874. This was published to record the folklore of the British Isles.[112] Stagl describes it as an 'immense parlour game among rural notables'.[113] It was a lineal descendant of the great collections of questions for travellers of the late eighteenth century.[114] The private or professional travels developed into what Richards calls the data pilgrimage or state nomadology.[115] The early nineteenth-century survey of Ireland is a particularly telling example of this.

General Charles Vallancey was posted to Ireland as a military engineer around 1765, under the threat of an impending French invasion. He was, Ó Danachair says, the first to turn an attentive eye and ear to the sayings and doings of the common people.[116] He was, however, opposed to the education of the Irish and referred to the Irish language as the 'jargon yet spoken by the unlettered vulgar'.[117] Abandoning his early liberalism he worked in the interests of the Empire fortifying the ports of Cork and Dublin and suppressing agrarian unrest in Munster. He blamed eighteenth-century confusion surrounding Irish mythology on the mistranslation of the Irish eponym *Éire*. Revealing the strong Orientalist tendency within English antiquarianism he argued that it did not mean Ireland but rather Iran or Persia. Vallancey was a contemporary of General William Roy, soldier, surveyor, antiquary and field archaeologist, 'the leading protagonist of a national cartographic survey and the virtual founder of the Ordnance Survey'.[119] A great number of these advanced their social status through involvement in what Edney calls the genteel, polite and amateur avocation of science, 'the genteel scholars, antiquarians, and natural historians formed a Learned Empire centred on London, polite society, and the Royal Society in particular'.[120]

Many were military men, William Mudge of the Ordnance Survey, James Rennell the Surveyor General of Bengal, Nevil

Maskelyne the Astronomer Royal and George Everest, superintendent of the Great Trigonmetrical Survey of India.[121] Everest, who was later knighted and had the world's highest mountain named after him, visited Thomas Larcom in Dublin in 1829. Sir Thomas Mitchell, Surveyor-General of New South Wales, and John Oxley, his successor, discovered and named places in their paths.[122] Similarly Augustus Pitt-Rivers and Thomas H. Huxley, William Crooke and Richard C. Temple in the nineteenth century were colonial figures-cum-anthropologists and sought patronage from the colonial government.[123]

Colby had been second-in-command to Mudge in Britain and he 'inherited the custom that the senior officers of the department should be primarily men of science'.[124] Surveyors were 'conscious of their professional activity being the bedrock of development'.[125] Colby was against the employment of civilian surveyors and had the backing of Marquis Wellesley, the lord-lieutenant of the day, who stated that the survey 'cannot be executed by Irish engineers and Irish agents of any description. Neither science, nor skill, nor diligence, nor discipline, nor integrity, sufficient for such work can be found in Ireland.'[126] The scientific aims of the survey, conceived of as an important national and imperial endeavour, would be executed most efficiently by the soldiers of that culture, the imperial army.

Habits of the People

The social and cultural interests of the survey serve as a counterpoint to the accuracy, discipline and skill of imperial science. While it is possible to map and to measure any particular townland without any recourse to the more abstract idea of culture, it is not as simple to describe its inhabitants when they speak a different language or behave in a manner that is unfamiliar. In a thirty-seven page pamphlet outlining the general sections of the memoirs Larcom produced the 'heads of inquiry' used by the survey. The most quoted passage of these headings directed surveyors to inquire about the 'habits of the people'. Although relatively neglected until now this is one of the crucial and central aspects of the survey. The word habit has interesting historical and contemporary meanings. It is derived from the Latin *habitus* or *habēre* meaning to have, to be, or to be constituted. In the nineteenth century it emerged as a catchphrase for the social, psychological and moral make up of individuals or groups.

Its senses extended from the ethnographic to the zoological describing the demeanour, disposition, character and behaviour of people and even insects.[127] In this sense a habit is a hallmark of civilization.

A habit was viewed almost in terms of a racial instinct. The habitual behaviour of a group in a particular environment was understood as contributing to their hereditary cultural and physical makeup.[128] Elite interest in habits in the eighteenth and nineteenth centuries was generally determined by reformative tendencies, 'the war against habits revealed the hierarchical relationship between the enlightened, rational people and their inferiors who indulged in unjustifiable habits and rituals . . . within the frame work of the great modern project, habits were seen as reprehensible and irrational . . . habits were viewed as "non-culture"'.[129] It was this perception that enabled Gilbert White to speak plainly about the 'sordid' ways of the Irish in the eighteenth century just as contemporary ideologues condemn the smoking, bonfires or fireworks of particular groups over others.

The colonial or imperial interest in exotics beyond national boundaries reflected practices in regard to urban or rural under classes within them. The homely vernacularity was, however, contrasted with a more exotic, more irrational external one. Unlike the underclass at home the colonial or imperial subject constituted an absolute other. The Empire was 'a British project, but it was the English, who defined its codes of belonging, and Englishness which were under attack when codes of conduct were forgotten'.[130] While the questions are detailed, specific and carefully organized using the latest techniques and theories available for statistical and social inquiry, there is an implicit implication of the Irish as outsiders. Many explicitly inquire after the remarkable and the peculiar:

> Habits of the people. Note the general style of the cottages, as stone, mud, slated, glass windows, one story or two, number of rooms, comfort and cleanliness. Food; fuel; dress; longevity; usual number in a family; early marriages; any remarkable instances on either of those heads? What are their amusements and recreations? Patrons and patrons' days; and traditions respecting them? What local customs prevail, as Beal tinne, or fire on St John's Eve? Driving the cattle through fire, and through water? Peculiar games? Any legendary tales or poems recited around the fireside? Any ancient music, as clan marches or funeral cries? They differ in different districts, collect them if you can. Any peculiarity of costume? Nothing more indicates the state of civilization and intercourse.[131]

It could be pointed out, as Fabian does in the reports and travelogues of similar contemporary scientific travellers in Africa, 'how little *déjà vu*, recognition as remembrance, was expressed when explorers made notes on the countenances, speech, daily work, seasonal worries, and so forth of African peasants'.[132] Everything appears utterly strange, absolutely other to the observer. The application of the questions and the criteria that created them served to furnish evidence of racial difference. It did not echo or replicate homely similarities, home remained a lonely Shangri-la of Empire. The existence of an equivalent popular culture within the empire was not referred to. As Hall says, 'colonial subjects were, and were not, the same as those of the metropole... colonial heathens were, and were not, more in need of civilisation than heathens at home. And there was no certainty that England was civilised.'[133]

Many of those employed in the survey of Ireland had grown up in an England without railways or police, the social structure of the English countryside had changed much less than expected. In 1801, two thirds of the population still lived in 'face-to-face village communities' and by 1851, at the other end of the Industrial Revolution, the largest sector of the work force was engaged in agriculture.[134] Some thirty years subsequent to the survey Andrew Lang, James Anson Farrer and Sir Edward Burnet Taylor, 'the father of anthropology and godfather of the anthropological school of folklorists' were fusing savage customs, prehistoric antiquity and social evolution into a new ethno-folklore that drew analogies between exotic peoples and the English. The structure of the survey was not conducive to admitting any analogies with the English.[135] If the English surveyors were reminded of home, if the Irish shared any aspect of lifestyle with the English, nobody was admitting it. Nobody noticed if the Irish wake differed from the English wake or the English wake was similar to the Irish pattern. This would seem to suggest that it was not actually about collecting a precious or endangered Irish folklore that was very similar to English folklore but rather a selective reordering of knowledge.

Taylor's view that there was scarce 'a hand's breath of difference between an English ploughman and a Negro of central Africa' would have been unconscionable for the civilized and civilizing surveyor of Ireland.[136] Such cultural relativism is almost entirely absent from the writings of the surveyors. The racial distinction rather 'provided the basis for drawing lines as to who was inside and who was outside the nation or colony, who were subjects and who were citizens, what forms of cultural or political belonging were possible at any given time'.[137] Folklore itself, as it was understood in the life of the survey, was one such basis for drawing such lines between subjects and citizens, surveyors and aborigines, science and tradition, civilization and barbarity.

The Field of Folklore

One of the most popular ideas about the survey is that it was the saviour of Irish folklore. There is a relationship between the survey and folklore, the nature of the relationship, however, is more complex than this view would suggest. It is true that direct comparisons can be made, for example, between the methods of the survey and those of the Free State's Irish Folklore Commission almost a century later. If the survey mapped the edge of the Empire, the commission mapped the symbolic heartlands of the burgeoning nation-state on the edge of Ireland. The commission was a lineal descendent of the survey and a culmination of nineteenth-century advances in European ethnological theory and methodology.[138] The heads of inquiry foreshadow Seán Ó Súilleabháin's *Handbook of Irish Folklore*. The memoirs or letters correspond with the diaries of the professional folklore collectors. The civil assistants and officers of the Royal Engineers became the folklore collectors themselves. The Irish scholars could be compared with the local research assistants of contemp-orary practice. The early statistical–ethnographic questionnaires mirror the more recent use of questionnaires by folkorists. Ó Súilleabháin acknowledged the pioneering aspect of this method-ology when he noted that:

> One of the greatest sources of information we have in Ireland is the Ordnance Survey books, which were made about a century ago by three men, John O'Donovan, Eugene O'Curry, and George Petrie. They went around about a hundred years ago and took down all the placenames of the country and recorded material of very great importance. But the greatest importance lies in the diaries kept by these men, because these diaries give the atmosphere in which the work was done. Now we are doing the same thing and are asking our collectors to keep a diary.[139]

Séamas Ó Duilearga, the head of the commission, wrote in regard to county Clare in 1929, three years after the founding of the Folklore of Ireland Society and a century after the survey, that he was the first 'collector of the tradition of the Irish speakers of Thomond since O'Curry and O'Donovan wrote their letters from Corofin and Ennistymon to the Ordnance Survey almost a century before'.[140] Commenting on the hankering sense of loss that O'Curry expressed on his father's death, Ó Duilearga says he was 'inspired by the same feeling . . . to go to Clare to see what I could find there a century later'.[141]

The terms fieldwork and field-diary themselves are derived from military operations; to be in the field was to be engaged in military operations when the field was the field of battle. The sense of field-book and fieldwork appeared first in 1616 and 1767 as borrowings from the natural sciences and the lexicon of surveying. Chesney says, 'by 1650, many developments in natural history and the pract-ical sciences had their origins in the acquisition of information for military, political and economic ends, which gave impetus to the accumulation of data and particularly promoted surveying and mapping'.[142]

The field-naturalist or field-observer of the eighteenth and nineteenth centuries were given their contemporary anthropological meaning by Bronislaw Malinowski in 1922 and Margaret Mead in 1935. There are further grounds for comparison: the survey followed the Act of Union just as the Irish Folklore Commission followed the establishment of the Irish Free State. Both embodied contemporary political ideologies and were institutionalizations of these. In official discourse the concept of tradition or folklore itself is a classification or genre of knowledge. It is sometimes given with one hand and taken away with the other. It is noble or ignoble, up or down, good or bad depending on the reigning political regime. It could be said that the survey, which essentially equated folklore with non-civilization, emptied the land of tradition while the commission refilled it.

George Petrie urged O'Donovan to 'get as much of everything as you can, manners, customs, traditions, legends, songs, etc'.[143] Instructions to the writers of memoirs, often adapted to each parish, outline a clear ethnographic methodology:

> In order to check and verify statements, inquiry respecting the same object should be made both of interested and uninterested persons; the character of each referee and the means he possesses of acquiring information being always kept in view. In any abstract questions, it would be well to ascertain the feelings of the higher and lower orders, of the clergy and laity. In all inquiries it is important to record with

minuteness any striking facts, which may bear on or elucidate the subject in question, adding the names of all the parties concerned and the place and date of the transaction.[144]

Incorporating a range of ideas from Smith to Malthus, from Quetelet to Sadler and combining the German sense of statistics with British inflexions of antiquarianism and ethnology, it continues to inquire whether the schools that the gentry sent their children to were English or Irish, what the nature of the teachers was, if education caused crime, whether there was any instruction in the Irish language and if it encouraged English language instruction, 'whether any taste for literature, the fine arts and speculation science exists among the higher orders', what the spirit of the people is taking their race into account, if crime is hereditary or racial, does the society hide criminals, is prison beneficial, how many idiots or lunatics are at large and what was the relationship between sex, age and crime'.[145]

Such schematic inquiry reflects Stocking's idea of the metaphorical extendibility of the categories of nineteenth-century social and cultural evolutionism. These reached across different lines of 'domestic life (woman, child), of socioeconomic status (labourer, peasant, pauper), of deviancy (criminal, madman), and of "race" (Celtic Irishman, black savage) – they all stood in a subordinate hierarchical relationship to those who dominated the economic life, who shared political power, or who most actively articulated the cultural ideology of mid-Victorian Britain'.[146]

The reference to the spirit of the people bespeaks the influence of German intellectuals such as Johann Gottfried von Herder and the brothers Jacob and Wilhelm Grimm whose, 'discovery of the popular was part of a movement of cultural primitivism in which the ancient, the distant and the popular were all equated'.[147] Custom and tradition were central and fieldworkers were instructed to inquire:

> Whether there are any fairy men, medical old women or similar rural practitioners in the parish, or whether there is a resort from the parish to any such in the neighbourhood. Whether any charms are used in reference to refractory churns, hens etc. Whether any cures are effected, by prescriptions in old Irish manuscripts, by washing at holy wells or other such ceremonies. What means are there for curing horses or other diseased animals?[148]

Handbooks, manuals and glossaries on rural society and agricultural practices had replaced the earlier reports.[149] The portrait of manners and customs was a 'remarkably stable sub-genre, which turns up in a wide range of discursive contexts'.[150] Like the memoirs or letters of the survey, manners and customs discourse usually appeared as an appendage to a superordinate genre. This was later translated into the Irish language by the folklore commissioners as *nósanna agus piseoga*.[151] The purposeful accumulation of knowledge and the ordering of information leads inevitably to the questioning of the 'assumption that the accumulation of knowledge by one people about another constitutes a process in which objective truth is revealed through the disinterested quest of learning for its own sake'.[152]

The ancient customs and curious vestiges encountered by the memoir-writers were seen unambiguously as impediments to progress and not cultural expressions in themselves. This discourse amounted at times to the compilation of a typology of relative racial and cultural inferiority. The important point is that the survey, apart from producing maps, also produced deep description and discourse. This discourse grew from the careful recording and translation of a comprehensive database of knowledge and information. In order to make sense it was not simply translated into English, this was an obvious necessity as well as a desire, but also into the cultural universe of hegemonic imperial science. The purpose of this was, as Pratt says:

> To incorporate a particular reality into a series of interlocking information orders – aesthetic, geographic, mineralogical, botanical, agricultural, economic, ecological, ethnographic, and to make those information orders natural, to find them there uncommanded, rather than assert them as the products/producers of European knowledges or disciplines.

In the nineteenth century it was precisely information that sustained the fantasy of Empire. This information was overseen by 'a sort of extended Civil Service recruited from Britain's dominant classes':

> These people were painfully aware of the gaps in their knowledge and did their best to fill them in. The filler they liked best was information. From all over the globe the British collected information about the countries they were adding to their map. They surveyed and they mapped. They took censuses, produced statistics. They made vast lists of birds. Then they shoved the data they had collected into a shifting series of classifications . . . this paper shuffling, however, proved to have a great influence. It required keeping track,

and keeping track of keeping track. It required some kind of archive for it all. Unquestionably the British Empire was more productive of knowledge than any previous empire in history. The administrative core of the Empire was built around knowledge-producing institutions like the British Museum, the Royal Geographic Society, the India Survey.

As a specialized department of knowledge-acquisition the survey embodied the imperial 'project of a unified and systematic comprehensive knowledge'.[155] The process of gathering, classifying and ordering information in taxonomies became statecraft, the archive became 'the collectively imagined junction of all that was known or knowable, a fantastic representation of an epistemological master pattern, a virtual focal point for the heterogeneous local knowledge of metropolis and empire'.[156]

It could of course be argued that archives are largely irrelevant to everyday life. The archive was not just illusory, however, it could intervene to shape the political definition of actual territory. In the case of Tibet in the India Survey 'the archive functioned both to imagine territory as representation and realise it as social construction. What began as utopian fictions of knowledge . . . often ended as territory. This system of representational order – an order of social imagination so powerful that it could, in effect, construct social reality – was responsible for fashioning the modern idea of Tibet.'[157] The uncivilized or primitive Ireland of the survey was not a *de facto* entity, it was a social and cultural construction. This construction became an objectified intellectual space. Similar processes of objectification and cultural reformulation are noted in the nineteenth-century census of India. Through ongoing processes of inquisition, 'colonial peoples begun to think of themselves in different terms, not only are they changing the content of their culture, but the way they think about their culture has changed as well . . . what has been unconscious now to some extent becomes conscious. Aspects of the tradition can be selected, polished and reformulated for conscious ends.'[158]

This is similar to the early stage in the life of folklore that Honko calls the arrival of the external discoverers of folklore who begin to notice, recognize or emphasize aspects of community life and convert them into folklore, 'the user of the object or custom, or the singer of the song, learns to view it as part of a tradition determined by an interest that has come from outside but that has begun to penetrate the community'.[159] A process of definition or classification of tradition or, more accurately, neo-tradition was begun. This process, although not unchallenged, situated or sited the community intellectually and physically as rustic and folksy, 'the folklorist of the 19th century sought messages from antiquity in the archives of the human mind ... with his assistance the nation became aware of its own history, language and literature, and the rising social class – the bourgeoisie – gained social legitimization as the descendant of peasant society'.[160] It was envisaged, however, that the history, language and literature would be English.

The result of this encyclopaedism was partly the production of proof of the existence of a primitive superstitious people. Pels says that 'these forms of identification, registration, and discipline emerged in tension and in tandem with technologies of self-control that fostered notions of cleanliness, domesticity, ethnicity, and civilization'.[161] Although examined more fully in the following chapters a few examples at the conclusion of this one will serve to illustrate and introduce this point. In the Parish of Urney in Donegal the people were described as dirty, 'many stow their potatoes under their beds and the manure heap is generally at the threshold of the door'.[162] This is echoed in the description of the Parish of Templecarn where a general want of cleanliness was recorded.[163] At the time English colonial officials in Jamaica, another theatre of conflict faced by Britain, were writing that 'all sense of propriety or decency, all morality and all cleanliness are utterly wanting'.[164] It becomes apparent that it is precisely the observer, and the observer alone, who is moral, clean, sober and decent.

The pilgrimage to Lough Derg, very much a part of contemporary culture, featured as an ancient superstition with the re-enforcement that the water of the Lough was 'boiled and turned into wine by the miraculous power of the priests'.[165] Lieutenant W. Lancey, writing in response to the questionnaire of the North West Farming Society in the Parish of Killymard, presents a typical mixture of ethnographic detail and ethnic stereotype 'their cottages are stone, thatched, some whitewashed, a few slated, consisting of 2 rooms with small glass windows and as usual not over clean'. He continues to describe the festival of May, *Bealtaine* in the Irish language, 'Bealtinne is still kept up. They make and drink plenty of whiskey, with which they associate almost everything in life, whether good or evil . . . their food and dress are of the usual very common kind, but it is said they are improving in civilization.'[166]

Many accounts could be described as reworkings of prefabricated stereotypes that exoticize, shock and reassure in turn, the affect could be summarized as 'wow, oh no, its okay'. Preoccupied with contemporary bourgeois ideas of industry and education the descriptions suggest that the Irish, if encouraged and reformed, could well be incipient English citizens. In the Parish of Kilbarron, Robert Ball remarks that, 'the morals of the lower order are manifestly improved within the last 20 years'.[167] This improvement is not understood in absolute or philosophical terms but in terms of a convergence to hegemonic English standards of civilization. Ready and willing to be improved the people in the Parish of Raymoghy are said to be 'very well disposed to be industrious but they have no encouragement'.[168] It is the idea of industriousness itself, shared here by a particular writer and a particular reader within a particular culture, that is taken for granted. By their nature a lifestyle of seasonal agricultural work or fishing afforded periods of inactivity and winter entertainments that were perceived as idleness by hardnosed industrialists, bourgeois moralists and hard-pressed surveyors.

When the committee on survey and valuation proposed in 1824 a government valuation of land and buildings in every Irish townland, it may not have envisaged that it would develop beyond trigonometrical observations to cultural observations. In doing this it shifted its parameters far beyond what the limelight could discern. In its favour it did have the advantage of the latest theories and methods of the burgeoning social, cultural and natural sciences of anthropology, ethnology, statistics, geography, geology and botany. In this regard alone it is an interesting example of the institutionalization of contemporary ideas, theories and methods of research that developed from the sixteenth to the nineteenth centuries. The survey marks a significant stage in the eventual appearance of Irish ethnography and its emergence out of English and European practices of natural history, statistics, travel, nation description and antiquarianism into a professional practice. Poised gingerly between the colonialist discourse of barbarism and the nationalist repossession of tradition the survey heralded a new chapter in the multilayered story of Ireland. While it charted some new ground it also went over much of the old ground.

NOTES

1. Justin Stagl, *A History of Curiosity: The Theory of Travel 1550–1800* (London, 1995), p.2.
2. Angélique Day, '"Habits of the people": Traditional Life in Ireland 1830–1840, as Recorded in the Ordnance Survey Memoirs', *Ulster Folklife*, 30 (1984), pp.3–17.
3. Jennifer Schacker, *National Dreams: The Remaking of Fairy Tales in Nineteenth-century England* (Philadelphia, 2003), p.61.
4. Fintan Cullen, *Visual Politics: The Representation of Ireland 1750–1930* (Cork, 1997), p.49.
5. Roy Foster, *Modern Ireland 1600–1972* (London, 1988), p.32.
6. George W. Stocking Jr., *Victorian Anthropology* (New York, 1987), p.234.
7. Thomas Hylland Eriksen and Finn Sivert Nielsen, *A History of Anthropology* (London, 2001), p.1.
8. Andrew Hadfield and John McVeagh (eds), *Strangers to that Land: British Perceptions of Ireland from the Reformation to the Famine* (London, 1994), p.63.
9. Malcolm Chapman, *The Celts: The Construction of a Myth* (New York, 1992), p.187.
10. Clare Carroll, 'Barbarous Slaves and Civil Cannibals: Translating Civility in early Modern Ireland', in Clare Carroll and Patricia King (eds) *Ireland and Postcolonial Theory* (Cork, 2003), p.68.
11. Chinua Achebe, *Home and Exile* (Edinburgh, Canongate, 2001), p.60.
12. John Van Maanen, *Tales of the Field* (Chicago, 1988), p.15.
13. Clare O'Halloran, *Golden Ages and Barbarous Nations: Antiquarian Debate and Cultural Politics in Ireland 1750–1800* (Cork, 2004), p.98.
14. Raymond Williams, *Keywords: A Vocabulary of Culture and Society* (London, 1976), p.57.
15. O'Halloran, *Golden Ages*, p.97.
16. Adam Kuper, *Culture: The Anthropologist's Account* (Cambridge, MA, 1999), p.25.
17. Williams, *Keywords*, pp.57–8.
18. Stocking, *Victorian Anthropology*, pp.9–11, 14–20.
19. Ibid., p.11.
20. Rev Michael O'Flanagan (ed.), *Ordnance Survey Letters 1834–1841*, 43 volumes typeset and bound in Boole Library, University College Cork (Bray, 1927–35), Fermanagh, pp.44–7. John O'Donovan, Eugene O'Curry, George Petrie, Patrick O'Keefe and Thomas O'Connor wrote these in the course of their fieldwork for the survey. For further information see Archival Sources. Hereafter OSL.
21. Catherine Hall, *Civilising Subjects: Metropole and Colony in the English Imagination 1830–1867* (Cambridge, 2002), p.215.
22. Lewis Perry Curtis Jr., *Apes and Angels: The Irishman in Victorian Caricature* (Washington, 1997), p.21.
23. Stocking, *Victorian Anthropology*, p.63.
24. Thomas Richards, *The Imperial Archive: Knowledge and the Fantasy of Empire* (London, 1993), p.71.
25. Stocking, *Victorian Anthropology*, p.64.
26. Fintan O'Toole, 'Going Native' in *Black Hole Green Card: The Disappearance of Ireland* (Dublin, 1994), p.56.
27. Stocking, *Victorian Anthropology*, p.243.

28. David N. Livingstone, 'Climate's Moral Economy: Science, Race and Place in post-Darwinian British and American Geography', in Anne Godlewska and Neil Smith (eds), *Geography and Empire* (Oxford, 1994), p.138.
29. Richards, *Imperial Archive*, p.14.
30. Ibid., p.15.
31. Ibid., p.13.
32. Ibid., p.21.
33. Henrika Kuklick, *The Savage Within: The Social History of British Anthropology 1885–1945* (Cambridge, 1991), p.6.
34. Elizabeth Tonkin *et al.* (eds), *History and Ethnicity* (London, 1989), p.12.
35. Johannes Fabian, 'Presence and Representation: The Other and Anthropological Writing', *Critical Inquiry*, 16 (1990), p.758.
36. Diarmuid Ó Giolláin, *Locating Irish Folklore: Tradition, Modernity, Identity* (Cork, 2000), pp.32–62.
37. Stagl, *A History of Curiosity*, pp.234–5.
38. Damien Murray, Romanticism, *Nationalism and Irish Antiquarian Societies 1840–80* (Maynooth, 2000), p.23.
39. Richard M. Dorson, *The British Folklorists: A History* (London, 1968), p.433.
40. Ibid., p.395.
41. Bernadette Cunningham, *The World of Geoffrey Keating: History, Myth and Religion in Seventeenth-Century Ireland* (Dublin, 2004), p.25.
42. John A. Murphy, *The College: A History of Queen's University College Cork* (Cork, 1995), p.2.
43. Linda T. Smith, *Decolonising Methodologies: Research and Indigenous Peoples* (London, 1999), p.8.
44. Murphy, *The College*, p.9.
45. Gillian M. Doherty, *The Irish Ordnance Survey: History, Culture and Memory* (Dublin, 2004), p.36.
46. O'Halloran, *Golden Ages*, p.185.
47. Ibid., p.173.
48. Ibid., p.161.
49. Ibid.
50. Daithí Ó hÓgáin, '*Béaloideas* – Notes on the History of a Word', *Béaloideas*, 70 (2002), p.83.
51. Gearóid Ó Crualaoich, *The Book of the Cailleach: Stories of the Wise-woman Healer* (Cork, 2003), p.3.
52. *Dictionary of the Irish Language: Based mainly on Old and Middle Irish Materials* (Dublin, Royal Irish Academy 1983 Compact Edition), p.303.
53. Ó Giolláin, *Locating Irish Folklore*, p.48.
54. *Dictionary of the Irish Language*, p.65.
55. Breandán Ó Buachalla, 'Foreword', in *Foras feasa ar Éirinn: the History of Ireland by Geoffrey Keating*, (London, 1987), p.8.
56. Marika Vicziany, 'Imperialism, Botany and Statistics in early Nineteenth-century India: the Surveys of Francis Buchanan 1762–1829', *Modern Asian Studies*, 20, 4 (1986), p.649.
57. Stocking, *Victorian Anthropology*, pp.15, 32.
58. Vicziany, 'Imperialism, Botany and Statistics', pp.648.
59. Mary E. Daly, *The Spirit of Earnest Inquiry: The Statistical and Social Inquiry Society of Ireland 1847–1997* (Dublin, 1997), p.10.
60. Birmingham (1835), Bristol and Glasgow (1836), Liverpool, Leeds and Ulster (1838).
61. Jacob Grimm, 'Circular Concerning the Collecting of Folk Poetry', in Alan Dundes (ed.), *International Folkloristics: Classic Contributions by the Founders of Folklore* (Oxford, 1999), p.3.

62. Arthur Geddes, 'Scotland's "Statistical Accounts" of Parish, County and Nation: c. 1790–1825 and 1835–1845', *Scottish Studies*, 3, 1 (1959), p.20.
63. Ibid., p.18.
64. David Englander and Rosemary O'Day, *Retrieved Riches: Social Investigation in Britain 1840–1914* (Aldershot, 1998), p.4.
65. John H. Andrews, *A Paper Landscape: The Ordnance Survey in Nineteenth-century Ireland* (Oxford, 1975), p.87. Reprinted in Dublin by the Four Courts Press in 2001.
66. Doherty, *The Irish Ordnance Survey*, p.38.
67. John O'Connor, *The Workhouses of Ireland: The Fate of Ireland's Poor* (Dublin, 1995), p.48.
68. Day, "'Habits of the people'", p.23.
69. Englander and O'Day, *Retrieved Riches*, p.6.
70. Ibid.
71. Doherty, *Irish Ordnance Survey*, p.23.
72. Andrews, *Paper Landscape*, p.145.
73. John McEvoy, Statistical Survey of the County of Tyrone; with observations on the means of improvement. Drawn up in the years 1801, and 1802, for the consideration, and under the direction of the Dublin Society (Dublin, 1802).
74. Stephen A. Royle, 'Irish Manuscript Census Records: A Neglected Source of Information', Irish Geography, IXI (Vol 11) (1978), p.113.
75. Niall Ó Cíosáin, 'Introduction', in *Poverty before the Famine: County Clare 1835: First Report from his Majesty's Commissioners for Inquiring into the Condition of the Poorer Classes in Ireland* (Clare, 1996), p.iii.
76. Ibid., p.iv.
77. Andrews, *Paper Landscape*, p.17.
78. Mathew Edney, *Mapping an Empire: The Geographical Construction of British India, 1765–1843* (Chicago, 1997), p.45.
79. *Transactions of the Kilkenny Archaeological Society* (Dublin: John O'Daly, 1853), p.100.
80. Peter Pels and Oscar Salemink, 'Five Theses on Ethnography as Colonial Practice', in Peter Pels and Oscar Salemink (eds), *Colonial Ethnographies*, a special issue of *History and Anthropology*, 8, 1–4 (1994), p.6.
81. Ibid., p.7. See also G. de Rohan-Csermak, 'La Première Apparition du Terme "Ethnologie"', *Ethnologia Europaea*, 1, 3 (1967), p.170.
82. Stagl, *History of Curiosity*, p.236.
83. Åke Hultkrantz, *General Ethnological Concepts* (Copenhagen, 1960), p.114.
84. Talal Asad, *Anthropology and the Colonial Encounter* (London, 1973), p.80.
85. Peter Pels, 'Professions of Duplexity: A Prehistory of Ethical Codes in Anthropology', *Current Anthropology*, 40, 2 (1999), p.104.
86. Fabian, 'Presence and Representation', p.758.
87. Pels and Salemink, 'Five Theses', p.9.
88. Stagl, *History of Curiosity*, p.68.
89. Pels and Salemink, 'Five Theses', p.9.
90. Stocking, *Victorian Anthropology*, pp.31, 49.
91. Stagl, *History of Curiosity*, p.241.
92. Hultkrantz, *General Ethnological Concepts*, p.139.
93. Ibid., p.135.
94. Dorson, *British Folklorists*, p.3.
95. Murray, *Romanticism, Nationalism*, p.5.
96. Dorson, *British Folklorists*, p.2.
97. Roger D. Abrahams, 'Phantoms of Romantic Nationalism in Folkloristics', *Journal of American Folkore*, 106, 419 (1993), p.16.

98. Kuklick, *The Savage Within*, p.36.
99. Stocking, *Victorian Anthropology*, p.54.
100. Murray, *Romanticism, Nationalism*, pp.5–6.
101. Ibid., pp.8–9.
102. Abrahams, 'Phantoms of Romantic Nationalism', p.10.
103. Stagl, *History of Curiosity*, p.289.
104. Murray, *Romanticism, Nationalism*, p.4.
105. Joseph Lennon, 'Irish Orientalism: An Overview', in Clare Carroll and Patricia King (eds) *Ireland and Postcolonial Theory* (Cork, 2003), p.129.
106. Donald W.A. Baker, *The Civilised Surveyor: Thomas Mitchell and the Australian Aborigines* (Melbourne, 1997), p.5.
107. Geddes, 'Scotland's "statistical accounts"', p.20.
108. O'Halloran, *Golden Ages*, p.53.
109. Ibid., p.46. O'Halloran is developing an idea of Thomas R. Trautmann from *Aryans and British India* (London, Berkley, 1997).
110. Stagl, *History of Curiosity*, p.293.
111. John W. Foster, 'Encountering Traditions', in John W. Foster (ed.), *Nature in Ireland: A Scientific and Cultural History* (Dublin, 1997), p.62.
112. van Maanen, Tales of the Field, p.15.
113. Stagl, *History of Curiosity*, p.294.
114. Ibid., p.295.
115. Richards, *Imperial Archive*, pp.19–20.
116. Caoimhín Ó Danachair, 'The Progress of Irish Ethnology 1783–1982', *Ulster Folklife*, 29 (1983), pp.3–17.
117. O'Halloran, *Golden Ages*, p.45.
118. Ibid., pp.45–7, 181.
119. C.W. Phillips, Archaeology in the Ordnance Survey 1791–1965 (London, 1980), p.2.
120. Mathew Edney, 'Mathematical Cosmography and the Social Ideology of British Cartography 1780–1820', *Imago Mundi*, 46 (1994), p.109.
121. Ibid., p.110.
122. John Atchison, 'Eton Vale to Bamaga – Place, Geographical Names and Queensland', *Queensland Geographical Journal*, 5 (1990), pp.1–27. "A very significant proportion of Queensland localities are named after colonial worthies and, perhaps, not so worthy".
123. Kuklick, *The Savage Within*.
124. Andrews, *Paper Landscape*, p.20.
125. Atchison, 'Eton Vale to Bamaga', p.10.
126. Andrews, *Paper Landscape*, p.21.
127. The Oxford English Dictionary, second edition prepared by J.A. Simpson and E.S.C. Weiner (Oxford: Oxford University Press, 1989), p.993.
128. Stocking, Victorian Anthropology, p.64.
129. Jonas Frykman and Orvar Löfgren (eds), *Force of Habit: Exploring Everyday Culture* (Lund, 1996), pp.8–9.
130. Hall, *Civilising Subjects*, p.25.
131. Mary Hamer, 'Putting Ireland on the Map', *Textual Practice*, 3 (1989), pp.184–201.
132. Johannes Fabian, 'Remembering the Other: Knowledge and Recognition in the Exploration of Central Africa', *Critical Inquiry*, 26 (1999), p.55.
133. Hall, *Civilising Subjects*, p.22.
134. Stocking, *Victorian Anthropology*, pp.5, 209.
135. Dorson, *British Folklorists*, p.199.
136. Ibid., p.194.

137. Hall, *Civilising Subjects*, p.20.
138. Ó Giolláin, *Locating Irish Folklore*.
139. Stith Thompson (ed.), *Four Symposia on Folklore* (Westport, 1953), p.6.
140. Séamus Ó Duilearga, 'Notes on the Oral Tradition of Thomond', *The Journal of the Royal Society of Antiquarians of Ireland*, 95 (1965), pp.134–5.
141. Ibid., p.135.
142. Helena C.G. Chesney, 'Enlightenment and Education', in John W. Foster (ed.), *Nature in Ireland: A Scientific and Cultural History* (Dublin, 1997), p.368.
143. OSL, Roscommon, Vol.2, p.127.
144. Angélique Day and Patrick McWilliams (eds), *Ordnance Survey Memoirs of Ireland* (Belfast, 1990–98), forty volumes, Vol.34, p.84. Hereafter abbreviated OSM.
145. Ibid., 34, p.84.
146. Stocking, *Victorian Anthropology*, p.231.
147. Peter Burke, *Popular Culture in Early Modern Europe* (Aldershot, 1994), p.10. Ó Giolláin, *Locating Irish Folklore*, pp.18–31.
148. OSM, 34, p.86.
149. Shahid Amin, 'Cataloguing the Countryside: Agricultural Glossaries from Colonial India', in Peter Pels and Oscar Salemink (eds), *Colonial Ethnographies*, a special issue of *History and Anthropology 8* (1994), p.35.
150. Mary L. Pratt, 'Scratches on the Face of the Country: or, what Mr Barrow saw in the land of the Bushmen', *Critical Inquiry*, 12 (1985), p.120.
151. The phrase could be translated back in Thomsian terms as manners and customs or manners and superstitions.
152. Ovidio Carbonell, 'The Exotic Space of Cultural Translation', in Román Álvarez and Carmen-África Vidal (eds), *Translation, Power, Subversion* (Cleavedon, 1996), p.83.
153. Pratt, 'Scratches on the Face of the Country', p.125.
154. Richards, *Imperial Archive*, p.4.
155. Ibid., p.13.
156. Ibid., p.11.
157. Ibid., p.16.
158. Bernard S. Cohn, 'The Census, Social Structure and Objectification in South Asia', in Bernard S. Cohn, *An Anthropologist Among the Historians* (New York, 1990), pp.228–9.
159. Lauri Honko, 'The Folklore Process', in *Folklore Fellows' Summer School Programme* (Turku, 1991), p.35.
160. Ibid., p.26.
161. Peter Pels, 'The Anthropology of Colonialism: Culture, History and the Emergence of Western Governmentality', *The Annual Review of Anthropology*, 26 (1997), p.165.
162. OSM, 39, p.179.
163. Ibid., 39, p.161.
164. Hall, *Civilising Subjects*, p.59. Valerie Kennedy, 'Ireland in 1812: Colony or Part of the Imperial Main? The "Imagined Community" in Maria Edgeworth's The Absentee', in Terence McDonough (ed) *Was Ireland a Colony? Economics, Politics and Culture in Nineteenth-century Ireland* (Dublin, 2005), p.276.
165. OSM, 39, p.160.
166. Ibid., 39, p.98.
167. Ibid., 39, p.77.
168. Ibid., 39, p.144.

CHAPTER THREE

Colonialism's Culture: Ethnography, Cartography and Translation

From the Latin *translātĭo* translation originally suggested transference from one place to another. It combines locomotion and the senses, attributes that go to the heart of curiosity and exploration.¹ Friel's play *Translations* dramatizes the complexity of cultural closure, or perhaps opening, in a colonial context. It certainly caught the conscience of the King as it led to a succession of articles about the nineteenth-century Ordnance Survey. Gailey sketched the ethnographic interest of the memoirs in 1982 and this was followed two years later by Day's further contextualization of this scheme. Boyne published an article in 1984 and a biography of John O'Donovan in 1987. In 1989 Hamer published the first serious critique of the project. In the 1990s Andrews defended the survey from what he saw as a misrepresentation of it.² The source of the disagreement in much of the debate is quite specific: the Anglicization or translation of Irish place names (see Chapter 7).³ In viewing the survey as a colonial or state ethnography and highlighting its essentially cross-cultural nature this work argues that this is not the only aspect of it that could be considered as translation.

As outlined in the last chapter social, political and cultural research became more systematic and extended its field of application from the sixteenth century up to the Act of Union with Britain in 1800. These years saw the emergence of sometimes newly denominated socio-cultural sciences such as folklore, ethnology, sociology, statistics, geography, archaeology, philology, cartography and political science (see Chapter 2).⁴ The survey was described as a peripatetic university and a wellspring of such emergent academic disciplines. There were no Catholic universities in Ireland when it was undertaken and many in the colonial regime were actively opposed to the education of the Irish. As ways of understanding, interpreting, enumerating and accounting for amorphous masses of people these disciplines are, in many ways, an inheritance of colonial governmentality, elaborate extensions of the negotiation of the relationship between imperials and their subjects.

The idea of the survey as colonial, state or revenue ethnography has not been fully considered until now.[5] Its role in the emergence of professional Irish folklore and ethnology has remained unclear. The relationship between methods and political tasks has been understated, as Pels says 'professional ethnography . . . may be better regarded as a specific offshoot of a wider field of colonial intelligence rather than . . . the fulfilment of an intellectual goal to which colonial ethnographies vainly aspired'.[6] The professionalizing disciplines believed that 'to avoid colonial struggle – race conflict, indigenous revolt – one should follow a colonial strategy based on anthropological knowledge and planning, in order to achieve the desired evolutionary progress cheaply and without bloodshed'.[7] In the survey the precision of mensuration and science, enveloped by the ideology of Empire, wrestles with the uncertain knowledge of another culture. The archive is discordant in so far as the rhapsody of the imperial emissary is sometimes contested by the mimicry of the subject scholar.

In the story of travel, survey and social research the disciplines of cartography, ethnography and translation form a triangle of representational theory and technique. The various linkages and breaks between the three codes became the foundation for a poetics of interpretation. Power, politics and language are three of the cardinal points of the cultural compass. In the who's who and whodunit of interpretation there are no truly innocent orientations, just opportune vantages and perspectives. There are no bystanders, just interested parties feigning disinterest. Neither are there any purely or objectively interested parties, just players feigning objectivity. Each standpoint is strategic and is locatable relative to others. Each position, each variant degree of interest is mapped. Niranjana, Venuti, Álvarez and Vidal point to the asymmetrical power relations between languages and the ways in which this is expressed in the partial and selective processes of translation.[8] In the end game of discursivity, translation, like ethnography and cartography, is a site of representation, power and historicity. Niranjana says:

> The context is one of contesting and contested stories attempting to account for, to recount, the asymmetry and inequality between peoples, races, languages. Since the practices of subjection/

subjectification implicit in the colonial enterprise operate not merely through the coercive machinery of the imperial state but also through the discourses of philosophy, history, anthropology, philosophy, linguistics, and literary interpretation, the colonial 'subject' – constructed through technologies or practices of power/knowledge – is brought into being within multiple discourses and on multiple sites.[9]

The metaphor of cultural translation unites many different ethnographic paradigms and schools of thought. One of the ethnographer's roles is that of the translator making order of chaos, making sense of other cultures, 'anthropologists are presented as semiotic tour guides, escorting alien "readers" in rough semiotic space'.[10] In the nineteenth-century disciplinary multiplex of the survey, translation is more than a component: it is a key. It is useful to consider therefore what translation might be and outline something of what it does. Rather than simply transferring a transparent meaning from one language to another, or from one culture to another, the interpreter often confers meaning, 'the object of translation offers up, not a text to be read, but latent significance to be *written*. The object of ethnography becomes a possible object of knowledge through the operation of written inscription, which makes another culture *known*.'[11] Beyond the pragmatic ideas of fidelity or accuracy there are more searching issues implicated in translation. On the bench of the colonial envoy it becomes something like a clearing house, a licensing agency for a nomadic culture.

Ethnography textualizes experience, belief, value, speech or behaviour. The texts produced become a corpus removed from immediate discursive situations. Taking a long view of this process the social sciences developed sophisticated analyses and theories of the rhetorical aspects of research. The self-critical stance of ethnography grew out of a keen awareness of the interpersonal nature of data collection. As a translation of experience into text it is recognized as an uncertain exercise coloured by many nontextual factors. Foremost amongst these was the culture and personality of the ethnographer and the repertoire of linguistic and literary devices available to him or her in the language in which the ethnography is written. Aunger says for example, 'there is no recognition in the ethnographic medium that the social and psychological reality of some far-off place and time is transformed into the mental representations of a reader through at least one intervening intelligence (the ethnographer's) and several instances of physical mutation (e.g., into patterns of ink on paper)'.[12]

It is understandable therefore that translation is an appropriate metaphor to approach the interplay of different cultural codes. In the context of the survey it is considerably different from the creative interpersonal flows of meaning that occur in all communicative acts. It doubles as the strategy of a hegemonic language to engage with a weaker subject one. It should be remembered that weaker here does not refer to current ideas about minority or lesser-spoken languages. The Irish language was not a minority or lesser-spoken language in the nineteenth century. The diminution of the status and significance of the language in the nineteenth century is itself rhetorical. Although Irish was widespread, commonplace and modern it was constructed as a survival of pre-English barbarity. There is an abiding sense of real meaning residing in the dominant or target language. Translation, specifically into the English language, is more than a desideratum, it is expressed evangelically as the inevitable fate and solemn edict of progress.

The *am Faoilteach* Problem

The frustrated efforts of nineteenth-century Scottish Gaelic scholars to fix local time categories and match them to the Gregorian calendar epitomize the search for real time and real space that Ardener calls the '*am Faoilteach* problem'. On considering the seasonal approximation of the Scottish Gaelic term and comparing it with the English month of February, the lexicographer Dwelly was forced to conclude that the old periods were movable. Of course they were only movable in relation to the standard English calendar.[13] A similar fixation with precise measurement or, in terms of translation, with equivalence, characterized the cultural work of the survey. The Irish language, the names of townlands, the culture as a whole, was not actually unsettled, they were unsettling however.

All scientific discourse works with temporal conceptions. Ideas about time are involved in any possible relationship between ethnographic or anthropological discourse and its referents.[14] The observer and the observed, the translator and the translated, the mapmaker and the mapped are removed from each other temporally. As distancing devices 'categorizations of this kind are used, for instance, when we are told that certain elements in our culture are "neolithic" or "archaic"; or when certain living societies are said to practice "stone age economics"; or when certain styles of thought are identified as "savage" or "primitive"'.[15]

The logic and coherence of the survey was English logic and coherence. The measurement, naming and inscription of people and land translated them out into a symbolic otherworld. They were subsumed within a dominative English language discourse and narrative. Edward Said's paradigm slips uncannily into place when Orientalism is replaced by Celticism. At times the survey's Celticist antiquarian discourse, influenced by the English model, bears an uncanny resemblance to Orientalist discourse. Said described this comprehensively as:

> A *distribution* of geopolitical awareness into aesthetic, scholarly, economic, sociological, historical, and philological texts; it is an *elaboration* not only of a basic geographical distinction . . . but also of a whole series of 'interests' which, by such means as scholarly discovery, philological reconstruction, psychological analysis, landscape and sociological description, it not only creates but also maintains; it is, rather than expresses, a certain *will* or *intention* to understand, in some cases to control, manipulate, even to incorporate, what is a manifestly different (or alternative or novel) world; it is, above all, a discourse that is by no means in direct, corresponding relationship with political power in the raw, but rather is produced and exists in an uneven exchange with various kinds of power, shaped to a degree by the exchange with political power (as with a colonial or imperial establishment), power intellectual (as with reigning sciences like comparative linguistics or anatomy, or any of the modern policy sciences), power cultural (as with orthodoxies and canons of taste, texts, values), power moral (as with ideas about what 'we' do and what 'they' cannot do or understand as 'we' do).[16]

In subjecting Irish culture to systematic translation and examination in the name of science, nation or Empire many antiquarians and Celticists shared a common intellectual paradigm with Orientalism. Many elements of the theory survive from the centuries-old classification and registration of exotic others. The criteria of scientific accuracy dominated in translations from old and middle Irish. These translations were directed at the arcane milieu of the discipline and shrouded in the etiquette of elite erudition. The scholar was trained in the encryption of footnotes, glossaries and commentaries. The footnotes in O'Donovan's translation of the *Annals of the Four Masters* in 1851 replace the text itself. The adherence to forbidding exactitude has been partly accredited to the influence of German scholars whom along 'with their compatriots working in

Oriental Studies, the emphasis was on exactness and the production of texts acceptable to other scholars'.[17]

O'Grady's model was 'explicitly Orientalist' and was carried into the twentieth century by poetizers, novelizers and mystics who, along with the scholarly translators, philologues, keltologues and folklorists undertook a rudimentary processing, editing, adapting and novelising of all things ancient, Irish or Celtic.[18] The ideas of evolutionary discourse or much twentieth-century historical, literary, archaeological, geographical, sociological or folkloristic discourse 'in the double movement of its quest to historicize the present and to resurrect the past may be seen, precisely, as a necrology practised on the living'.[19] The vital link to a living language and culture is artificially broken.

Outlining the alienating experience in translations from Arabic, Jacquemond says that 'the non-professional reader . . . is soon rebuked by its harshness, its radical strangeness, and its lack of appeal and learns to satisfy himself with the second-hand knowledge he is provided with through the Orientalist's writings'.[20] This kind of translation presents the Irish, or the Arab, as exotic, mysterious or 'as beautiful and elusive as any aspect of Irish culture'.[21] Being beautiful and elusive the problems faced by the living language and culture can be overlooked or even seen as unrelated. The ethnographic antiquary or surveyor becomes cartographer, cryptographer and translator. The elite English language discourse is authorized to rummage amongst the ruins, to decipher the runes and to render the mysteries intelligible in the futuristic idiom of science and reason. The Irish past is a puzzle better left to intrepid archaeologists and philologists to try to put back together in English.

In this way translation slowly but surely becomes the prerequisite for intelligibility. The source language is constructed as Neolithic, prescientific or unsettled while sense, if not intelligence itself, resides in the dominant language. Cultural elements are removed from their habitat of meaning and rendered novel, curious or antiquated. They are rendered so because to admit to their easy intelligibility, to those in whose culture and language they are found, would break the magic spell of discovery. Novelty, curiosity and peculiarity are the catch cries of the colonial antiquary or collector, not qualities of language or culture. The reader is guided, 'by the authoritative hand of the omniscient Orientalist-translator who is 'trained to decipher the otherwise unfathomable mysteries of the Orient'.[22] It needs to be said perhaps that these cultural elements were neither curious nor novel; indeed many continue to be part of everyday life in the present.

To date many interpretations of Ireland are dominated by fashionable and commoditized analyses of only a handful of literary works in the English language. The translation by a colonial or imperial government of key elements of a people's culture begs an important question. Who is translating what and why are they doing it? The translation is first and foremost authorized by the colonizer. The indigenous language to be translated, the source language, is to be translated because it is an impediment to progress or civilization. If translation theory recognizes 'the need to examine in depth the relationship between the production of knowledge in a given culture and its transmission, relocation, and reinterpretation in the target culture' it must also look beyond ethnocentric reflections upon elitist poesis.[23]

The difference between writerly or literary translations or translations in everyday communication, and translation as ordinance, lies in the prescriptive, appropriative and instrumentalist nature of the latter. The power relations underpinning the contact between the two languages are basic to the understanding of the survey. These relations have remained sketchy and fuzzy. Translation fuelled an ascendant canon of representation in which there was a suppression of realpolitik. There was a clinical and crucial elision of the asymmetric relationship between the languages and cultures involved. Comparison was not an option; Irish was the argot of the aborigine, dwarfed by the lordliness of English and its privileged position in the grand imperial design.

If every definition is a space, each redefinition becomes a space for meaning or a sounding board for further new meanings.[24] This is the very dynamo of creative process itself but when it is allied to, and embedded in, a colonial regime the cartographic, ethnographic or translative affect is erasure and re-inscription. Translation redefines as it lenites the rough edges of otherness and eclipses the symbolic heart of the originary space. Through translation a cultural expression travels from the source language, it is interpreted and reconstructed as a new text in a kind of interlanguage and is finally incorporated in the target language. Acts, artefacts, utterances, notions, tropes, concepts, beliefs, values and feelings are appropriated by the empowered language and made part of its own repertoire. They may not cease to exist in the source language but they are symbolically nullified, socially restricted and culturally marginalized. Indigenous elements of culture are transformed by the powerful apparatus of knowledge in the receiving culture that represents it on its own terms through a dialectic of exoticization–naturalization of the other.

The others are not so much distinctive as they are incompatible, deviant or recalcitrant. They must therefore be domesticated or homogenized in the other language. The translator as ethnographer becomes a mediator who transplants 'the source text in a language different from its own, and in a linguistic and cultural habitat alien to its original environment'.[25] Cartography and ethnography were political practices in the nineteenth century that relied upon translation 'to bring back the cultural other as the same, the recognisable, even the familiar' and, as Venuti says, 'this aim always risks a wholesale domestication'.[26] The survey attempted to bring back the Irish as the same, looking and sounding familiar in the English language.

O'Donovan's role as etymologist and orthographer was that of a cultural operator 'entrusted with monitoring, organising, or negotiating the comparative statuses of each system. The tensions that exist between them can never be suppressed, only channelled or reconverted into various options.'[27] His options were limited, however, by the fact that he was under the command of the superintendents of the survey. He learned that his posting, was 'within an empire whose heartbeat lay . . . in London'.[28]

Translation and the protean processes of hybridization or creolization were also part of nineteenth-century cultural process in Ireland witnessed by the surveyors. These processes reinforced the commitment to standardization and translation. The Irish language, imprinted with the character and traits of the people, was described as unsettled, unattached, off standard and nomadic. Richards' idea of the nomadic includes 'knowledges that resist bureaucratic codification . . . peoples that defy national concentration . . . armies that resist defeat . . . entire societies that inhibit the formation of power centers'.[29] Far from exemplifying the free and open migration of meaning across languages or cultures, the survey could be likened to a cultural rite of passage. The place, the name and the language remained as mute anachronisms, vestiges of barbarous inarticulacy, of the world before English civilization. Translation was a spiritual, moral and cultural deliverance. People translated themselves to sell things, buy things, make sense, to be like their betters, to avoid stigma, to be free and happy, to leave behind what was said to be past, to live the imperial dream. They were of course free to do this; they were not as free not to do it.

In Australia nostalgic ethnocentrism influenced the naming of places, co-opting, neutralizing or sanitizing the unfriendly nowhere of colony. It is true that in the survey the vast majority of the place names were not imposed, they were not always translated in the strictest sense of the word, but they were sanitized and neutralized as they were Anglicized, settled or laid down in English orthography. The officers of the Royal Engineers consulted only those considered to be of sound judgement and social status, the former largely being determined by the latter. For his part O'Donovan urged the superintendents of the survey to resort to Irish speakers.

Acutely conscious of the culturally specific aspect of the names he believed that 'no person should be allowed to meddle with those names except one acquainted with the whole circle of Irish lore'.[30] The names were more than convenient labels, they were expressions of a wider indigenous universe of discourse. The British military experienced many difficulties in the field as a result of their lack of Irish. Lieutenant Fenwick expressed his frustration in County Roscommon, 'I am constantly at my men about these matters, but many of them are Presbyterians from the North who care not for Eremites and Friars, White, Black and Grey with all their trumpery' (see Chapter 7).[31]

In confronting a construction of the Irish past, that may be neither past nor Irish, the contemporary theory of translation illuminates the theory of ethnology. Translation, like cartography or ethnography, attempts to catch and hold culture within its own perspective. At the same time it poses as an actual or factual presentation of something that already exists. If translation provided a place in history for colonized people worldwide it is also true that, 'the history of contact between, say, an African people and a colonial power is likely to be very differently conceived by members of the two groups concerned'.[32] With the advent of the survey an English Ireland and the English Irish are mapped, translated and defined. As Niranjana says, 'what is at stake here is the representation of the colonized, who need to be produced in such a manner as to justify colonial domination, and to beg for the English book by themselves'.[33] The proverbial horse was brought to the water.

Ethnography, Cartography, Translation

Ethnography, cartography and translation fit into each other like Russian dolls both as symbols in themselves and image creators. In the nineteenth century they were the pseudo-scientific equivalent of story or narrative. They constructed worlds utilizing a repertoire of figures and words. They translated experience, imagination, observation, sense and sentiment into an imperial teleology, a providential vision of reality, part of a narrative of a promised world. Proceeding from an ethnographic understanding of culture allows for a re-enlivening of some of the questions raised by the fieldworkers themselves during the course of the survey.

If culture approximates to everyday, shared meanings through which different groups express themselves, it is clear that one group's meaning is not the same as another's. Culture is also defined by meanings that are not shared. It is important to consider that it is by first of all constructing it as being like this or being like that, that the frontier science of Empire appropriates the language, the word and the story of the other. It takes on the character of a symbolic quarrel in which the other's beliefs, behaviour, dress, cuisine or habits, become mere figures. What emerges is not so much a transparency as it is a negative, an unreal and unrealized image, fixed in time and space, a silent shadow cast by the past and projected into the future by the visionary technology of science and power.

The emergence of relativism in ethnography, associated with the work of Geertz, Clifford and Marcus, marked a move away from domineering monologues to more hetroglossic or multivocal interpretations.[34] The patronizing propensity, the platitudes and certitudes of disciplines partnered in power and their concomitant socio-political implications, was highlighted. This began with the recognition that the only fact is that the facts do not speak for themselves. Facts are the colours in the kaleidoscope of consensus. Whatever way they appear to change they end up in harmony. The realization that 'ethnographic data are not "given" but "made"' and the pretension 'that *a* reality . . . be representative of *the* reality' was not only the posture of authority but the posture of the author.[35] Fetishized tradition or traditionalized knowledge, understood on the whole as fragmented and disintegrated, was pasted into the copies of the surveyors to make up a scrapbook of fantasy and nostalgia for the kindergarten of humanity and civilization. Tradition was the fairy in the compass, the miraculous mote in the eye of the naive native.

It is not exactly what the survey contains that is relevant, but rather how it is contained. There is certainly a modicum of ethnographic information in the memoirs and letters, the maps may be accurate or useful and the Anglicized or translated place names were based on a sound methodology. The point is this, however, each representation is dependant upon an interlocutor and that interlocutor is in turn locatable in a particular linguistic and discursive home. Translation is often the fulcrum upon which the relative weight of different cultures is balanced; it can sway in one direction or another. Each interpretation takes place in a specific social and historical context that informs and structures it.

Viewing the survey from the perspective of the theory of translation enables a retracing of important stages and processes in its execution. The idea of Irish knowledge must be offset from the idea of knowledge about Ireland. The idea of Irish folklore, understood as pristine, present and correct, must be offset from the idea of the folklore of Ireland. The distinction is a fine one, the latter is often data suspended on the webs of colonial archives and discourse. One of the questions therefore concerns the creation of knowledge about Ireland in general and the positing of a putative Irish alterity. The survey constructs a physical, social and ethnographic image of Ireland that is an image of the extension of colonial power. In much the same way as Said's Orientalism or Mudimbe's Africanism, the survey's Celticism, itself an inflexion of English antiquarianism and ethnology, was appropriative and instrumentalist. This was conducted within a colonizing structure that enveloped the physical, human and spiritual aspects of experience and represented 'the domination of physical space, the reformation of *natives'* minds, and the integration of local economic histories' into the Empire's perspective.[36] Ideas about colonialism or imperialism can be vague and loose at times and their use requires further qualification.

Ireland and the Empire

The main difference between Ireland and a proper colony, in Leersen's opinion, is that Ireland was a kingdom with a parliament of its own.[37] The Irish Parliament enjoyed nominal independence during the period between 1782 and 1783. He continues to say:

> In many other respects (the expropriation of the land, its distribution among immigrant settlers, the social establishment of these settlers as

a new ruling class, the submission of the population both in social and economic terms, their employment as providers of cheap labour, the subordination of the country's internal economic interests to those of the 'mother country') Ireland did have the character of a colony.[38]

Colonial expansion coincided with the emergence of a modern capitalist system of economic exchange. This reinforced the use of the subjugated people as a resource that the burgeoning economy of the colonizer exploited. The relationship between the colonizer and the colonized was ordered into a rigid hierarchy of difference. Writing of the intimately related categories of class and race Gibbons argues that Ireland was 'a colony whose subject population was both "native" and "white" at the same time'. The otherness of the Irish was disconcerting because it did not immediately lend itself to obvious racial divisions.[39]

Ireland was readily Africanized and Indianized by writers such as Robert Southey or Samuel Taylor Coleridge who were contemporaneous with the survey.[40] The view of Ireland as a colony is based upon the 'construction and naturalization of an unequal form of intercultural relations' that arose from a protracted period of asymmetric political and economic entanglement.[41] Under such conditions 'the idea of the colonial world became one of a people intrinsically inferior, not just outside history and civilization, but genetically predetermined to inferiority. Their subjection was not just a matter of profit and convenience but also could be constructed as a natural state'.[42] The looseness or vagueness of concepts like colonialism can be converted by colonialist discourse itself into useful tropes that characterize other cultures as weaker or inferior.

The British Act of Union with Ireland in 1800, like the India Act of 1784 and Canada Act of 1791, was part of a process of imperial consolidation, formalization and rationalization following the American War of Independence. It tidied up the constitutional anomaly of the separate parliament in Dublin that provided a counterpoint to the power of Westminster. Following the uprising of 1798 the British government was increasingly aware of the dangers within the Empire. Earlier colonialist discourse on the barbarism of the Irish tended to re-emerge during crises of legitimacy, 'what, in 1788, had been unremarkable antiquarian dabbling in the material culture of the past, now assumed a dangerous contemporary dimension. Native Irish rebels and their weapons had re-emerged from the page of history.'[43] Under conditions of relative colonial normalcy the tendency was to politicize the ethnographic while in times of crisis the tendency was to ethnographicize the political.

William Pitt, the Prime Minister of Britain, said in the House of Commons in 1799 that the roots of the problems in Ireland lay 'in the situation of the country itself – in the present character, manners and habits of the inhabitants – in their want of intelligence, or in other words their ignorance'.[44] The Union, which came into effect on 1 January 1801, created 'a new English-dominated "British Empire in Europe"'.[45] In the House of Lords Lord Clare advocated enlightened government for Ireland through perpetual martial law. He had promised to make the Irish as tame as cats and on his death people threw dead cats at his coffin.[46] The new imperialism was motivated by 'a vindictive patriotism'. Britain, having lost America, was determined to extend its economic control through forced expansionism. British public opinion was hostile to colonization, then associated with the slave trade, but favourable to an imperialism that fostered English interests.[47] With the Act of Union British imperialism re-embraced its old colonial familiar. In the fifty years following the Union 'the discrepancy between the values of British society and the means employed to preserve colonial domination over Ireland grew steadily more apparent'.[48]

Expressed in prison reform, the Poor Law and the first Reform Act of 1832 emerging humanitarianism in Britain was not extended to Ireland, a fact that points to a sharp disparity between values and behaviour. Most tenants in Ireland had the status of serfs, rents were revised at the whim of a landlord and they were evicted lawfully without reimbursement. Landlords, whose titles were enshrined in the survey, exercised a form of absolute power. The Union placed the Irish in a liminal and contradictory position, 'they were national subjects incorporated into the nationstate through parliamentary and economic structures and given the title of citizens, but they remained a colonised and alien population, were denied certain fundamental rights of citizenship, and continued to be constructed as culturally, religiously and racially other'.[49]

The Union also removed trade barriers between Ireland and Britain and permitted British industrialists, suffering a depression in the 1820s, to dump goods onto the Irish market. Cleary says that 'after the Union in 1800 Ireland was constitutionally integrated into the most advanced industrial capitalist economy of the time' and that this was 'constrained within a colonial relationship mediated through London'. This evolved out of a 'bastardized variety of colonial feudalism' exacted by landlords upon the tenantry.[50] The description of the Irish, in the survey's memoirs in particular but also in the letters of the Irish fieldworkers, was not wholly innocent or politically neutral.

The newspapers and journals of the time, the *Times*, *Edinburgh Review*, *Punch* and *Blackwoods*, showed sympathy towards the British poor yet fostered a racialist antipathy towards the Irish. The customs and manners discourse of the civilizing mission of the Empire was employed frequently to highlight 'the abysmal ignorance and adherence to superstition' in Ireland.[51] The descriptions in the survey were not significantly different from those in political pamphlets, broadsides, travel descriptions, diaries, plays, novels, newspapers, journals, magazines, parliamentary debates, reports, and official and private papers. They were the mainstay of a rationalizing, differentiating discourse in the service of imperial power and control. This discourse explained and justified itself in civilizing what the young Benjamin Disraeli described in 1836 as 'this wild, reckless, indolent, uncertain, and superstitious race'.[52]

Ethnographic detail can also be an aid to ideological persuasion through which science is politicized and politics scientized. The political and rhetorical nature of tradition is evident here, whether it is valorized or rescued in the survey, it is first and foremost a category of the survey. It is a double-edged sword handed to the subject culture. It conceals and classifies diverse resistances and negates oppositional claims to authority. When politics revealed itself to the surveyors in the form of informants who were hiding from the sheriff, fieldworkers who were mistaken for tithe proctors or preachers, the presence of Daniel O'Connell, it was always carefully controlled and managed.

Said distinguished colonialism from imperialism: '"imperialism" means the practice, the theory, and the attitudes of a dominating metropolitan centre ruling a distant territory; "colonialism", which is almost always a consequence of imperialism, is the implanting of settlements on distant territory'.[53] The Civil and Down Surveys of the seventeenth century were proper colonization, the confiscation and settlement of land by an exogenous group. Ireland may not have been a geographically distant territory and while plantation, although considered after the famine, was not a practice of nineteenth-century imperialism in Ireland, the Union was an annexation of what Young calls 'England's first and always exceptional colony'.[54] For Said the survey parallels circumstances in

India, Algeria or Palestine in terms of 'the silencing of their voices, the renaming of places and replacement of languages by the imperial outsider, the creation of colonial maps and divisions also implied the attempted reshaping of societies, the imposition of foreign languages and systems of education, and the creation of new élites'.[55]

The survey epitomized post-Union consolidation and economic expansion, 'in one sense, colonialism functions as a part of global capitalist expansion at an early stage, when the non-Western world functioned primarily as a source of materials and labour or as a human warehouse for those individuals (criminals, Irish nationalists, displaced agricultural workers) left behind by that expansion'.[56] It was an extension of imperial power emanating from the metropolitan centre. Colonialism is not solely about money or the exercise of raw political power, however, it is also about culture. Cohn argues that 'the bulk of social and cultural anthropological fieldwork has been done in colonial settings. In a very real way the subject matter of anthropology has been the study of the colonized'.[57] The Empire saw itself as the superintendent of all knowledge, it 'produced and published so much information – linguistic studies, orientalist tracts, ethnographies, geographical reports – that only a large corporate or governmental entity could possibly possess the resources necessary to comprehend Britain's comprehensive knowledge'.[58]

Young says that colonialism is a practice while imperialism is a concept. Colonialism in the past represented a diverse set of practices varying from direct rule to indirect rule, assimilation or imposition. There are both settled and unsettled colonies.[59] The different meanings of imperialism and colonization continually overlap and intersect each other. Colonization was also a cultivation of intellect, a battle for hearts and minds, an ethnographic entanglement, a race for knowledge that required knowledge of race. The Empire combined political economy and ethnography in the management of culture. The Englishman was engaged in 'bringing his superior know-how and his science to other "inferior" populations . . . they represented Science and Technology and . . . this knowledge enabled the societies they subjugated to realize progress. And to accede to civilization.'[60] It has been said that the British Empire represented the highest state of social organization. In its attempt to attain comprehensive knowledge it imagined 'a not-too-distant future when all species would be identified, all languages translated, all books catalogued'.[61]

The Act of Union was a belated climax to the seventeenth-century strategy of plantation, safeguarding the official state religion and preventing Ireland being accessible to enemies like Spain or France through so-called Irish treachery. By this time Ireland's economy was almost fully integrated with that of Britain. By 1800 more than 85 per cent of all Irish exports went to Britain.[62] An important part of this consolidation was that Ireland would be opened up to a programme of Anglicization. This was articulated as a desirable part of liberation, emancipation and development. In this the survey had an important role to play, Judd says:

> Ireland after the Union was not simply an increasingly well-integrated part of the United Kingdom. It was also a colony, and an abominably treated one at that. Victorian liberals did not need to travel, like their modern equivalents, to India or Africa to witness abject poverty and degradation or to see examples of man's inhumanity to man. They merely had to cross the Irish Channel.[63]

The Maoris, the peoples of the Cape and the Chinese of Hong Kong benefited more from the conscience of the imperial administrator. It could be argued that Ireland was disadvantaged in comparison to the so-called proper colonies. In Ireland taxation contributed several million pounds per annum to imperial burdens such as the upkeep of a sizable military, naval establishment and a paramilitary police force. Formal colonies were exempt from such taxes. Ireland was governed directly by a Lord-Lieutenant or Viceroy whose powers were ill defined.

Unrest in Ireland provoked measures of repression and coercion that would have been unthinkable in Britain. Fitzpatrick says, 'in these respects, Ireland was not only exceptional within the United Kingdom but akin to a colony, efficiency in government being valued above the liberty of the subject and the sanctity of property'.[64] The imperial administration, whether grim or benevolent, viewed the Irish as a separate and subject population. If physical colonization lessened as the nineteenth century wore on, Anglicization and the construction of racial difference in discourse intensified and diversified. The duality of this discourse was closely paralleled in British discourse on the peoples of Asia, Australasia and Africa.

While it is undeniable that 'the nineteenth-century Irish might elect to play the parts of colonials (whether deferential or resentful), metropolitans, or colonizers' it is too easy to obviate the various processes of colonization that led to this situation.[65] These distinctions can sometimes be dependent upon the position of the writer. If Ireland was not, technically, a colony then was it a colonizer? This is a question of identification, of the 'definition of enunciative space'.[66] It matters little to the colonized whether it is colonialism proper or some deviant form that is responsible for their condition. Only the pedants among the colonized would opt to live in a proper colony. A cultural interpretation of colonialism on the other hand suggests something more intangible and non-indictable, and easier to obviate. The fault may even be attributed to the colonized themselves as when it is suggested that the Irish simply decided to stop speaking the Irish language as a lifestyle choice. This argument assumes a person, firstly denied education in their own language, can switch languages even before they know or have come in contact with another one. It also assumes, as already mentioned, that the Irish language is no longer spoken which is quite simply wrong. In the nineteenth century violent colonization was replaced by 'colonization through the mediated instrumentality of information (the operations of a post-modern world)'.[67] The survey is evidence of the role played by ethnographic information in colonialist discourse.

The sometimes old world image or even romantic notion of colonialism, although itself a deceptive disguise, is far less dramatic when it is qualified by the evidence that it did not always succeed and did not always go away. The emergence of postcolonial nation-states in the wake of Empire wasn't always an 'ethically and politically proper readjustment of the wrong of colonization'.[68] In many cases nationalizing states could be viewed as inheriting the intellectual apparatus of Empire. In striving for uncertain legitimacy official state nationalist discourse often reiterated colonialist discourse. Sometimes cloaked in obscure Celticist or Orientalist language, nineteenth-century English antiquarian notions, translated into Irish, served to remap revolutionary Ireland, symbolically cordoning the homely aborigines in *Gaeltacht* reservations.[69]

In complex and subtle ways evolutionary ideas about the survival of language, superstition, customs or rites came to define resistance to the colonialist regime that created them in the first place. The mountaineers and cottiers that first appeared in colonialist discourse were privileged in the newly independent state. At first glance this appears to be have been justified and while it may also have been necessary the duplicitous nature of this discourse was not unchallenged. Renowned Irish novelist Máirtín Ó Cadhain was acerbic in his criticism of what he saw as the fatalistic and fetishist folklorization of Irish speaking areas. Seán Ó Ríordáin, the foremost poet in the Irish language, and the one who most eloquently enunciates the postcolonial moment in Ireland, engaged in a constant critique of the cultural politics and discourse of the southern Irish state.[70] Although it has been overlooked, the more specific question of the role and status of the Irish language itself within the survey is relevant to the definition of the concept of colonialism or imperialism. The prevalence of the Irish language at the time – it could be described as being in its heyday – emphasizes the relevance of translation theory to it and raises many questions about the nature of the survey itself as well as its later reception and registration.

The Rhetoric of Empire

Ireland was predominantly Irish speaking at the time of the Union and during the life of the survey.[71] Of a population of five million in 1801, four million were Irish speaking. This did not change significantly in the following decades. In fact there is some evidence that the Irish language flourished due to a significant population increase up to 1831.[72] The idea that this is not significant in the interpretation of nineteenth-century Irish culture is perhaps unique. It has meant that pivotal aspects of postcolonial studies in Ireland, or at least those that arise from the contemplation of the position of the Irish language in both the nineteenth and twentieth centuries, have been ignored. Any literature with postcolonial pretensions that manages to overlook a colonized language and culture is seriously lacking in credibility. It may even suggest that this literature is in itself part of colonialist discourse since the elimination of the Irish language is a colonial inheritance. The clarity, efficacy or proficiency of the survey itself partly evolves from this elision. If the Irish language is removed then the singular reality of the universe of evolutionary discourse in English is unchallenged. Here the equally simple and no less profound symbolic problem of the Irish language place name is emphasized. The giant may stumble when a meagre pin pierces his heart.

Many nineteenth-century antiquaries testified to the extensive use of Irish both outside of large towns and within. A protestant minister estimated in 1815 that there were two million Irish speakers in Connaught alone. Edward Wakefield wrote that, with the exception of Leinster, Irish was ordinarily spoken throughout the country and that the economically disadvantaged knew no English. The London Hibernian Society, concerned with spreading the Holy Scriptures in Ireland, politely suggested that it 'demanded the most serious consideration'.[73]

Doherty says that 'English was the language of the *arrivistes* in the eastern seaboard, in towns, and in plantation estates in Munster, Ulster and the Midlands. It was the language of the ascendancy and of government, while Irish was the language of the rural poor.'[74] The letters of the survey offer evidence of the language in all but a few counties that had been previously planted. Ó Muraíle says that the letters come from a country that was Irish speaking to an unprecedented extent. There was never before, or has never been since, as many people speaking the Irish language at the same time.[75]

Even if this was not the case the survey's attention to the Irish language and culture has a marked symbolic dimension. From its privileged position as receptor, interpreter and translator it was English that was to accede to linguistic and cultural dominance. It reached backwards into the reservoir of the language to subject it to a process of Anglicization. While it measured the physical contours of the land it fundamentally altered the linguistic and symbolic environment. The increasing linguistic duality of Ireland and the hegemony of the English language as the symbolic filter of all things Irish was the leitmotif of nineteenth-century Ireland. Kiberd describes the survey as 'one of a number of modernizing experiments conducted in the colonial laboratory that was Ireland in the mid-nineteenth century'.[76] The reconstitution of physical and intellectual space by the survey was an act of ethnographic translation or colonization. The absenting of the Irish language is not merely an oversight but one of the defining characteristics of colonialist discourse in Ireland.

Whether or not statistics reflect the extent or knowledge of the Irish language does not erase its relevance. Many of them were accumulated in similar projects to the survey itself and should not be seen as unproblematic. It is surely sufficient to point to the strategic construction of the presence or absence of the language and to question the construction itself. The language was not marginal, it was marginalized. The statistics are enmeshed in a perspectival struggle to amass the property of discourse and to legitimate or negate oppositional claims to the truth. As Derrida says 'one should never pass over in silence the question of the tongue in which the question of the tongue is raised'.[77] Here the British Empire and the English language was culture-giver, the voice of authority over which alternative voices were inaudible.

The salience of much of Memmi's or Fanon's writings on the situation of the colonized can be sensed, the physical and mental estrangement as well as the academic, antiquarian or folkloristic archaizing of the living language and tradition. Memmi says, 'possession of two languages . . . actually means participation in two physical and cultural realms. Here, the two tongues are in conflict; they are those of the colonizer and the colonized.'[78] It is also significant that bilingualism itself is often wrongly assumed to mean an absence of Irish and that the invigorating processes of cultural hybridization are themselves overlooked. It is as if knowledge of the English language precludes knowledge of any other language or that cultural adaptation or accommodation implies discontinuity rather than continuity. It is striking that almost any inflexions of Irish appear as stains on the forward-looking face of English civilization. It is as though the threshold of civilization can only be crossed over once. There is no going back, no crisscrossing, no shared space.

MacLeod points out that 'science has no nation; but nations have science . . . whatever their motives and means and whatever their allegiances to economic doctrine, western Europeans came to be united on the principle by which colonies served as plantations or primary producers for the trade and manufacturing industries of the metropolis. Within this context grew, by the early nineteenth century, a diverse range of rational projects which we collectively label "colonial science".'[79] Whyte loosely defines this as a phase in which scientific work is carried out by residents of the territory in question who see themselves as dependent on the scientific tradition of the metropolis.[80] This was carried out mainly by members of the ascendancy, membership of societies like the *Irish Archaeological Society* (1840) largely comprised of protestant clergy, the gentry and urban middle classes.[81] The nascent disciplines of Irish academia were themselves both Anglicized and Anglicizing.

Little attention has been given to the heterogeneous nature of colonizers and their interactions with local people and what are often emergent knowledge traditions. Thomas says 'colonial culture . . . includes not only official reports and texts related directly to the process of governing countries and extracting wealth, but also a variety of travellers' accounts, representations produced by other colonial actors such as missionaries and collectors of ethnographic specimens'.[82] This census-like knowledge enumerates, classifies, hierarchizes and locates the subject population creating a charter for intervention.

Accepting that the colonial situation can make the maker as much as it does the made, the survey can be viewed as a 'socially transformative endeavour that is localized, politicized and partial, yet also engendered by longer historical developments and ways of narrating them'. Such projects are both a culmination and combination of diverse and convergent forces from the exploratory and religious to the administrative and commercial, 'missionaries and explorers engaged in trade, and settlers who were no doubt preoccupied with running plantations or trading might also have been correspondents of the Royal Geographical Society, who occasionally dabbled in ethnology or geology'.[83]

Critiques of colonialism reached a moment of new reckoning, its illusory image of permanence, pervasiveness and power has faded to reveal its workshop of signs, practices, axioms, aesthetics and tropes. Colonial power, while it is invidious and unimagined, is not absolutely omniscient or omnipotent, 'colonial projects of knowledge were frequently incoherent and contingent, were profoundly heterogeneous, were discontinuous and often haphazard'.[84] Running counter to this approach to colonialism is the problematization of the colonial archive and its use in the present to construct academic discourses, historical, ethnographic and literary, that place them in teleological narratives of state formation.[85] As Spurr says 'imperialism has survived the formal ending of colonial rule, but so has colonial discourse'.[86] Neither the survey nor colonialist discourse are history, each are redolent of, and resonant within, contemporary cultural discourses.

Colonialist discourse or the rhetoric of Empire in the nineteenth century in particular meant the imposition of specific knowledge, discipline and value upon dominated groups. It included 'everything from legal statutes to memoranda, from newspaper accounts to novels, from telegrams to poetry'.[87] It worked to constitute reality for those represented and those representing them. It was a system of knowledge and beliefs about the world within which acts of appropriation took place, were interpreted and justified. It largely evaded and still evades discussion of economic or political exploitation, 'rather it conceals these benefits in statements about the inferiority of the colonized, the primitive nature of other races, the barbaric depravity of colonized societies, and therefore the duty of the imperial power to reproduce itself in the colonial society, and to advance the civilization of the colony through trade, administration, cultural and moral improvement'.[88]

Sometimes dated to the publication of Said's *Orientalism* in 1978, colonialist discourse analysis attempts to understand and deconstruct colonial discourse 'and the exercise of colonial power through discourse'.[89] Said recently commented that he was drawn to Irish culture by precisely the same interests that led him to write *Orientalism*.[90] Bhabha says that the predominant strategy of colonialist discourse is the creation of a space through the production of knowledge in terms of which surveillance is exercised. Its objective is to 'construe the colonized as a population of degenerate types on the basis of racial origin, in order to justify conquest and to establish systems of administration and instruction'.[91] The survey was not an isolated project; it was contemporary with the establishment of the Irish Constabulary, called Peelers after their founder Robert Peel, and the National Education system, and it deployed both force and education in the mapping of a new Anglicized Ireland. The rhetoric of reform, improvement and progress was matched with a threat of punishment for the non-compliant.

Colonial knowledge collection, 'the act of assembling the colonial archive, created and imposed categories and had a part in shaping what it was that was notionally "simply" being reported'.[92] The central problem in relation to the survey as a colonial archive is that:

> The colonial archives on which we are so dependent are themselves cultural artefacts, built on institutional structures that erased certain kinds of knowledge, secreted some and valorized others . . . we cannot just do colonial history on our given sources: what constitutes the archive itself, what is excluded from it, what nomenclatures signal at certain times are themselves internal to, and the very substance of, colonialism's cultural politics.[93]

Leersen wrote that the survey foreshadowed later folklore commissioners and echoes the praise lavished on the survey for 'salvaging . . . a great deal of local lore and learning'.[94] The depiction of the survey as the salvager of a dying culture also risks repeating one of the key tropes of colonialist discourse. At least it is also possible to view it as one of a number of strategies that were employed in the early nineteenth century to quantify, categorize, inquire into and inscribe the indigenous culture. The surveyors 'can claim to have contributed to the cultural heritage of the societies they study by a sympathetic recording of indigenous forms of life . . . but they have also contributed, sometimes indirectly, towards maintaining the structure of power represented by the colonial system'.[95]

As the survey engages with a different cultural knowledge it tends to categorize it as a ragbag of incoherent ancient traditions. It casts itself in the role of the heroic rescuer of it and appropriates it within it own hierarchy of worth, value and usefulness. This emphasis on the other's knowledge as tradition or traditional is problematic. The language of tradition was often utilized as a strategy to smooth over the more overt and explicit power relations:

> Ethnographic compendia such as 'tribes and castes' volumes, famine commission reports, judicial records, and so forth – often record encounters that occur as colonial disciplines are imposed and colonial knowledge amassed. In these documents, colonial ethnography and characterizations of 'tradition' and 'custom' are often deployed, I suggest, precisely at those junctures where, in those encounters, the illusion of consent is most severely put to the test, as opposition to those disciplinary procedures or ethnographic inquiries is recognized. Supposedly uniform 'custom' and 'tradition' are then made to serve as explanations for intransigence that would supposedly evaporate if the native 'superstition' could only be overcome.[96]

The emergence of ethnological discourse in the nineteenth century and contemporary ideas about progress and industry promoted a scheme:

> In terms of which not only past cultures, but all living societies were irrevocably placed on a temporal slope, a stream of Time – some upstream, others downstream. Civilization, evolution, development, acculturation, modernization (and their cousins, industrialization, urbanization) all terms whose conceptual content derives, in ways that can be specified, from evolutionary Time. They all have an epistemological dimension apart from whatever ethical, or unethical, intentions they may express. A discourse employing terms such as primitive, savage (but also tribal, traditional, Third World, or whatever euphemism is current) does not think, or observe, or critically study, the 'primitive'; it thinks, observes, and studies *in terms* of the primitive. Primitive being essentially a temporal concept, is a category, not an object, of Western thought.[97]

Lotman draws attention to the location of cultural typologies in language saying that 'one's own culture' is considered to be the only one. To this 'is contrasted the lack of culture of other groups . . . from the point of view of that culture which is accepted as the norm and whose language becomes the metalanguage of that cultural typology, the systems which are opposed to it appear not as other types of organization, but as non-organization'.[98] In the survey English is the language of description in which the Irish culture is typologized. It is not an archive of rescued Irish culture, rather it is a document defining Irish culture as one that needs rescuing. It can be rescued in what is rhetorically construed as a more effective, more widespread, more sophisticated and more civilized language. The simple solution to the problem of Irish is English.

Ethnography, cartography and translation are consonant with each other as well as reflecting actual processes in the execution of the project itself. In terms of nineteenth century socio-cultural evolutionism all three were cultural and political moves in the scrabble of language and power. As part of the repertoire of power each of these theories and methods produced particular social and cultural constructions embedded in culturally specific perspectives. All three relied upon quasi-scientific methods and theories that underpinned colonialist or imperial ideology. The far-from-factual notions of objectivity, neutrality, transparency, fidelity, accuracy or authenticity were central to this ideology. Behind these notions the ethnographers, translators or cartographers are themselves illuminated in their diplomatic disguises and cultural camouflage. They become visible and recognizable as actors whose role was to order culture and enculturate order.

It is not to say that all surveyors were animated by identical motives and acted mechanically upon them. The persuasiveness of such perspectives created the impression in the observer that the other culture was in actual fact unsettled, unfixed or unpredictable. These are all intentions, effects or tendencies that are characteristic of colonialist discourse. The knowledge of the colonized people, marginalized as tradition, antiquity or folklore, remains subterranean. In this sense the map is a diagram of domination and the archive an imperial inventory of the exploitable at a time when capitalist expansion was allied to cultural assimilation. The artefact in the museum is a counterpoint to the commodity on the market.

The colonial regime was linked to an administration of local gentry, those whom the nineteenth-century writer William Carleton called the squandry, united in the common mission of neutralizing and sanitizing indigenous culture. The obscurantist Celticist discourse on civilization rarefied and esotericized the Irish language to supersede it with English. This discourse secreted the real economic and political benefits of power while highlighting native recalcitrance in a reactionary rhetoric of tradition. It was often this very rhetoric that outlived the original context of its production in later reconfigurations of colonialist discourse.

NOTES

1. Justin Stagl, *A History of Curiosity: The Theory of Travel 1550–1800* (London, 1995), p.2.
2. John H. Andrews, *A Paper Landscape: The Ordnance Survey in Nineteenth-century Ireland* (Oxford, 1975); Andrews, 'Notes for a Future Edition of Brian Friel's Translations', *The Irish Review*, 13 (1992/93), pp.93–106; Andrews, '"More suitable to the english tongue": The Cartography of Celtic Placenames', *Ulster Local Studies*, 14, 2 (1992), pp.7–21; Gillian Doherty, *The Irish Ordnance Survey: History, Culture and Memory* (Dublin, 2004); Brian Friel, John H. Andrews and Kevin Barry, 'Translations and a Paper Landscape: Between Fiction and History', *Crane Bag*, 7, 2 (1983), pp.118–24; Brian Friel, *Translations* (London, 1981); Alan Gailey, 'Folk-life Study and the Ordnance Survey Memoirs', in Alan Gailey and Daithí Ó hÓgáin (eds), *Gold under the Furze: Studies in Folk Tradition* (Dublin, 1982), pp.150–64; Patricia Boyne, 'Letters from the County Down: John O'Donovan's first Field Work for the Ordnance Survey', *Studies*, (Summer 1984), pp.106–16, and also *John O'Donovan (1806–1861): A Biography* (Kilkenny, 1987); John Paddy Browne, 'Wonderful Knowledge: The Ordnance Survey of Ireland', *Éire-Ireland*, (Spring 1985), pp.15–27; Angélique Day, '"Habits of the people": Traditional Life in Ireland 1830-1840, as Recorded in the Ordnance Survey Memoirs', *Ulster Folklife*, 30 (1984), pp.22–36; Mary Hamer, 'Putting Ireland on the Map', *Textual Practice*, 3 (1989), pp.184–201; Gillian Smith, '"An Eye on the Survey": Perceptions of the Ordnance Survey in Ireland 1824–1842', *History Ireland*, (Summer 2001), pp.37–41.
3. John H. Andrews has strongly defended the Ordnance Survey from what he has termed its 'accusers'. He views translation as a gesture of respect, recognition and acknowledgement. For further reading see Andrews, 'Notes for a Future Edition of Brian Friel's Translations' and 'More suitable to the english tongue', and also Friel, Andrews and Barry 1983. For an appreciation of Andrew's contribution to historical geography see Kevin Whelan, 'Beyond a Paper Landscape – John Andrews and Irish Historical Geography', in Kevin Whelan and F.H.A Aalen (eds), *Dublin City and County: From Prehistory to Present* (Dublin, 1992), pp.379–424.
4. Stagl, *History of Curiosity*, p.2.
5. Andrew West, 'Writing the Nagas: a British Officer's Ethnographic Tradition', in Peter Pels and Oscar Salemink (eds), Colonial Ethnographies, a special issue of *History and Anthropology*, 8, 1–4 (1994), pp.55–89.
6. Peter Pels, 'The Anthropology of Colonialism: Culture, History and the Emergence of Western Governmentality', *The Annual Review of Anthropology*, 26 (1997), p.167.

7. Ibid., p.163.
8. Tejaswini Niranjana, *Siting Translation: History, Post-Structuralism, and the Colonial Context* (Berkeley, 1992); Lawrence Venuti (ed.), *Rethinking Translation: Discourse, Subjectivity, Ideology* (London, 1992); Venuti, *The Translator's Invisibility: A History of Translation* (London, 1995); André Lefevere (ed.), *Translation, History, Culture: A Sourcebook* (London, 1992); Román Álvarez and Carmen-África Vidal (eds), *Translation, Power, Subversion* (Clevedon, 1996); Lance Hewson and Jacky Martin, *Redefining Translation: The Variational Approach* (London, 1991).
9. Niranjana, *Siting Translation*, p.1.
10. Gísli Pálsson, 'Introduction. Beyond Boundaries', in Gísli Pálsson (ed.), *Beyond Boundaries: Understanding, Translation and Anthropological Discourse* (Oxford, 1993), p.1.
11. Kadiatu Kanneh, '"Africa" and Cultural Translation: Reading Difference', in Keith A. Pearson, Benita Parry and Judith Squires (eds), *Cultural Readings of Imperialism* (London, 1997), p.275.
12. Robert Aunger, 'On Ethnography: Storytelling or Science?', *Current Anthropology*, 36, 1 (1995), p.110.
13. Edwin Ardener, 'The Voice of Prophecy: Further Problems in the Analysis of Events', in Malcolm Chapman (ed.), Edwin Ardener: *The Voice of Prophecy and Other Essays* (Oxford, 1989), pp.135–43.
14. Johannes Fabian, *Time and the Other: How Anthropology Makes its Object* (New York, 1983), p.28.
15. Ibid., p.30.
16. Edward W. Said, *Orientalism: Western Conceptions of the Orient* (London, 1978), p.12.
17. Michael Cronin, *Translating Ireland: Translation, Languages, Cultures* (Cork, 1996), p.134.
18. John W. Foster, *Fictions of the Irish Literary Revival: A Changling Art* (Syracuse, 1987), p.19.
19. Bernard McGrane, *Beyond Anthropology: Society and the Other* (New York, 1989), p.111.
20. Richard Jacquemond, 'Translation and Cultural Hegemony: The Case of French-Arabic Translation', in Lawrence Venuti (ed.), *Rethinking Translation: Discourse, Subjectivity, Ideology* (London, 1992), p.149.
21. Jeffrey Gantz, *Early Irish Myths and Sagas* (London, 1981), p.17.
22. Jacquemond, 'Translation and Cultural Hegemony', p.150.
23. Álvarez and Vidal (eds), *Translation, Power, Subversion*, p.2.
24. Edwin Ardener, 'The Voice of Prophesy: Further Problems in the Analysis of Events', in Malcolm Chapman (ed.), *The Voice of Prophesy and other Essays* (Oxford, 1989), p.149.
25. Omar S. Al-Shabab, *Interpretation and the Language of Translation: Creativity and Convention in Translation* (London, 1996), p.24.
26. Lawrence Venuti, *The Translator's Invisibility: A History of Translation* (London, 1995), p.18.
27. Hewson and Martin, *Redefining Translation*, p.154.
28. John Atchison, 'From Eton Vale to Bamaga – Place, Geographical Names and Queensland', *Queensland Geographical Journal*, 5 (1990), p.13.
29. Thomas Richards, *The Imperial Archive: Knowledge and the Fantasy of Empire* (London, 1993), p.20.
30. John B. Cunningham, 'The Letters of John O'Donovan in County Fermanagh: Dogs, Turkeycocks and Ganders', *Ulster Local Studies*, 14, 2 (1992), p.27.

31. Patricia Boyne, 'Letters from the County Down: John O'Donovan's First Field Work for the Ordnance Survey', *Studies* (Summer, 1984), p.107.
32. John Beattie, *Other Cultures* (London, 1966), p.24.
33. Tejaswini Niranjana, 'Translation, Colonialism and the Rise of English', in Svati Joshi (ed.), *Rethinking English: Essays in Literature, Language, History* (Delhi, 1994), p.124.
34. Clifford Geertz, *Works and Lives: The Anthropologist as Author* (Stanford, 1988); James Clifford and George E. Marcus (eds), *Writing Culture: The Poetics and Politics of Ethnography* (Berkeley, 1986); Robert Aunger, 'On Ethnography: Storytelling or Science?', *Current Anthropology*, 36, 1 (1995), pp.97–130; Christie Beverly J. Stoeltje, 'The Self in "field": A Methodological Concern', *Journal of American Folklore*, 112, 444 (1999), pp.158–82.
35. Johannes Fabian, 'Presence and Representation: The Other and Anthropological Writing', *Critical Inquiry*, 16 (1990), p.762.
36. V.Y. Mudimbe, *The Invention of Africa: Gnosis, Philosophy, and the Order of Knowledge* (London, 1988), p.2.
37. Joep Leerssen, *Mere Irish and Fíor-Ghael: Studies in the Idea of Irish Nationality, its Development and Literary Expression Prior to the Nineteenth Century* (Cork, 1996), p.387.
38. Ibid., p.388.
39. Luke Gibbons, *Transformations in Irish Culture* (Cork, 1996), p.149.
40. Tim Fulford, 'Romanticism and Colonialism: Races, Places, Peoples 1800–30', in Tim Fulford and Peter J. Kitson (eds), *Romanticism and Colonialism: Writing and Empire 1780–1830* (Cambridge, 1998), pp.38–9.
41. Bill Ashcroft *et al.* (eds), *Key Concepts in Post-Colonial Studies* (London, 1998), p.46.
42. Ibid., p.47.
43. Clare O'Halloran, *Golden Ages and Barbarous Nations: Antiquarian Debate and Cultural Politics in Ireland, c.1750–1800* (Cork, 2004), p.124.
44. Robert Kee, *Ireland: A History* (London, Abacus, 1982), p.16.
45. Denis Judd, *Empire: The British Imperial Experience, from 1765 to the Present* (London, 1996), p.40.
46. Benedict Kiely, *Poor Scholar: A Study of William Carleton* (Dublin, 1972), p.88.
47. Marc Ferro, *Colonization: A Global History* (London, 1997), pp.13–14.
48. Richard N. Lebow, *White Britain and Black Ireland: The Influence of Stereotypes on Colonial Policy* (Philadelphia, 1976), p.25.
49. Amy E. Martin, '"Becoming a Race Apart": Representing Irish Racial Difference and the British Working Class in Victorian Critiques of Capitalism,' in Terence McDonough (ed.) *Was Ireland a Colony? Economics, Politics and Culture in Nineteenth-century Ireland* (Dublin, 2005), p.190.
50. Joe Cleary, 'Misplaced Ideas? Colonialism, Location and Dislocation in Irish Studies', in Claire Connolly (ed.) *Theorizing Ireland* (New York, 2003), pp.102–3.
51. Lebow, *White Britain and Black Ireland*, p.40.
52. Ibid., p.61.
53. Edward W. Said, *Culture and Imperialism* (London, 1993), p.8.
54. Robert J.C. Young, *Postcolonialism: An Historical Introduction* (Massachusetts, 2001), p.16.
55. Edward Said, 'Afterword. Reflections on Ireland and postcolonialism', in Clare Carroll and Patricia King (eds), *Ireland and Postcolonial Theory* (Cork, 2003), pp.178.

56. Gregory Castle, *Postcolonial Discources: An Anthology* (Massachusetts, 2001) p.503.
57. Bernard S. Cohn, 'The Census, Social Structure and Objectification in South Asia', in Bernard S. Cohn, *An Anthropologist Among the Historians* (New York, 1990), p.224.
58. Richards, *Imperial Archive*, p.112.
59. Young, *Postcolonialism*, pp.16–19.
60. Ferro, *Colonization*, p.20.
61. Richards, *Imperial Archive*, p.39.
62. Judd, *Empire*, pp.41–2.
63. Ibid., pp.43–4.
64. David Fitzpatrick, 'Ireland and the Empire', in Andrew Porter (ed.), *The Oxford History of the British Empire, Volume 111, The Nineteenth Century* (Oxford, 1999), p.495.
65. Ibid., p.520.
66. Young, *Postcolonialism*, p.19.
67. Richards, *Imperial Archive*, p.23.
68. Colin Graham, *Deconstructing Ireland: Identity, Theory, Culture* (Edinburgh, 2001), p.83.
69. The term *Gaeltacht* refers to the present Irish-speaking areas within the state that are supported by government grants for home improvements, industry or cultural projects. There is a sense in which this was equally a linguistic and ethnographic mapping by an English-speaking Ireland.
70. Stiofán Ó Cadhla, *Cá bhFuil Éire? Guth an Ghaisce i bPrós Sheáin Uí Ríordáin* (Baile Átha Cliath, 1998).
71. Reg Hindley, *The Death of the Irish Language* (London, 1990), p.8.
72. J.E. Caerwyn Williams and Máirín Ní Mhuiríosa, *Traidisiún liteartha na nGael* (Baile Átha Cliath, 1985), p.312.
73. Máiréad Nic Craith, *Malartú Teanga: an Ghaeilge i gCorcaigh sa Naoú hAois Déag* (Bremen, 1993), pp.14–33.
74. Doherty, *The Irish Ordnance Survey*, p.141.
75. Nollaig Ó Muraíle, 'Seán Ó Donnabháin, "An Cúigiú Máistir"', in Ruairí Ó hUiginn (ed.), *Scoláirí Gaeilge, Léachtaí Cholm Cille*, XXV11 (Maynooth, 1997), pp.11–82.
76. Declan Kiberd, *Inventing Ireland: The Literature of the Modern Nation* (London, 1995), p.614.
77. Jacques Derrida, 'Des Tours de Babel', in Rainer Schulte and John Biguenet (eds), *Theories of Translation* (Chicago, 1992), p.219.
78. Albert Memmi, *The Colonizer and the Colonized* (London, 1990), p.173.
79. Roy MacLeod, 'On Science and Colonialism', in Peter J. Bowler and Nicholas Whyte (eds), *Science and Society in Ireland* (Belfast, 1997), p.4,
80. Nicholas Whyte, *Science, Colonialism and Ireland* (Cork, 1999), p.7.
81. Damien Murray, *Romanticism, Nationalism and Irish Antiquarian Societies 1840–80* (Maynooth, 2000), p.8.
82. Nicholas Thomas, *Colonialism's Culture* (Cambridge, 1994), p.16.
83. Ibid., p.97.
84. Roger Knight, 'Colonialism and its Forms of Knowledge', a review of Bernard S. Cohn's work of the same title, *Postcolonial Studies*, 1, 3 (1998), p.437.
85. John Beverly, 'Theses on Subalternity, Representation, and Politics', *Postcolonial Studies*, 1, 3 (1998), pp.305–19.
86. David Spurr, *The Rhetoric of Empire: Colonial Discourse in Journalism, Travel Writing, and Imperial Administration* (London, 1993), p.5.
87. Castle, *Postcolonial discourses*, p.502.

88. Ashcroft et al., *Key concepts*, pp.42–3.
89. Homi K. Bhabha, *The Location of Culture* (London, 1994), p.67.
90. Said, 'Afterword', p.177.
91. Bhabha, *Location of Culture*, p.70.
92. Knight, 'Colonialism and its Forms of Knowledge', p.439.
93. Frederick Cooper and Ann L. Stoler (eds), *Tensions of Empire: Colonial Cultures in a Bourgeois World* (Berkeley, 1997), p.11.
94. Joep Leersen, *Remembrance and Imagination: Patterns in the Historical and Literary Representation of Ireland in the Nineteenth Century* (Cork, 1996), p.102.
95. Talal Asad, 'Introduction', In Talal Asad (ed.), *Anthropology and the Colonial Encounter* (London, 1973), p.17.
96. Gloria G. Raheja, '"The Ajaib-Gher and the Gun Zam-Zammah: Colonial Ethnography and the Elusive Politics of "Tradition" in the Literature of the Survey of India', *South Asia Research*, 19, 1 (1999), p.32.
97. Fabian, *Time and the Other*, p.17.
98. Juri M. Lotman, 'On the Metalanguage of a Typological Description of Culture', *Semiotica*, 14, 2 (1975), pp.97–123.

CHAPTER FOUR

The Bonnet and the Brogue: Race, Place and Language

The word memoir, from the French *aide-mémoire*, referred to technical aids to memory. These were souvenirs like stones, sticks or knots which commemorated things or events in the experienced world. Such objects lent themselves to collection, quantification, arrangement, addition and subtraction and were the precursors of writing.[1] Their purpose was to preserve the impressions of travellers and store the information of surveyors and they were the first step in the removal of knowledge from its bearer.[2] Stagl says that writing breaks the spell of the immediate and constitutes an artificial memory.[3]

Evolving into the *ars memorativa* or the mnemonics of classical rhetoric they represented the imagistic remembrance, recall, documentation, ordering and marshalling of human knowledge as in museums, archives, manuscripts and books.[4] The memoir also became an elite literary autobiographical genre in the seventeenth, eighteenth and nineteenth centuries when diary keeping became fashionable. Cuddon relates this to the cult of anthropocentric humanism in which people explored and analysed themselves in detail leading ultimately to autobiography.[5]

The authoritative aspect of the memoir is reflected in its nineteenth-century sense of a learned essay comprising the proceedings or transactions of learned societies. A memoir is an official report, note, record or memorandum that mixes sourced information with personal detail. The term also suggests the descriptive prose that accompanied maps or the writings of travel-explorers. Early military and estate maps were often accompanied by written descriptions. Vallancey provided military reports with his maps of the south, the Down Survey supplemented its maps with prose 'often with notable literary panache'.[6] William Petty, Daniel Augustus Beaufort, George Sampson and William Shaw Mason were all producers of prose as well as maps. The memoir was a compilation of encyclopaedic knowledge and had become a high art by the middle of the eighteenth century.

Cartographers of high repute published memoirs to accompany their larger and more original maps, laying out in minute detail the many sources and the manner of their reconciliation. Indeed, a map's accompaniment by a memoir was a sign of the cartographer's pretension that the map ought to be considered as a cartographic landmark. Through his memoir, a cartographer assured his public that a map was based on the best available sources and he displayed his own conscientiousness and intellectual virtuosity.[7]

The memoir authorized the map just as the memoirist was the author of the knowledge. From the outset the survey had ambitions beyond the mapping and measurement of boundaries. Conceived of as statistical accounts or remarks in nineteenth-century terms the memoirs were also ethnography.

Natives, Aborigines and Mountaineers

In the memoirs of the survey the people are variously referred to by race, religion, class or situation. The Irish are referred to as the Roman Catholic inhabitants, the lower order of people, the country people, the inhabitants, the aborigines or the Irish Aborigines, the natives, the cottiers, the labourers, the vulgar, the peasantry, the mountain people, mountaineers or the old Irish. Interestingly this is often the case not just for the officers of the Royal Engineers but also for the writers of the Antiquarian or Topographical Department. While adopting the terminology of their superiors in the survey the latter, however, referred to their compatriots and their language at other times as the ancient families, the Hibernian tongue, the Milesian Irish or Erse (the Irish language). If they were not already explicitly derogatory in the nineteenth century, even outside of elite English-speaking circles, some of the terms have become so today. The term mountaineer was used as a term of contempt in the English of Louth for example.[8]

Peasant is derived from the Latin *paganus*, heathen or pagan, and the French *païsant* or *païs* meaning a canton or country district.[9] The *Declinatio Rustica* of the thirteenth century defined the 'six declensions of the word peasant' as villain, rustic, devil, robber, brigand and looter, and in plural wretches, beggars, liars, rogues, trash and infidels.[10] It emerges in the sixteenth century with its early sense of the lowest rank, rustics, bumpkins, brutes, yokels or clowns.

In a general sense it meant the intimate relationship between farm, family and community, tilling the land and living in the country, those for whom agriculture was a way of life and not a business. It is further undermined by its use as a term of abuse suggesting uneducated or common.[11] The French *paisant* became *padhsán* in Irish and continues in English orthography and speech as pysawn, used to refer to any unseemly person. As is the case with many residual terms it has been largely outdistanced, its ontology or rationale eroded, constructed as it was from 'images of preindustrial European and colonial rural society. Informed by romantic sensibilities and modern nationalist imaginations. These images are anachronisms.'[12] Its use here is obviously dictated by its use as a category or classification of the survey itself.

In referring to housing, habits, language, cultivation, situation, religion, education or morals in general, distinctions were drawn between the Irish and the Scots. Those referred to as Scottish in the memoirs were treated in a more sympathetic manner than the Irish and were portrayed as being more civilized to a degree. At other times they were viewed in terms of a provincial gaucherie, having 'developed specific limited colonial characteristics . . . but not others'.[13] This distinction is illustrated by John McCloskey's statistical report from Derry:

> In the Ballymullins district the Irish language is universally spoken. English is also known to all, but they are attached to their own tongue and are disposed to give all their confidence to the stranger who addresses them familiarly in it. Too much of that confidence was reposed in Robert Wetherall, a reputed prophet born in Aughlish in the last century. Among his predictions were the beheading of Lewis XVI by his own subjects, the union of the 'bonnet and the brogue', meaning the Scotch and Irish.[14]

Direct references to language are brief and sometimes excluded altogether but they highlight the inherently cross-cultural nature of the survey. It is useful to look at the way in which the memoir-writers depicted the Irish as well as those they referred to as Scottish. At times it is unclear whether the writers are actually writing about Scottish people or about the accent with which English was spoken. The references are terse at times, Lieutenant G.A. Bennett wrote that, 'there are no prevailing customs, and the language spoken is in general tolerably good English with a Scotch accent, as is usual in the north'.[15] The implication is that this dialect of

English is inferior to that to which the writer is socially attached. As was the case in the writing of Maria Edgeworth: 'moral rectitude is embodied in correct grammar and pronunciation'.[16] Most memoir-writers valorized the use of English and their reports tried to reassure where possible that it was in fact spoken in a particular area.

In Co Tyrone Lieutenant W. Lancey records that, 'the Valleys are filled with Protestants, chiefly of the Scottish churches, the mountains are tenanted by Catholics. English is spoken by all but the vernacular tongue of the highlands is Irish, but is on the decrease.'[17] In Co Tyrone Rev Gilbert King replies to a query of the North West Farming Society in 1823 that 'they all speak English, but in the mountain districts Irish is a good deal spoken. Their manners, from not having had much intercourse with the upper classes, are short and blunt.'[18]

The manners of the people are linked to language and they are considered often under the heading of improvements. The use of English is understood explicitly as a positive sign of civilization and progress. Often the reference boils down to G. Downes' hopeful note suggesting that the 'Irish language is on the decline'.[19] On the other hand J. McCloskey's statistical report speaks of promising bilingualism:

> The English language has come into very general use, except among the most unfrequented recesses of the mountains where it has not yet supplanted the native tongue. The latter is understood probably by all: it is certainly spoken by most of the people, and there may be seen not only Catholic manuals but also the Book of Common Prayer in the Irish language.[20]

The evidence of the memoirs suggests that the Irish language was present except in the most extensively planted areas. The linguistic-racial distinction, often drawn between the Protestant English-speaking people and the Catholic Irish-speaking people, may not have been as simple as some memoirists suggest. In the mainly Presbyterian parish of Drumachose, Co Derry a general will to attain knowledge of the Irish language was noted by J. Butler Williams:

> The parishioners are very anxious to obtain books from the Irish Bible Society. They have also a wish for some acquaintance with the Irish language, as they feel their ignorance of it highly inconvenient, not only in their intercourse with some parts of the county, but also

on visiting other counties to purchase goods. In the markets where Irish is spoken those unacquainted with the language are regarded as foreigners, and to cheat them is considered a praiseworthy deed. The wish to acquire that language prevails in all the surrounding parishes.[21]

Apart from the obvious Irish-English division there is evidence of further linguistic difference even amongst the part of the population generalized as Scottish. McCloskey reports from the parishes of Banagher, Bovevagh and Dungiven in Co Derry that:

> The greater part of the people speak the English language, but the dialect abounds in Scotticisms. In Boevagh and the western part of Banagher the Presbyterians retain the broad Scotch accent of their ancestors . . . the Irish language is much spoken: among the mountains are some women and children who can scarcely speak a single word of English.[22]

Created from fragments and field-notes intended for publication, current notions of English literary taste and style were a part of the production and consumption of the memoirs. The magical effect of the fluency of the memoirs is to make the observed translucent. This is reminiscent of what Venuti calls the illusory effect of translation, the way in which a translation is made to appear original from the translator's effort 'to ensure easy readability by adhering to current usage, maintaining continuous syntax, fixing a precise meaning, the effect of which is to conceal the numerous conditions under which the translation is made, starting with the translators crucial intervention in the foreign text'.[23]

In a sense there is no text, no culture prior to the intervention of the translator or interpreter. Ireland is textualized as it transcribed, translated and presented. The Ireland of the memoirs is an Ireland mediated by the intellectual, ideological and elite literary standards of the English language and colonialist discourse. Apart from language the descriptions of places are also permeated by racialist assumptions.

Home Pleasing Scenery

The word landscape was originally a technical term used by painters and was borrowed from the Dutch *landschap*. A landscape wasn't just any old place, it was a place that reminded the observer of a painting. The nineteenth century saw the increased intervention of humans into what was previously considered part of God's creation 'in the form of science, agricultural improvement, and the industrial revolution'.[24] The growth of cities led to increased tourism and an educated philosophical or romantic attraction to the rural landscape. A visual culture of description emerged that was related to 'the craft of empirical representation. The prevalence of maps, map-like representations, and pictures with a "realistic quality" are expressive of this visual culture.'[25]

In the eighteenth and nineteenth centuries the aesthetic of the picturesque had begun to influence English and European culture in general. The term picturesque was originally derived from the Italian *pittoresco* that referred to any subject that was thought to be paintable.[26] In the eighteenth century the scenery of Ireland was eulogized as romantic and beautiful. At the end of that century tourism was flourishing and 'publishers recognised the trend and began issuing special travellers' handbooks for Ireland similar to those on offer for tourists on the European grand tour. These directories increased in number after 1800.'[27] The fieldworkers of the Antiquarian or Topographical Department often had to compete with tourists for accommodation in the course of their work. The memoirs, and to a lesser extent the letters, blend the ethnographic and statistical quest for information with the literary style of the traveller's gazetteer and the artistic eye of the painter.

In 1833 a number of the civil assistants who had been sketching hills for the proposed one-inch map of Ireland were employed in obtaining information for the memoirs. Andrews argues that their 'eye for country' was a necessary qualification.[28] Scholar and artist George Petrie was employed by the survey from 1833 to 1846 as were his pupils William Frederick Wakeman and George du Noyer. Their sketches and paintings reflected the elite picturesque aesthetic or taste in much the same way as the popular English genre of travel writing epitomized by Mr and Mrs Samuel Carter Hall.[29] The gaze of the memoir-writer was often directed by the conventions of the picturesque. Brett says that this way of viewing or looking at the land 'was embedded in social class and education to a very marked degree'.[30] Edney calls attention to the way in which the British looked at South Asia:

> If the view at hand (at eye?) failed to meet the established Picturesque norms, then the observer could alter (that is, cure) the view in its graphic representation until it agreed with those norms. Overall, the

various 'natural' gazes replicate the power relations inherent in social discipline. The act of observation embodies the observer's physical (military, political, gendered) power over the observed; it also embodies a moral power in that it defines and creates and normalizes the observed.[31]

Parallels have been drawn between the conventions of the picturesque and early forms of ethnographical description. The picturesque became 'a powerful framing device for the way in which non-Western cultures came to be perceived, represented, and colonized'. There is a tension between the convention of romantic description and the ethnographic information associated with observation, empirical record making and experimentation.[32]

Under the heading of Topography or Artificial State one of the inquiries concerned the general appearance and scenery of an area. Far from being neutral verbal articulations of what was engraved on the map, responses to this heading often reveal more about the observer than they do about the observed. Edney says, 'the Picturesque gaze was ... an inherently elitist and class-based aesthetic; it was an educated taste. Its possession was the hallmark both of gentlepersons, who were confident of their social position and their natural right to own and to shape the land and its products, and of social climbers . . . who asserted their right to such status.'[33]

J. Hill Williams writes that the town of Tanderagee in Co Armagh presented 'a uniform appearance' while Tanderagee Castle was 'well planted and tastefully laid out'.[34] The notion of the picturesque is closely associated with what the survey called the state of improvement. This spoke of the productivity of an area, the relative degree of settlement or even the visible signs of law and order. Large estates and gentlemen's seats feature prominently in these descriptions. Lieutenant C. Bailey notes that Castle Dillon in Loughgall, the residence of Lieutenant General Sir Thomas Molyneux who was himself a contributor to the memoirs, 'stands in the centre of an extensive demesne containing a great deal of wood and ornamental plantation, with a large and picturesque lake which adds much to the beauty of the place'.[35]

The artificial creation of the picturesque is always noted and singled out for praise, as is evidence of cultivation. Lieutenant Bailey describes the parish of Tynan, Co Armagh as 'rich and well cultivated, with several gentlemen's seats containing a good deal of wood and ornamental plantation. The southern portion of the parish consists of a poorer and less productive soil, with very few trees or hedgerows to improve the scenery. It has a bleak and cold appearance.'[36] The 'home pleasing' scenery is most praised, the marginal or flat landscapes inhabited by the Irish are described as forbidding by some writers. Features of the local economy or farming are seldom recognized and never related to the season or time of the year. T.C. Hannyngton and James Boyle are impressed with what they view as the English neatness, industry and civilization in the parish of Ballymartin, Co Antrim:

> In the north east district the appearance of the country, though not actually interesting, is pleasing and agreeable ... the farmhouses indicate in their construction and appearance an idea of comfort and neatness and of industry and civilisation on the part of their occupants ... in the other portion of the parish the country is but partially cultivated and broken in, and the system of agriculture less improved. The houses and cottages are by no means neat nor compact in their appearance ... Lord Templetown has recently put down a large quantity of Spring planting on Lyle's Hill. This will, when it begins to show, greatly relieve the bareness of the scenery.[37]

The memoir-writers scour the landscape for signs of civilization. In cases where they determine none to exist the references are very brief and general. In the Grange of Ballyrobert, Co Antrim, James Boyle reports that 'the appearance or scenery of the grange is not generally interesting, owing to the almost total absence of planting ... towards the south of the grange (a district rather retired and unfrequented) the aspect of the country becomes gradually less pleasing and finally bleak and bare ... the country is destitute of planting. Dry stone fences prevail.'[38] In the memoir for the parish of Kilbroney in Co Down, written by J. Hill Williams, there is a good example of the memoir-writer as tourist or traveller carrying us with him on a reverie through the landscape:

> On coming to Rostrevor from Hilltown the road is excessively bleak and uninteresting, the view being confined by the proximity of high bare mountains on each side, until on arriving within a mile of the town of Rostrevor the contrast offered to the view is striking and delightful. From a barren and cold region, the traveller had passed at once from the contemplation of the rich and wooded valley at the foot of which the town is built, with the wooded mountains and the bay (then appearing like a lake) with the passing sail scattered over its tranquil surface. From the Quay to the Wood House, the road to Kilkeel, cut out from the foot of the mountain and a little elevated

above the sea, is completely shaded over by the foliage of the trees on each side thus forming a cool and agreeable walk in summer, the beauty of which is enhanced by numerous turns in the road forming vistas of foliage.[39]

From the bleak, bare, barren and cold area the memoirist is surprised by the contrast he is presented with. The wooded hillside suggests civilization, it is a cultural landscape rather than a natural one, the sea appears pleasing because it looks like a lake. The final excitement is the journey through the trees that twist and turn entertaining the observer with an optical phantasmagoria. This could be likened to the subjective visual experiences created by the optical devices of the mid-nineteenth century that were initially intended for scientific observation but quickly became forms of popular entertainment. These included the thaumatrope, the Faraday Wheel, the phenakistiscope, the diorama, the stroboscope and the zootrope.[40] The landscape is a stimulus to the observer's subjective contemplation.

In Aghalurcher, Co Fermanagh, the landscape described by Lieutenant J. Greatorex is picturesque and painterly, 'the scenery and general appearance of the country immediately surrounding Lisnaskea is very beautiful and interesting . . . through out the parish the artist might find many spots upon which he could dwell with pleasure, and it is only to be regretted that so fine a district is not made more of, by planting and improvements'.[41] The Ireland of the memoirs is not only painterly but writerly as well. In the memoirs colonialist discourse blends with genteel English letters and taste. The miscellaneous papers attached to J. Stokes' memoir for Ballynascreen in Co Derry note that, 'planting is much wanted: it appears indeed in many places, but the generation is yet to arrive which shall avail itself of the many capabilities of the country that shall invite well-tutored art to lend her helping hand to exhibit and adorn the countless beauties "that nature's boon pours forth perfuse, o'er hill and dale and plain"'.[42]

The tree was the very sign of civilization to the observer and always suggested improvement, plantation and richness. In Longfield, Co Tyrone J. R. Ward's draft memoir highlights their absence reporting that 'there is little wanting but trees to render it highly picturesque'.[43] M. M. Kerkland, writing from the parish of Slanes, Co Down, recoils in horror from the view presented to him noting 'a sickening want of variety and tameness in its appearance. The coast is bleak and rocky without being picturesque, and there is not a tree to be found from one end of the place to the other'.[44] As Fabian says, in regard to the explorers of Africa, the recognition of people, practices or institutions was suppressed while occasionally familiar features from a particular cultural landscape penetrate the cordon of memory.[45] For some memoir-writers the tree was evidence of Englishness and reminiscent of home. The parish of Camlin, Co Down had 'the appearance of a thickly wooded country, more resembling a rich English than an Irish district'.[46]

To this can be contrasted the vernacular stone wall boundaries of other areas. Lieutenant W. Lancey writes in the statistical report for the parish of Desertegney, Co Donegal that, 'the road to Dunree Fort passes to the right of the best land in the parish and the unvaried monotony of stone walls and the almost total absence of trees . . . is relieved by Mr Harvey's house and small plantations on the coast'.[47] Lieutenant E.W. Durnford writes that the parish of Toomregan, Co Fermanagh was in dire want of an Englishman:

> The scenery of this parish owes very little to art, but nature has been profuse, affording every variety of hill and dale from the lofty Slieve Russel to the plains skirting the river a little above Ballyconnell. Added to these are several land loughs placed in good situations and tempting, as it were, spirited men to improve upon the scenery of nature . . . what would many English gentlemen in England give for such desirable situations? Here they are neglected and serve only to gratify the eye by presenting a contrast to the usual barren appearance of the country.[48]

The scenery was not passive but part of an ideological vision that permeated all aspects of the survey. This is suggestive of the idea of a natural physiognomy, the notion that each country, county or parish had its own particular type of landscape.[49] In the nineteenth century civilization was understood as resulting partly from the environment and climate. The barbarous features in the landscape reflected the barbarous features of the people. The dry stone walls, bogs and mountains were Irish and the trees, gardens and flowers were English. J.B. Williams writes from the parish of Tamlaght Finlagan, Co Derry, that:

> Several of the cottages between Roe Park and the parish of Bovevagh have little ornamental gardens before their doors, with trailing plants against the walls. These are neatly trimmed and are indeed an improvement on the usual accompaniment of a pig and a dunghill,

while the fragrance of the flowers indicates to the stranger that the proprietor is capable of appreciating the value of the English word 'comfort'.[50]

One obstruction to improvement noted by Lieutenant Lancey in the parish of Desertegney in Co Donegal was what he calls the peasants' interference with gardens and their use of timber: 'no plantations can be expected to thrive where the people cut the young trees as soon as they are fit for their purposes. Even peas and beans can scarcely be grown for depredations constantly committed on gardens, and the few persons who are willing to improve are overawed by the peasants'.[51]

The British officers seldom allow themselves the luxury of reflection on their roles as outsiders or foreigners. Writing from the town of Enniskillen, Lieutenant John Chaytor refers to himself as a stranger. He writes that the view from the distance 'cannot fail to arrest the eye of the traveller (especially approaching from the south or west) as a scene of rare imagery, conveying the romantic idea of a town swept off *terra firma*, and as it were floating upon the surface of the waves'.[52] As he draws uncomfortably closer, however, its appearance is described as very different and the officer is repelled:

> The town is neither lighted nor watched, and from the broken and very irregular state of the pavements it requires more than ordinary caution to walk them at noonday without stumbling! What then must a stranger encounter, who enters the town on a winter's night with no other guide to his way than the faint gleam of a shop window, or perchance a dull lamp at the door of an inn? Surely these alone are sufficient to stamp an unfavourable impression and at once bespeak the supineness of the corporation or the sordid habits of the people.[53]

The accounts of a *terra nullius* or empty landscape were a characteristic of the travel writing of the eighteenth and nineteenth centuries and the descriptions in the memoirs are similar. The accounts given under the heading of the general appearance and scenery present an unpopulated country anticipating or awaiting improvement. In reality the population of Ireland had never before, or since, been as numerous. It is as if nature, as Spurr says:

> Stands for an empty space in the discourse, ready to be charged with any one of a number of values: nature as abundance, as absence, as original innocence, as unbridled destruction, as eternal cycle, as

constant progression. These meanings, which lie layered in the discourse, fill in one for the other as the occasion requires.[54]

One small and brief exception is from Ligar, Bleakly and Stokes' memoir for the parish of Dunboe, Co Derry where 'on a summer's evening, as the sun sets behind the Inishowen mountains and leaves the rocks and all around in a softened light, with his last rays reflected by the wet sand, a scene presents itself, the solitary grandeur of which is only occasionally broken by groups of figures moving along the beach or emerging from the lengthened shadows of the cliffs'.[55] Even at a distance the people seem almost to trespass upon an otherwise idyllic scene or lend it a romantic aspect.

There is a persistent and recurrent concern with education, libraries, reading rooms, schools, churches and chapels. Their presence or absence is noted and attributed to the race, class or religion of the people. In the parish of Leckpatrick, Co Tyrone, George D. Mansfield stated in reply to the queries of the North West Farming Society that 'habits of decency and order are observable amongst the children of the poor who are educated. The one is the rational accountable being, the other the idle mischievous and inconsiderate creature who, from want of proper employment, is constantly engaged in either what is useless or wicked.'[56] The rationale of the memoirists could be explained by class as much as by race. These were often interrelated categories in evolutionary discourse. The dominant morality was a middle- or upper-class English one pursuing radical change in the idiom of improvement, reform and progress.

As Kucklick says 'racial characteristics were not only determinants of culture but also were themselves determined by all aspects of individuals' and peoples' cultures: wilful behaviour, traditional practices, material possessions, and natural surroundings'.[57] The confidence of such assertions in relation to Ireland is qualified by the fact that rural life in Britain as experienced by the surveyors or sappers themselves, in spite of the Industrial Revolution, did not differ radically from the one they were describing in Ireland. The difference, whether subtly or bluntly stated, was that the particular discursive idiom of the survey attributed what it described as the failings and shortcomings of the Irish to racial immaturity or inferiority.

In the parish of Aghadowney, Co Derry, Stokes, Ligar and T. Fagan recount that 'there has been much moral improvement from the general introduction of schools into the parish. This has manifested itself by increased cleanliness, temperance, punctuality of attendance at public worship, and a continually growing intelligence and love of information.' Intelligence would result from practising the proper religion, behaving properly, being properly hygienic, speaking the proper language and acquiring the proper information.[58] 'Adopted during the eighteenth and nineteenth centuries by the aristocracy and the bourgeoisie, by clergy, officialdom, and industrialists', the clock symbolized the new standardized concept of time and 'confronted the traditional peasant view and use of time'.[59] In the Grange of Mallusk, Co Antrim in 1838 it served for James Boyle as testimony to the progress of the people that 'each house possesses a clock. Most of the better description of farmers have jaunting cars, and most of the young men have watches.'[60] In the parish of Skerry, Co Antrim Lieutenant R. Boteler recorded that the clock was referred to as 'she' while a party was held 'at the setting up of her'.[61]

Conspicuous governmental or colonialist concerns are revealed from time to time. The presence of a military regiment in Enniskillen is commended for contributing to its state of improvement. On the other hand so-called ancient customs are noted for their potential to obstruct improvement. Vernacular gender relations are sometimes sketched, men drink too much whiskey while women drink too much tea.

Generally, however, the Irish culture is seen as an obstacle or something to be corrected. In the parish of Aghanloo, Co Derry, Ligar and Stokes write that 'the game of card playing, cock-fights at Easter, Christmas and All Hallow Eve with hurling are diminishing from the gradual increase of seriousness in the general character of the parishioners'.[62] There is a brief reference by J. Stokes to local resistance in the parish of Ballynascreen, Co Derry to the Draper's Company, 'it is remarkable that there is an obstruction to improvement derived from the inhabitants themselves. The Irish inhabitants view with much suspicion the exertions of the company.'[63]

A Peaceful and Industrious People

While the attitudes of the writers are sometimes subjective or personal they are generally responsive to prefabricated inquiries and directives and tend to vary very little. As a result an almost formulaic, or at least repetitive, pattern of description emerges. In the areas that are designated as Scottish the reason, very often explicitly stated, for the improvement of an area is given as colonization. It is unfailingly recorded as an improvement in terms of morality, religion, education, language, culture and road building.

In the parish of Ballymartin, Co Antrim, James Boyle and T.C. Hannyngton write that, 'there is not any local record or tradition which can throw any light on the history of this parish previous to its having been, in common with all the surrounding districts, colonised by the Scots in the sixteenth and early part of the seventeenth centuries . . . the colonisation of the parish of this period laid the basis for the several improvements which have since taken place in it'.[64] These improvements are also attributed to Lord Templetown, the local landlord. In the same parish, one in which there was no Catholic landowner, the people are presented as being naturally given to education, 'there is not any of the neighbouring parishes in which there is more taste for information or anxiety for education than in this . . . the introduction of schools in this neighbourhood has been almost coeval with its colonisation by the Scots and the establishment of Presbyterian congregations'.[65] The population are described as 'an intelligent, enlightened, moral and industrious people, having cleanly and comfortable residences, nicely whitened and surrounded by little clusters of trees . . . the people are in a prosperous state, and seem alone intent upon industry'.[66] When a memoir-writer records, as was often the case, that 'there is nothing peculiar to them', it is shorthand for the absence of traditional practices that could obstruct improvement or acculturation to English norms. The distinctive Scottish traits noted in this area contradict the memoir-writer's earlier assertions and transgress the strict racial coding often employed:

> They have many pithy saws and proverbs, and their dialect, idioms, customs and manners are purely Scottish and by no means pleasing. Their manners, even when intending civility, are far from being courteous, but their ideas and principles are generally honest and manly. They are more than a little stubborn. Improvements are usually looked on by them as innovations, and suggestions as interfering. It is difficult to persuade them to change.[67]

The descriptions of the settled populations are sometimes similar to the descriptions of the natives wherever the local culture seems vibrant. James Boyle wrote of the Grange of Ballyrobert, Co Antrim:

> Their accent, idioms and customs are strictly Scottish. Their manners are uncourteous and anything but agreeable or civil. They are inquisitive, but at the same time incommunicative and are very jealous and suspicious. They are by no means bright in their intellects, nor, speaking generally, intelligent in their conversation . . . they are rather superstitious, their creed in this respect being similar to the people of the neighbouring parishes.[68]

The following passage of Boyles' follows a formulaic pattern and similar passages occur frequently in reference to this sector of the population. The memoir-writer attributes improvement explicitly to colonization:

> The colonisation of the country by a peaceful and industrious people, the establishment of religious congregations and regular clergy, the subsequent introduction of schools, the construction of good roads and the enjoyment of property in the peace and security which ensued the few troublesome years after their settlement, have been the causes of all the subsequent improvements.[69]

Obviating the agrarian serfdom of the nineteenth century the presence of a gentleman is always mentioned as proof of improvement. In the parish of Aghalurcher, Co Fermanagh Lieutenant Greatorex notes:

> The Scotch and English settlers brought habits of industry and cleanliness with them that in this parish are very apparent . . . the late Sir Henry Brooke by his attention to the habits and comfort of his territory went far towards affording his dependants an opportunity of rising above the general humiliating state of the Irish farmer . . . there do not appear to be any obstructions to improvement, there being no legal disputes or uncertainties about rights, about land or boundaries, nor are there any ancient customs tending to obstruct improvement. For it is not to be expected, as nearly the whole of this district was forfeited at the close of the rebellion in 1641, that any of the ancient families exist at present, at least not upon the lands of their forefathers, and it therefore follows that the new settlers soon laid aside any of the then ancient laws or customs which fell of course into desuetude by degrees.[70]

In the town of Lisburn 'the civilised manner of the people' is attributed by George Scott to the settlement in the area of French Protestant refugees.[71]

Advanced aspects of material culture, the use of technology, techniques of cultivation and a particular kind of spatial organization are attributed to the Scottish areas. In the parish of Glenavy, Co Antrim, James Boyle differentiates systems of farming and agricultural techniques:

> Their system of farming is anything but Irish, as potatoes are generally planted and moulded by the plough. Threshing machines are numerous, the hedgerows neatly trimmed, the lanes shady, the fields large and well laid down and squared ... the implements of husbandry are of an improved description: good iron ploughs driven by reins, and large farming carts drawn by draught horses of a suitable breed. All these circumstances, with the aspect and scenery of the country, give it more the appearance of a rich English district.[72]

This relatively advanced agricultural economy and organization is extended to their social and emotional life by J. McCloskey in his statistical report on Ballynascreen, Desertmartin and Kilcronaghan in Co Derry where:

> The disposition of the Scotch is cold but not unsocial; their manners reserved but not impulsive; their affections mostly bounded by the family circle but not irresponsive to the call of benevolence; their habits frugal, yet capable of softening occasionally into generous relaxation; and in cool perseverance, calculating prudence and command of temper they deserve to be taken as models by the Irish.[73]

In contrast the Irish are impulsive, emotional, irrational and intemperate, their character and personality, like the landscape or houses they occupy or the language they speak, is rough, plain, unEnglish and uncivilized.

From the Midst of Smoke and Filth

The portrait of the Irish differs from the depiction of the Scottish and their lifestyle is presented as evidence of pre-colonial chaos. Lieutenant George A. Bennett comments wryly on the local names in the parish of Drumgath, Co Down that reflect a vibrant vernacular language and culture:

> A very prevalent name in the parish, particularly towards the west, is that of McConvills formerly McConwell. The different branches of

them are more generally known through the parish by such distinctive appellations as the following: Jack Caul, Charle's Jack, the Knocks, the Roashawens, the Warities, the Butts, the Arts, the Bawns, the Stockas, the Oin Paddoins, the Pawricks, the Gonghrees, the Petheronels, the Tonyanthonys and the Nealanthonys.[74]

Similar to the examining, disciplining, and improving features of the British observations of India the memoir-writers often name and shame those whom they considered the most troublesome natives. In the parish of Loughguile, Co Antrim, Lieutenant Greatorex describes the mountain people under the heading of progress of improvement:

> Roman Catholics are to be found in almost every part of the parish but particularly in its mountainous districts, which are almost exclusively occupied by them. The wild valley of Glenbush is entirely occupied by Roman Catholics. The names principally found are those of McAuley, Quin, Murray, McAlester, Casey. They are a wild, lawless, uncivilised race and frequently create disturbance at the neighbouring fares. Some families, particularly the Murrays, are renowned cattle thieves.[75]

The memoirs occasionally take on the character of surveillance and provide intelligence on the local population. The accounts of the places locally called the Creagh and the Loup in Co Derry are examples. J. Boyle pens the following report on the parish of Artrea, Co Derry under the early improvements heading:

> The earliest settlement which took place in this country is said to have been in this parish. It seems, however, to have been partial, as in some of its districts the inhabitants are almost exclusively Roman Catholics and probably native Irish. Such for instance is the district along the lake called the Creagh, which is very backward in civilisation and inhabited by the families of the McErlanes, Diamonds and Ritchies, all of whom are Roman Catholics, the former particularly numerous and very troublesome. This is the least civilised district in the parish, the people being rather lawless and ignorant and their habitations small, dirty and comfortless.[76]

In the area referred to as the Loup the same author similarly lists certain families by their surnames:

> The most prevalent names are those of O'Neill, Scullion, McVey (who are to be found in every townland) and Devlin. These are said

to have been the original possessors of the soil. The McVeys and Scullions were at one time very troublesome, and until within the last 4 years the Loup was a very unsafe and uncivilised part of the country. Now, owing to the exertions of the clergy, but particularly to those of the agent of the Magherafelt estate, on which it is situated . . . this part of the country is much improved and the inhabitants, though less enlightened, are as to conduct little inferior to those of the rest of the parish.[77]

The memoirs mix calculation, ethnographic detail and quasi-scientific terms. Using technical terms borrowed from Scotland like cluster or clachan, families are presented as tribal clans. In the parish of Dungiven C.W. Ligar relates that, 'three-fourths of the inhabitants of the Beneda glen is of the clan McCloskey and are distinguished by surname of "Hamish", "Ruadh", "Fadh". The population is very dense and there may be frequently seen 3 generations together in one house sleeping together in the same bed of straw, accompanied by the pig who, however, has a corner to himself.'[78]

In the Beneda glen in the same county other families are named with added references to their vernacular denominations. Boyle wrote that Portlee in the parish of Drummaul, Co Antrim was, 'almost exclusively inhabited by Dowds, and in other townlands the most prevalent names are those of McCann, McAteer, Fenton, Granny (sometimes pronounced Grant) and McAuley, all of whom are Roman Catholics and apparently descendants of the aboriginal inhabitants'.[79]

If the tree was a sign of Englishness, the thatched house was a sign of Irishness. The word cottage came untranslated into the Irish language in the nineteenth century giving local names like *Páirc na gCottageí* or the field of the cottages in Ring, Co Waterford.[80] In the Irish language these were often simply called houses. In English the term cottier or cottager was used to describe the Irish in general. The Irish or aborigines of the memoirs are typically of the lower class of labourers and they become part of what Frykman and Löfgren call the moral problem of cleanliness, 'during the nineteenth-century, people in the countryside were exposed to a systematic barrage of propaganda aiming to teach them better morals and hygiene, to bring them up to be citizens pure in word, thought, and deed'.[81] What they call the pride of the farm, the dung heap, became the focus of this clean-unclean principle and the elaboration of cultural and racial boundaries. For the surveyors and memoir-writers the thatched house and the dung heap signalled Irishness. Rather than representing the everyday social and economic conditions they were evidence of racial inferiority. Lieutenant Greatorex's report on the parish of Aghalurcher, Co Fermanagh is typical:

> The cottiers and labourers live in the most wretched way in mud huts, in many instances without either window or chimney, with, in almost every case, an offensive heap of liquid manure at the very entrance to their miserable abodes. Still they appear happy and contented, and it is really surprising to see what healthy stout children issue from the midst of smoke and filth, brought up by the parents almost in a state of nudity in all weathers and entirely fed upon potatoes.[82]

In such descriptions there is no account of the taxation placed on glass or how tenants on estates were reluctant to display any signs of prosperity in the fear that the landlord or his agent, or his agent's agent, would raise the rent on them.

The Irish are presented as the poorest, least civilised, worst dressed, most troublesome, most superstitious and dirtiest of the inhabitants. The degree of dirtiness increases or decreases according to the presence or absence of the Irish. In the parish of Artrea, Co Derry, J. Boyle writes that 'it may be remarked that the richer, better cultivated and more fertile parts of the parish are inhabited by the settlers . . . they seem to be more independent in their circumstances, to have larger farms, better houses and greater idea of and taste for comfort than the Irish, who are not quite so daring or independent'.[83] The same writer in the parish of Dunaghy, Co Antrim writes, 'that the houses of the Roman Catholics are less neat, cleanly and of an inferior description to the Protestants. They are also less educated and more careless in their persons'.[84]

The observations of the writers can be called ethnographic in so far as this was an elite genre in nineteenth century Britain designed to find evidence of the evolution of man or justification for colonization. The discourse of the memoirs is distinctly class-bound but also functions as an ethnocentric or normative catalogue of race. The class-bound description is often subsumed within the superordinate description of race. Notwithstanding the questionable presentation of certain livelihoods or lifestyles as dirty, dirt itself was not seen as a result of poverty or exploitation but rather as a national trait. The replies to the queries of the North West Farming Society in the 1820s, written by Lieutenant R. Stotherd, for the parish of Clogher, Co Tyrone, is an exception:

> There is very little order, cleanliness or neatness in general to be found either in the houses of the poor ... the turf stack often approaches within a few yards of the door and thus intersects the view and stops the currency of the air. The yard in front of the house is full of the odour of the cow house and stable, for they are often built in the very front and sometime adjoining the dwelling house. The lanes and approaches to the house are narrow, rough and filthy in the extreme. Within no order is visible: you may see pigs and fowl eating in the kitchen and everything dirty and confused ... the potatoes at meals are thrown out in a basket and so laid on the table or on a stool, and the whole family gather round, master, mistress, children and servants in a mass, and eat out of the basket without knife, fork or any other appendage at meals. A man who can give his daughter in marriage 50 pounds or 100 pounds will live in this manner.[85]

The manners and custom discourse is inquisitive, curious, imagistic, ethnographic, evangelistic or antiquarian in turn but as Brantlinger says 'there might be many stages of social evolution and many seemingly bizarre customs and "superstitions" in the world, but there was only one "civilisation", one path of "progress", one "true religion"'.[86] Andrew Hammond's replies from Donegal to the questionnaire of the North West Farming Society illustrate the missionary zeal of the civilizing colonizer:

> The difference is very observable in the state of the peasantry. Formerly the children of the poor were either strangers to education: now we behold the blessings of education, of cleanliness, of decency, of regularity and, above all, the glorious light of the Gospel of our Lord, spreading their united influence over their native land. And it is a cause of great rejoicing to me to behold many worthy characteristics among the high and great who are anxious to promote the good cause and whose anxiety proceed from the true principle of good, namely that of love to God and a desire of bringing glory to his name.[87]

It could be said, paraphrasing Brantlinger, that Ireland, like Africa, 'grew "dark" as Victorian explorers, missionaries, and scientists flooded it with light, because the light was refracted through an imperialist ideology that urged the abolition of "savage customs" in the name of civilisation'.[88]

The manners and civilization that are valorized are English manners and civilization. The ethnographic details thrown up in observations and descriptions are shrouded in colonialist discourse. This is illustrated in the memoir for the parish of Duneane, Co Antrim, written in part by James Boyle and G.W. Hemans:

> In those districts inhabited by the Irish the appearance of their cottages and farms is characteristic of their own habits and dispositions. The houses are slovenly and comfortless and the ground partially cultivated, the fields being enclosed with crooked and insecure ditches, while the people are dirty and careless in their persons, unsettled and irregular in their dispositions, and the passing generation ignorant and unenlightened.[89]

Occasionally there are signs of the trope that gave rise to the loaded image of the Irish as noble savages in Anglo-Irish literature. On the one hand they are open and lively and on the other suspect and wild, enemies to civilization and improvement. Chapman outlines this anomalous position in the representation of the Irish and the evolution of the cross-disciplinary discourses that are urban, educated, rational and rooted in the English language.[90] In a statistical report on the Co Derry parishes of Ballynascreen, Desertmartin and Kilcronaghan, J. McCloskey states that 'the most striking features in the character of the natives are a social disposition, generous affections, a natural vivacity, an open hearted frank deportment . . . there are shades yet darker in the portrait. Their thoughtlessness, their irritability, their improvidence, their stoical indifference to personal and domestic convenience, and the apathetic acquiescence of the poor.'[91]

The parish of Faughanvale in the same county is described by the Irish scholar John O'Donovan, reiterating colonialist jargon, as having 'a half-ruined appearance. Sundry heaps of dirt, with lazy curs sleeping in the sun, houses old and dingy, and a neat fresh signboard to one of the whiskey shops also give it a genuine Irish character'.[92] This may or may not be tinged with irony but there is no recognition here that such circumstances, in other times and in other words, were familiar to O'Donovan in his own experience and upbringing. This is a question to which I will return later.

Manly Sports by Daylight

It has been claimed that the survey rescued Irish folklore but the Irish in the memoirs are categorized by their regular attendance at fairs, patterns and festivals, drinking, cock-fighting and the funeral cry. In Donacavey, Co Tyrone:

> The Beal fire blazes on St John's Eve, but even this ancient custom is gradually dying away. The numerous saints' days are regularly kept by the Roman Catholics and the idle and dissipated are never in want of an excuse for quitting their work ... look at the crowds of idlers who frequent the fairs throughout Ireland with no object but amusement. How much valuable time is lost? The Irish, I have frequently heard it said, are willing to work. I say that they are most unwilling; when at work they will do double the quantity of work of another man in the same time, with the view of having an idle day afterwards.[93]

The calendar customs and vernacular culture that would become the proud archive of professional Irish folklorists one hundred years later are described as going up in smoke in a grand pagan-style Saturnalia. In emphasizing what seem to be the most obviously savage or uncivilized customs in nineteenth-century evolutionary terms the following description from the parish of Dunboe, Co Derry, by Ligar, Bleakley and Stokes is typical. The details are too sketchy and superficial to be of any real use in terms of ethnographic description and it is likely that more could be found through contemporary research:

> Among the Irish population intemperate drinking at funerals and indulgence in the savage pastime of cock-fighting has been greatly diminished through the influence of the priesthood, but nightly dances and gatherings for mere amusement are still too frequent among a population not sufficiently educated to indulge in them safely, and the substitution for such customs of manly sports by daylight under the direction of moral and respectable persons would be of great advantage to the population generally.[94]

Through the careful combination of literary and visual representation the memoirs are partially a fabrication that Kanneh says has 'particular significance for ethnographic practice. Oral, or "traditional" cultures offer themselves up for visual de-coding in the drama of landscape and environment; in the visibility of bodies and faces; in the contours of masks and art forms.'[95] What results from the reconstruction of the other is sometimes a new document about the observer; it is less a window than it is a mirror.

Generalized information is presented as empirical truth with little or no reference to informants and virtually no sustaining detail. The observable is mixed casually with the unobservable and ultimately authorized by the status, race or position of the memoir-writer. In the parish of Tullyrusk, Co Antrim, Boyle, Scott and Hannyngton write:

> It may be stated with impartiality that the Roman Catholics are inferior in civilisation to the rest of the population. They do not dress so neatly or well or with the same taste. They are less cleanly in their habitations and live more poorly. They attend wakes for amusement and keep up cock-fighting. They are not so generally educated and they are inferior in their circumstances, their farms being small and a greater proportion of them engaged as labourers or servants. But at the same time they are fully more hospitable and generous, open and frank in their manners than their neighbours and possess most of the good traits of character peculiar to Irishmen, still, among aliens, retaining much of their native natural love of fun, their hospitality, pugnacity and improvidence.[96]

The Irish are written at, redefined and re-inscribed, often through the invisible processes of translation. They are childlike and innocent, warlike and devious in turn. This is premised upon the 'intermediate myth that is always used abundantly by all official institutions, whether they are the Assizes or the periodicals of literary sects: the transparence and universality of language'.[97] There is no difference or alterity on the other side, there is no other place, there is no other culture, there is no other language.

What the colonial ethnography of the survey purposefully uncovers is a subcultural nation expressed by the deviant, curious and savage aberration of tradition. Such ambivalent descriptions are part of a rich metaphorical complex of urban English language intellectualism in which the Celtic fringe is configured from the eighteenth century, in this there is little between 'those who find a favourable opinion of the Gael within this complex, and those who dip into it to find materials for derision'.[98] The statistical or ethnographic interest of the memoir-writers is framed in the contemporary Celticist or evolutionary ethnographic discourse of the English language.

The Two Races

Although the agricultural practices of the English countryside did not differ greatly from those in Ireland, the material culture of the Scottish and the English are described as being more advanced. In the parish of Drumachose, Co Derry, J. Butler Williams writes that 'good Scotch and English implements of husbandry have at length superseded the old Irish ones, particularly the Balteagh spade or "fack"'.[99] The differences are observable to the naked eye, in the parish of Aghanloo, Co Derry, Ligar and Stokes note that:

> The Irish and Scotch population differ or agree as follows: in style, comfort etc., of cottages. It is to be observed, however, that among the farmers the Scotch considerably exceed the Irish population in numbers. The latter are chiefly in about the Churchland. The style and comfort of the Scotch cottages is the best; all differences between the exterior of the houses.[100]

Race, religion and class are interchangeable, to be a labourer or cottier or Roman Catholic is also to be Irish. In the parish of Dunaghy, Co Antrim, Lieutenant J. Chaytor relates that 'the houses of the Roman Catholics are less neat, cleanly and of an inferior description to the Protestants. They are also less educated and more ignorant and more careless in their persons'.[101] In the parish of Dungiven the cleanliness or dirtiness of the place is not attributed by C.W. Ligar to a different form of social organization, or to the relative wealth or poverty of the people, but to race:

> The bad appearance of a house always depends not so much on the worldly condition of the individual who owns it, as on the circumstance of his being Scotch or Irish. If the latter, he will appear to be poorer than he really is by neglecting the roof, never whitewashing the wall. The same sluttish feeling extends to his farm to such a degree that in a deep glen near Dungiven, one side of which is inhabited by the Scotch and the other side by the Irish, the difference between the farms of each race can be strikingly observed in the superior neatness of the Scotch fences, the greater luxuriance of their fields, the superior air of comfort of their houses.[102]

The memoir of Dungiven, Co Derry offers an example of the extent to which the rational, reformative urge of the nineteenth century, exemplified by the survey, was prepared to go in inspecting and examining the people. The same writer records that an inspector 'marked every man's house "clean" or "dirty", according to the state they found in it. Their lists accordingly furnish accurate data from which to calculate even mathematically the proportion of sluttish feeling among them.'[103]

Thomas Jervis, an evangelist involved in the British survey of India, called Colby's system for the survey of Ireland 'a Fellenberg School on the grandest scale ... Promptitude, regularity, and cheerfulness are the essentials of such [a] system, and here their effects are displayed in every individual of the establishment as in the parts of a steam engine.'[104] The steam engine was an invention of the early nineteenth century. The system of education outlined by the Swiss Philipp Emanuel von Fellenberg was intended to bring the lower classes closer to the upper classes. The imperial reformers, improvers, moralists and educators walked in the footsteps of the colonizer bringing science, enlightenment and English to the country and the people. Although similar processes were going ahead internationally within colonies and emergent nation-states, indeed in the home of the Empire itself, in Ireland they were imbued with a distinct sense of racial supremacy.

In the Grange of Ballyscullion, Co Antrim, under the heads of social economy and early improvements, J. Boyle distinguished between what he unambiguously calls the two races:

> The western side of the grange of Ballyscullion is inhabited by a race who seem to be the descendants either of the aborigines of the country or of the earliest settlers in it, while the higher but eastern district is occupied almost exclusively by the descendants of the Scottish settlers of the 17th century ... the difference between the 2 races is still very striking in almost every respect, the Irish being uncivilised, turbulent and riotous, poor in their circumstances, slovenly in their persons and habitations, and not very industrious; while the Scots are industrious and orderly, comfortable in their stations and remarkably neat in their habitations.[105]

What is lacking here, according to the memoir-writer, is 'a resident gentleman, who could, by his station influence or example, control or improve them, or a landlord who would pay attention to their comfort or character'.[106] Hemans and Boyle's memoir for the parish of Duneane, Co Antrim, points to the detrimental consequences of the lack of contact with the Scottish or English or strangers in general:

> The march of improvement has been less rapid . . . this is perhaps to be attributed to its having been less generally colonized by Scottish or English settlers and to its being a more retired district, little frequented by strangers, and those districts of it along its coast inhabited by people who are said to have almost wholly subsisted upon fish and to have neglected the cultivation of the ground.[107]

Along with racial, religious and class differences there are further examples of how the memoirists created their typologies. Further to the clean-unclean principle mentioned above, a piece of rare detail given by J. Bleakley from the Co Derry parish of Ahanloo shows that shoes and dress in general were worn ritualistically on special occasions only, 'the Irish not so respectable as they are much poorer and have no taste for neatness: if the females can have a cotton or stuff gown for going to mass or to market, with a cloth or cotton shawl and cap of thick muslin with red, pink or blue ribbons . . . when going to market they often walk barefoot, and when near the towns wash their feet in a pond or stream and put on their stockings and shoes'.[108]

Medical conditions and pathology are also explained in racial terms. Boyle and Hemans in their memoir for the parish of Duneane, Co Antrim allude to the prevalence of scrofula amongst the native Irish:

> The most prevalent disease is scrofula, from which few of the lower class (the native Irish) along the coast are free. This is probably caused by the practise of intermarrying in the same family, for instance the McErlanes seldom intermarry out of their own clan. This is also customary with several other families and it may be remarked there are more idiots or half-idiots along the shore, but particularly near Toome, than in any other rural parish in the country.[109]

Faintly echoing the centuries-old colonialist writings of Gerald of Wales, incestuous sexual behaviour amongst the Irish leads also to disease. John Barrett writes to the North West Farming Society that the people in the parish of Inishkeel, Co Donegal, suffered from 'what is usually called the King's Evil; the prevalence of that disorder I conceive to arise from their intermarrying in the same district'.[110] What Fabian calls the game of physiognomy is played out recruiting science in support of 'moral judgements of persons, mostly to classify traits and types'. As in many similar descriptions in the accounts of explorers physiognomy was associated with character, race and class.[111]

C.W. Ligar in the parish of Dungiven used physiognomy to differentiate the people, 'the physiognomy of the clan McCloskey can be perceived by any stranger who will walk up the glen on Sunday, when they are all crowding to chapel at Dungiven'.[112] The same writer recorded that the women 'are all short with bodies clumsy and ill-made. Their faces are round and fat with small eyes, often set in an oblique manner' while the men 'are strongly made and broad shouldered. The countenance is Irish, with a well-proportioned nose and small eyes.'[113] The Scottish on the other hand 'are not quite so stunted as the native Irish. Neither have they that air of wildness to be seen in some individuals of the other race.'[114]

In one of the most universal of racial stereotypes the Irish are described as drinking too much alcohol. In the parish of Desertlyn in Co Derry, Ligar recounts that 'some say that the law recently enacted to permit publicans to sell spirituous liquors from the hours of 7 in the morning until 11 at night on weekdays, and from 2 until 11 on Sundays, is quite sufficient to demoralise a whole nation, but especially the Irish'.[115] The Irish are described in the memoir of the parish of Lissan by J. Stokes as being the best servants 'as they are more submissive than either Protestants or Presbyterians'.[116] The repetitious and formulaic nature of many of the descriptions suggest that some of the memoir-writers followed questionnaires to the letter and applied the criteria employed in them across the board according to the denomination, race or class in question. The final example comes from the area of Desertmartin in Co Derry. Written by Boyle it exemplifies the general treatment of the subject:

> It may be probable that the most permanent and important change which took place in the morals or habits of the people was at the time when the county was apportioned by James I to the London companies. About one half of this parish is included in that portion of the county allotted to the Draper's Company and this comprises its southern and more mountainous district, most of which is still occupied by the descendants of the native Irish, who seem to have retreated before the invaders of their country. The lower and more fertile parts are occupied by a much more independent, comfortable and enlightened class, who are all either of English or Scottish decent. [117]

Both culture and colony are derived from the Latin root *colonia* or *colo* suggesting tillage, cultivation, husbandry or agriculture. In this sense to colonize is to inhabit, pursue carefully, cultivate and care for the country. The memoirs, the hallmark of proficiency in nineteenth-century mapmaking, contain all previous meanings. The land and the people are distinguished and evaluated as pre-colonial (aborigine) and colonial (settlers). While colonization in this sense is a fundamental human imperative, turning bare nature into culture, here it is also a question of contested and contesting cultures.

The colonization of some areas by Scottish settlers is seen as proper cultivation and is contrasted to what are described as uncultivated, uncolonized areas where the Irish lived in anarchy. To colonize was also to enlighten, educate, reform and improve on both nature and culture. Colonization re-inhabits what it maps as an empty space implanting culture. A social, religious, cultural and racial typology emerges in the memoirs as a supplement to the maps. Different areas, or areas constructed as different, were mapped in evolutionary or progressivistic terms of stages of development or civilization. The memoirs simultaneously present and encode knowledge to achieve a largely rhetorical or discursive effect. The basis of this is ideological, the reportage on the Irish language is typical, its existence being accompanied by careful caveats that it was declining. The language itself, and the underlying process of translation and interpretation, remains invisible behind a fluent idiom of evolutionary discourse in the English language.

In the role of enlightened, culture-donning colonists the memoir-writers produced a statistical ethnographic gazetteer of Ireland blending perfunctory ethnographic detail with the trope of the picturesque. Strategy supported style in an exploration of Irish sublimity that produced pleasing portraits of colonial contentment (colonization as art) or debased images of indigeneity (colonization as moral necessity). The woods, lakes, trailing plants, flowers, peas, beans, trimmed hedges, shady lanes and gardens of the settlers are contrasted with the dry stone fences, pigs, dung heaps, turf stacks, scibs, crooked ditches, dogs and cottages of the Irish.

The watch, clock and jaunting car evince advancement towards industriousness, cleanliness, decency, temperance, religion and the English language. Ireland of the memoirs is peopled by wild, lawless, ignorant, dangerous, lazy, incestuous, unhealthy and uncivilized aborigines. It is darkened by colonial discourse. An objectification of the values and aesthetics of the colonizer accompanied this darkening.

NOTES

1. Justin Stagl, *A History of Curiosity: The Theory of Travel 1550–1800* (London, 1995), pp.6–7.
2. Ibid., p.15.
3. Ibid., p.18.
4. Ibid., pp.110–12.
5. J.A. Cuddon, *A Dictionary of Literary Terms* (London, 1979), p.64.
6. John H. Andrews, *A Paper Landscape: The Ordnance Survey in Nineteenth-century Ireland* (Oxford, 1975), p.144.
7. Mathew Edney, *Mapping an Empire: The Geographical Construction of British India, 1765–1843* (Chicago, 1997), p.98
8. Alf Mac Lochlainn, *Farasbarr Feasa ar Éirinn* (Dublin, Coiscéim 2005), p.27.
9. *Oxford English Dictionary*, (2nd edition) p.402.
10. Teodor Shanin, 'Short Historical Outline of Peasant Studies', in Teodor Shanin (ed.), *Peasants and Peasant Societies* (London, 1987), p.467.
11. Raymond Williams, *Keywords* (London, 1976), pp.231–2.
12. Michael Kearney (ed.), *Reconceptualizing the Peasantry: Anthropology in Global Perspective* (Colorado, Colorado Uni Press1996), p.1.
13. Bill Ashcroft *et al.* (eds), *Key Concepts in Post-Colonial Studies* (London, 1998), p.48.
14. Angélique Day and Patrick McWilliams (eds), *Ordnance Survey Memoirs of Ireland* (Belfast, 1990–98), 40 volumes, Volume 30, p.35. Hereafter abbreviated OSM.
15. OSM, 1, p.65.
16. Valerie Kennedy, 'Ireland in 1812: Colony or Part of the Imperial Main? The "Imagined Community" in Maria Edgeworth's The Absentee', in Terence McDonough (ed.) *Was Ireland a Colony? Economics, Politics and Culture in Nineteenth-Century Ireland* (Dublin, 2005), p.266.
17. OSM, 5, p.12.
18. OSM, 5, p.134.
19. OSM, 28, p.17.
20. OSM, 31, p.123.
21. OSM, 8, p.84. J. Bleakly and C.W. Ligar contributed to this draft memoir also.
22. OSM, 30, p.11.
23. Lawrence Venuti, *The Translators Invisibility: A History of Translation* (London, 1995), p.1.
24. Eric Hirsch and Michael O'Hanlon (eds), *The Anthropology of Landscape: Perspectives on Place and Space* (Oxford, 1995), p.6.
25. Ibid., p.8.
26. David Brett, *The Construction of Heritage* (Cork, 1996), p.40.
27. Andrew Hadfield and John McVeagh (eds), *Strangers to that Land: British Perceptions of Ireland from the Reformation to the Famine* (London, 1994), p.238.
28. Andrews, *Paper Landscape*, p.153.
29. See Samuel Carter and Anna Maria Hall, *Halls Ireland: Mr and Mrs Hall's Tour of 1840* (London, reprinted 1984 [1841]).
30. Brett, *The Construction of Heritage*, p.40.
31. Edney, *Mapping an Empire*, p.53.
32. Hirsch and O'Hanlon, *Anthropology of Landscape*, p.11.
33. Edney, *Mapping an Empire*, p.62.
34. OSM, 1, p.3.
35. OSM, 1, p.80.

36. OSM, 1, p.129.
37. OSM, 2, p.5.
38. OSM, 2, p.20.
39. OSM, 3, p.29.
40. Jonathan Crary, *Techniques of the Observer: On Vision and Modernity in the Nineteenth Century* (London, 1996), p.109.
41. OSM, 4, p.10.
42. OSM, 31, p.116.
43. OSM, 5, p.129.
44. OSM, 7, p.121.
45. Johannes Fabian, 'Remembering the Other: Knowledge and Recognition in the Exploration of Central Africa', *Critical Inquiry*, 26 (1999), p.57.
46. OSM, 21, p.67.
47. OSM, 38, p.27.
48. OSM, 4, p.128.
49. Hirsch and O'Hanlon, *Anthropology of Landscape*, p.11.
50. OSM, 25, p.92.
51. OSM, 38, p.27.
52. OSM, 4, p.50.
53. Ibid.
54. David Spurr, *The Rhetoric of Empire: Colonial Discourse in Journalism, Travel Writing, and Imperial Administration* (London, 1993), p.168.
55. OSM, 10, p.51.
56. OSM, 5, p.125.
57. Henrika Kucklick, *The Savage Within: The Social History of British Anthropology 1885–1945* (Cambridge, 1991), p.86.
58. OSM, 22, p.12.
59. Jonas Frykman and Orvar Löfgren, *Culture Builders: A Historical Anthropology of Middle-class Life* (New Jersey, 1996), p.15.
60. OSM, 2, p.112.
61. OSM, 13, p.94.
62. OSM, 11, p.11.
63. OSM, 31, p.13.
64. OSM, 2, p.5.
65. OSM, 2, p.7.
66. OSM, 2, p.8.
67. OSM, 2, p.11.
68. OSM, 2, p.23.
69. OSM, 2, p.29.
70. OSM, 4, p.10.
71. OSM, 8, p.8.
72. OSM, 21, p.88.
73. OSM, 31, p.122.
74. OSM, 3, p.20.
75. OSM, 13, p.64.
76. OSM, 6, p.14.
77. Ibid.
78. OSM, 15, p.16.
79. OSM, 19, p.53.
80. An Cumann Logainmneacha, *Logainmneacha as Paróiste na Rinne Co. Phort Láirge* (Dublin, 1975), p.23.
81. Frykman and Löfgren, *Culture Builders*, p.175.
82. OSM, 4, p.12.

83. OSM, 6, p.14.
84. OSM, 13, p.15.
85. OSM, 5, p.59.
86. Patrick Brantlinger, 'Victorians and Africans: The Genealogy of the Myth of the Dark Continent,' in Henry Louis Gates Jr., (ed.), *'Race', Writing and Difference* (Chicago, 1986), p.185.
87. OSM, 39, p.56.
88. Brantlinger, 'Victorians and Africans', p.182.
89. OSM, 19, p.103.
90. Malcolm Chapman, *The Gaelic Vision in Scottish Culture* (Montreal, 1978), p.18.
91. OSM, 31, p.122.
92. OSM, 36, p.6.
93. OSM, 5, p.72.
94. OSM, 11, p.109.
95. Kadiatu Kanneh, '"Africa" and Cultural Translation: Reading Difference', in Keith A. Pearson et al. (eds), *Cultural Readings of Imperialism* (London, 1997), p.276.
96. OSM, 21, p.138.
97. Roland Barthes, *Mythologies* (London, 1973), p.44.
98. Chapman, *Gaelic Vision*, p.18.
99. OSM, 8, p.88.
100. OSM, 11, p.10.
101. OSM, 13, p.15.
102. OSM, 15, p.15.
103. OSM, 15, p.14.
104. Edney, *Mapping an Empire*, p.313.
105. OSM, 19, p.4.
106. OSM, 19, p.5.
107. OSM, 19, p.101.
108. OSM, 11, p.13.
109. OSM, 19, p.102.
110. OSM, 39, p.71.
111. Fabian, *'Remembering the Other'*, p.58.
112. OSM, 15, p.20.
113. Ibid.
114. OSM, 15, p.21.
115. OSM, 31, p.45.
116. OSM, 31, p.106.
117. OSM, 31, p.58.

CHAPTER FIVE

Old Rhymes and Rags of Legends: Tradition and Civilization

The pioneering folklorist in the English language, Crofton Croker, collected a story in 1825 about a land surveyor called Ahern who attempted to fathom the Black Hole of Knockfierna but was 'drawn down into it and never heard of again'.[1] In this story it could be said that the subject fights back, the land swallows the surveyor, enlightened science is magically consumed by the landscape. The nineteenth-century Irish landscape through which the surveyors moved with the confidence and privilege of imperial science was assumed to be passive. The public's perceptions of the work of the surveyors are often presented as superstition, the awe and wonder that was only to be expected of archaic tradition in the face of futuristic science.

C.W. Ligar writes in the parish of Dungiven, Co Derry that, 'it is no uncommon thing to meet a Catholic who is capable of believing that the little Schalwalder [Schmalkalder or Smalkalder?] compasses used on the survey of Ireland are actuated by imprisoned fairies. One such instance occurred near Donald's Hill. Another was in Upper Cumber, where they believed the individual bearing it to be capable of foretelling future events.'[2]

This speaks of the science of folklore as well as of the folklore of science. Imprisoning it in their instruments the surveyors pressed vernacular magic into the service of science. In the course of their work they inevitably trespassed in areas popularly associated with the supernatural world. The necessity to take sightings from hilltops or from prominent features on the landscape located them on the numinous webs of the ancestral cognitive map. In the popular imagination the subdued fairies, not unlike the country labourers themselves, worked for the survey orientating and guiding them along the invisible boundaries of the local landscape.

As frontiersmen of science the surveyors were the nomadic practitioners of the new magic of modernity. In making their observations, however, they were themselves observed, interpreted and located within vernacular fields of knowledge. In order to make sense it could be said that the vernacular was modernized but the modern was also vernacularized. The surveyor, compass in hand, was a wise man with the second sight or special supernatural power known as *fios* in the Irish language. While being subjected by the survey to the correction and emendation of science the landscape was in fact intimately known and any movement on it, as assuredly as a fly enters a spider's web, was immediately obvious and open to interpretation.

This recalls Sahlins' idea that 'every reproduction of culture is an alteration, in so far as in action, the categories by which a present world is orchestrated pick up some novel empirical content'.[3] Just as King George of England could represent *mana* for the Hawaiian chief, contact between cultures can have unforeseen effects. The unpredictability of people's actual use of signs can mitigate the assumed universality of evolutionary scientific ideas, 'meaning is risked in a cosmos fully capable of contradicting the symbolic systems that are presumed to describe it'.[4] In enacting its own scientistic vision the survey, much like Captain Cook in Hawaii, was subsumed by the Irish through creative understanding or misunderstanding in their own mytho-practical terms.

To return momentarily to Crofton Croker's story, the imaginative world of the survey is dwarfed by the magnitude of another imaginative world, exemplified by the Black Hole, in which it finds itself diminished or lost. The fixated nomadic surveyor of Empire appears on the Irish horizon tilting at windmills much like Don Quixote.

Spying Out the Nakedness of the Land

Arriving untranslated into the Irish language the sappers make a brief appearance in an extempore Connemara song, '*Óra mhíle grá, nach ard é Cnoc an Chaisil, Is chuaigh na Sappers ar a bharr*'.[5] Belonging to a genre that is racy and Rabelaisian in tone this droll verse simply says that the hill of *Cnoc an Chaisil* is very high and that the sappers climbed to the top of it. It is a mock heroic image of the British soldiers perambulating the homely countryside in their drilled drudgery and labour. The mockery is aimed at the dramatic intensity of the officially orchestrated endeavours of a foreign military to map a land that is already as well known as the back of the proverbial hand.

It exhibits the playfulness, confidence and mimicry of vernacular culture poking fun at the grand imperial designs of the survey. This was not a landscape devoid of meaning, understanding or resonance but a place that was known, peopled, cultured, familiar, named and already mapped in hearts and minds; it was home. In this landscape the sappers are a chain gang imprisoned by the very focus of their attention. They are scientific-military emissaries of Empire imparting an unwanted gift of English culture wrapped in chains and paper. Selling the idea to the native was the work of discourse, they first needed to learn what it was and why they needed it, but they did not always accept it.

In the nineteenth century mapping retained political overtones and the Spring Rice Commission that instituted the survey discussed the possibility of opposition. Many private surveyors and engineers were consulted regarding their own experiences in the field, William Edgeworth and David Aher recounted the opposition they had met with. Trigonometrical stations and the equipment of the survey were under police guard. Smith says that, 'as well as cases of suspicion, muted resentment and passive resistance, there were more serious incidents of sabotage and violent conflict. Boundary Surveyors, Royal Engineers and civilian assistants were threatened or attacked on various occasions. Ordnance Survey records describe assaults on staff although they do not usually discuss the cause of disturbances.'[6] The officers, civil assistants and fieldworkers were observed while they observing.

J. Hill Williams recounts in the parish of Kilbroney, Co Down that 'every person seems to have the most decided objection of giving information relative to any part or portion of the place'.[7] From the parish of Drumachose in Co Derry, J. Butler Williams reports that:

> Many of the parishioners were greatly alarmed when the engineers first came into the country to make piles at the tops of the hills for the purpose of triangulation. They viewed with dismay the strange-looking men erecting (as they thought) redoubts on the most commanding situations, and actually imagined them to be emissaries of some formidable national enemy, coming to 'spy out the nakedness of the land' and to mark the point from which the country below could be most easily commanded by artillery. They conceived that the trigonometrical station on the top of Keady was the spot from whence a fire was to be opened on the lowlands in that neighbourhood, particularly those in Drummond.[8]

What may have been a real fear based on previous experience of the British military, or a concealed threat of opposition, is recounted as superstition in the face of progress and science. The sight of imperial soldiers struck terror into the hearts of many people in the late-eighteenth and early-nineteenth century. The uprising of the United Irishmen in 1798, just twenty years before the survey, was followed by horrific reprisals, torture, mutilation and murder. Lord Clare had said in Dublin that he would make the people as tame as cats.[9]

The rhetorical use of tradition folklorized resistance and functioned to preserve the illusion of consent. The implication was that 'it was only "superstition" and unenlightened "tradition" that prompted opposition'.[10] Aspects of the other's knowledge, particularly adversarial interpretations, were dismissed as naive misconceptions. The survey of India employed a language of tradition whenever it met resistance. Framing it as superstition or tradition versus science, this language decontextualized and depoliticized rivalry and denied the other culture agency. The others did not resist, they just misunderstood.

Giving new resonance to Richards' idea of the data pilgrimage, a striking example of science superseding superstition took place at Tummock, in Desertmartin parish, Co Derry where the survey used a *carn* or pile of stones that pilgrims visited to perform stations as an Ordnance Survey station.[11] The sacred and ritualized circumambulation of the station or pattern of the vernacular supernatural world is replaced by the profane ritualized circumnavigation of the country in the scientific world of Empire. Both are ritualized enactments that embodied equally abstract ideas and ideals. The ambivalent reception was not reserved for the British officers of the Royal Engineers.

O'Donovan and his colleagues in the Antiquarian or Topographical Department frequently encountered resistance or reticence. He records that the people in Dungiven were very shy 'in giving their names because they are afraid that they might be wanted for the "service of war" or some other plan of the Government'.[12] O'Donovan notes in Drogheda that 'the people are not willing to give information, suspecting it may be connected with Tithe affairs'.[13] Tithe proctors in the area were regularly beaten with sticks, a fact that quite understandably made the fieldworkers both nervous and cautious.

O'Connor and O'Keefe write from Ardee that 'everyone upon whom we called actually refused to give us any information, as a few tithe-processes were served the week before to some persons in these parishes'.[14] Not only was co-operation withheld at times, but also misinformation was given. This was the case with the Gentlemen's Stewards that Richard Griffith employed in the area of Navan.[15] In Co Kilkenny the boundary surveyor William Jones reports that the people would not 'give the least information'.[16] Poles used by the survey were often removed and in one case observers were attacked. In this instance Andrews guesses that the people considered the survey an attempt to suppress agrarian disturbances. On Bantry Commons in Co Wexford the police were called to protect field parties against local people who were angered with their local landlord.[17]

Such instances of resistance to the work of the survey have been overlooked as irrational in favour of a more heroic depiction like the one in a newspaper report of 1828. This painted an idealized picture of the people of Glenomara, Co Clare helping the engineers to build a trigonometrical station. It depicts them climbing the mountain in a great crowd with flutes, pipes, violins, and young women bearing laurel leaves. They insisted, however, on naming the station O'Connell's Tower.[18] O'Connell was the *bête noire* of the Tories in England at the time. O'Donovan wrote in 1837 that the Tories cited references from the Book of Revelations about the Apocalypse to prove that O'Connell was the devil.[19]

This ambivalence manifested itself not only in the field but also within the survey itself. From time to time the names chosen by O'Donovan for townlands displeased the gentry.[20] In other areas the local people disputed that areas marked as townlands were such at all but rather comprised of several townlands. O'Donovan himself frequently questioned the accuracy of Griffith's boundaries and complained of 'roguery and imposition'. On one occasion he was greeted with suspicion by people who took him to be a preacher.[21] Erroneous names were given intentionally and O'Donovan suggested the extreme solution that landlords, parsons, priests and farmers should be made to agree and swear the truth upon oath.[22]

Corporal Berry and other officers of the British Royal Engineers took umbrage with Thomas O'Connor for accusing them of careless name taking.[23] O'Donovan at one point availed of the opportunity to remind Larcom that the surveyors were not the first topographers or antiquaries in Ireland. O'Curry spoke more openly of the 'collective folly and stupid intellect of the Empire'.[24] Such instances point to an incoherence in what otherwise appears to be a monolithic colonial project. Resistance to the survey was not only blunt and external but also subtle and internal. It proceeded at times as both a complementary and contradictory rivalry that was political, religious, linguistic, economic and racial.

Abominable Amusements

Some approaches to the survey show how it reflected tradition as something concrete that was out there, encountered in the field, when it was more a categorization of the knowledge of a particular race. Many of the aspects of Irish vernacular culture that would be considered folklore proper a century later were purposefully highlighted as signs of primitiveness or an index of inferiority. Pattern or festival of the local parish priest, fair, fairy, funeral, hurling, dancing, storytelling and superstition formed a lexicon of euphemistic referents reserved for the Irish alone, the primitive, uncivilized or aboriginal race. They emerged as the stigmata of indigenousness increasingly constituted as a quasi-criminal subculture. The popular, interpreted as potential belligerence, becomes unpopular while the unpopular, interpreted as manifest gullibility, becomes popular.

The letters frequently echo the discourse of the memoirs and the Scottish tend to be presented as more enlightened than the Irish whom O'Donovan, while in Cavan, calls 'stupid numbskulls'.[25] James Boyle writes of the people of Ballymartin, Co Antrim in 1838:

> They are not quite so superstitious as the people of the neighbouring districts, but still most of the old and many of the young people firmly believe in ghosts, fairies, and enchantments. Many will swear to have seen the fairies, the devil (in the shape of a black dog or pig) and the wraiths of their friends before death. Respect is paid to old forts – even the forts are beginning to feel the sad effects of the march of knowledge.[26]

From the Grange of Ballywalter in the same county J. Bleakley notices that, 'there are at present no ancient hawthorns in the Grange of Ballywalter. There were many old hawthorns but they are all destroyed long since, as that superstition does not prevail in this part at present which is so prevalent in other districts, chiefly owing to its inhabitants being Presbyterians and somewhat enlightened and rather intelligent'.[27] As is the case in the memoirs, difference is equated with place, religion, language, race or physical appearance.

In a passage strongly reminiscent of the memoirs, and suggesting a shared paradigm, O'Donovan writes that:

> There is a remarkable difference between the countenances of these Scotch families and those of the Irish, so that they can be very easily distinguished. The Scotch have long pale faces with large mouths. The Irish, excepting those that are nearly starved, have round jolly faces and a look expressive of little care. They are always ready to joke, pun and tell extraordinary stories; the Scotch are more meditative, but always more stupid.[28]

The Scottish are solemn while the Irish are elf-like and mischievous. For O'Donovan many of the narratives concerning names of places are, 'the production of an ignorant unenlightened mind, that rests satisfied with the sounds of foolish words employed in setting forth improbable reports'.[29] The absence of tradition is just as significant as its presence and amusements are generally thought to result from idleness. If people were busy there would be no folklore.

The casual combination of work and leisure in the popular calendar is evidence of an anarchic unregulated society. In the parish of Carnmoney, Co Antrim, James Boyle relates in 1839 that 'the taste for amusement has with the people . . . declined as the pressure of the times has increased . . . now their ability either as to time or means is much more confined than formerly'.[30] The perceived lack of tradition in the areas designated as Scottish is sometimes accredited to their recent introduction to the country. Boyle observes this in Carnmoney where he says, 'there is scarcely a tradition in the parish. This is not much to be wondered, when it is remembered that but 2 centuries have elapsed since their ancestors first settled in the country, but it is rather surprising that scarce a farmer can tell how even his father or grandfather came into possession of the farm on which he dwells.'[31]

In the townland of Sheeptown in Newry, Co Down, the civil assistant R. Martin reports a pattern at Crown Bridge in his statistical remarks, 'a pattern did prevail to lately twice a year (Easter and Whitsuntide), but the difficulty of earning their living are burying such abominable amusements'.[32] In Enniskillen, Co Fermanagh, local pilgrimages are attributed by Lieutenant John Chaytor in his statistical report to 'the superstitious simplicity of the peasantry'.[33] Holy days, like fairs, are seen as worthless assemblies. In the parish of Toomregan, Co Fermanagh, Lieutenant Greatorex relates that 'holy days have lately been decreased and it were well they were entirely so, as they are only considered by the people as opportunities for assembling to drink and consequently to riot or to settle some long existing quarrel between factions'.[34]

The fact that large gatherings were viewed with fear in a time of political and agrarian unrest is veiled. The imperial rhetoric is, by its very nature, always resistant to indigenous representations. In the 1834 statistical report on the parish of Inishmacsaint, Co Fermanagh, Lieutenant P. Taylor enters that 'monthly fairs, the bane and demoralization of nine-tenths of the Irish population, are held in Derrygonneely on the 24th of each month'.[35] These fairs, Lieutenant Greatorex writes, were 'attended by numbers for pleasure, and they rarely end without some fighting or quarrelling occurring'.[36] In the parish of Clogher, Co Tyrone, Lieutenant R. Stotherd observes that 'fairs are held on every third Saturday and on the 6th May and 26th July. The traffic is the same as on market days. The last mentioned is called a "spolien" fair, and people amuse themselves on that day by stuffing themselves with mutton and mutton broth.'[37] Lieutenant W. Lancey's statistical report of 1834 for the parish of Clonmany, Co Donegal, following Larcom's heads of inquiry to the letter, invokes a typical image of an Irish area:

> Cock-fighting, hurling and dancing are declining. The game of common is still practiced. They frequent St Columb's Well at the Gap on the 8th of June and drive cattle into the sea, and on St John's Eve they light bonfires and drive cattle through them. They cry at funerals but do not hire persons called "keenie". The priests are endeavouring to get rid of as many superstitions as possible.[38]

The lament for the dead, often called the Irish cry by the surveyors, is a recurrent sign of the indigenous culture. In the parish of Artrea, Co Derry, the writers observe in 1836 that 'the Irish cry, which until within 8 or 10 years was yelled at the funerals of Roman Catholics, is now given up and hymns are now sung at their wakes and funerals by 8 or 10 singing boys who wear white shirts or surplices and sing in Chapel on Sundays'.[39] Conversely the absence of the keeny, the lament for the dead, also called the *taisch*, *tásg* or wake, is a sign of improvement. In the parish of Donagh, Co Donegal, Lieutenant W.E. Delves-Broughton writes in 1834 that:

> Wakes are usually visited by youths of both sexes. Among the lower orders, if the person who dies be very young or very old, a great deal

of merriment is expected to take place; consequently they amuse themselves with a variety of sports. But if the person who dies be respectable, then much gravity and decorum is observed. Sometimes matrons who are adept in the art of crying over their Catholic friends exert their ingenuity in reducing their wild plaintive notes into a kind of musical humming over the corpse of the deceased, and are so tenacious of them that their friends must drag them from the mournful scene.[40]

In its interest in, and sympathy for, the detail and meanings of the behaviour described, this account is exceptional. The more usual format adhered to is that of a taxonomic list of the racial characteristics of the Irish. The survey is not so much a storehouse of tradition as it is a process of cultural classification. There are references to curing priests, elf-stones, cunning women, distillation, legends and narratives but they are minimal and cloaked in colonialist rhetoric. Rather than being descriptive they are superficial and prescriptive and are best understood as belonging within a wider socio-political discursive framework.

Hagiology

The interest in vernacular discourse or storytelling became the defining characteristic of professional folklore and ethnology in twentieth-century Ireland. In the survey narratives are described in negative terms and are generally thought to be 'too ridiculous to be worthy of insertion'.[41] In the parish of Layd, Co Antrim, Lieutenant Chaytor relates in 1832 that there were 'many old persons who do little but recite tales and stories about ghosts, fairies, enchantments and the wonderful doings of Ossian, Fin McCoul and a great many other giants'.[42]

From the parish of Drummaul, Co Antrim, James Boyle reports in 1838 that 'their legends and traditions are not nearly as numerous or interesting as might have been expected . . . there is not a single tradition in the parish which in the slightest degree bears either on its history or that of the family connected with it. The only traditions are absurd stories relative to ghosts and the banshee Neeny Roe.'[43] In the latter report the writer seems to expect people to relate a formal history of the parish. In the neighbouring parish of Loughguile, Lieutenant Greatorex observes in 1833 that, 'their legendry tales are uninteresting and worthless, as they only relate to fairies, enchantments etc., in which they devoutly believe'.[44]

John McCloskey's statistical report on the Co Derry parishes of Banagher, Bovevagh and Dungiven gives a more considered and objective description of the much loved and respected vernacular narrative genre known as *Fiannaíocht* that glowingly recalls the exploits of the heroes Fionn and the Fianna:

> There is yet to be found some old persons in these parishes who retain fragments of poems ascribed to Ossian and tales of the old tales of Erin. Indeed, the names of Fion MacComhal, Gaul, the Son of Morni, Cuchullin, Ossian, Oscar and Fillan are 'familiar in their mouths as household words'; even the names of Fion's domestics have been transmitted to fame; and to this day the Irish cowherd dignifies his faithful dog with some of the names which these poems represent the ancient heroes of his country to have assigned the companions of their sports. The writer was lately treated with the recital of a poem, describing a combat between Cuchullin and Conloch, which, even from the rude translation of the unlettered reciter, appears to possess the true pathos of the Ossianic poesy. It would be desirable that some person who is versed in the language should rescue these fragments from the oblivion to which they are fast hurrying, through the decline of the seanachie profession – it need hardly to be added that these seanachies are indignant of the attempts of Macpherson and his coadjutors to deprive 'green Inisfail' of the voice of Cona.[45]

John McCloskey, the author of this report, differs from other writers and is aware of the contemporary Celticist antiquarian interest in 'fragments of ancient poetry'. The construction of a superstitious and thus primitive race and culture is not uncontested. O'Donovan refers dismissively to the interest of the romantic ascendancy antiquaries as hagiology or fairyology. He writes of the fairies:

> They have been looked upon by some as the souls of the departed, and by others as some of the fallen angels who expect still to be recalled to their original place of bliss, and therefore, are sometimes kind to mortals. The generality of the peasantry however, especially those who know nothing of theology or theogony, look upon the fairies in the same way as the Mahomedans do upon their genii, and this was probably the pure Pagan Irish belief before the Christian Religion introduced the idea of fallen angels and human souls wandering on this earth to be purged of their sins by the storms, the rains and the lashes of ariel demons. It is a pity that that little fairy elf, Crofton Croker, is not better acquainted with our ancient Irish

MSS., to give us the old history of the fairies of Ireland, for he seems to be well acquainted with their modern habits.[46]

O'Donovan's more sceptical or critical approach is undoubtedly a reaction to the burlesque or picturesque depictions of fairies, elves and giants in an increasingly folksy popular literature in the English language. This literature generally emptied the narratives of storytellers' potentially subversive messages. As a 'fairy book' Croker's 1825 *Fairy Legends and Traditions of the South of Ireland* was considered as second only to Edgar Taylor's *German Popular Stories* from 1823, which was an early translation of the Grimms' fairy tales.[47] It was around this time that the English word 'fairy' came to refer to the vernacular or ancestral supernatural world denoted in the Irish language by the older and completely unrelated term *sí*.[48]

An unusual singing competition is reported in the parish of Aghadowey in Co Derry where:

> In the still calm evenings of summer, usually in the moonlight evenings of July, a singing party from each county met on opposite sides of the River Bann and sang, standing at the waters edge, against each other. Judges were appointed and small prizes given. As the assembly was of young people and held after sunset, it was discountenanced by the religious part of the parishioners. The last one was held in the summer of 1817. The music was Irish and they sang alternately.[49]

In the various statistical accounts and reports customs and habits discourse prevails and tends to configure itself in a formulaic repetitive style from parish to parish with a final note of condemnation or rejection added. To what extent editorial or supervisory work was responsible for this must remain speculation. The following note by James Boyle and G.W. Hemans from the parish of Duneane, Co Antrim, exemplifies this:

> A fair, or rather a pattern, at Toome on Easter Monday is well attended by both sexes, but it has frequently been the scene of riots and sometimes of bloodshed. Wakes are much frequented as places of amusement – they sometimes burn fires on St John's Eve and the Ribbonmen occasionally observe Patrick's Day by walking in procession. Secret Societies prevail and it is said that nocturnal meetings are not infrequent. All sects, but particularly the Roman

Catholics, are superstitious and place implicit belief in fairies [and] the cures performed by the waters of Cranfield Holy Well. Their traditions concerning it, St Patrick, ghosts, banshees are still innumerable, but too absurd to be worthy of notice.[50]

Popular culture and political unrest are synonymous with each other and are recounted as though the security of an area depended upon the relative strength or weakness of tradition or custom in it. Contemporary vernacular culture, the everyday ethnographic reality of nineteenth-century Ireland, is notable mainly by its absence. The vitality and dynamism of tradition is defused in allusions to vestigial survivals or ancient beliefs. Its indefinite, unpredictable or even threatening nature leads to its constant redefinition as marginal, superstitious or quasi-criminal behavioural. This is reinforced by its tendency to reject or redefine its own position vis-à-vis the defining discourse.

The contemporary oral traditions of the people, although featured intermittently as appendages to the main descriptions, are interpreted and presented from an English perspective. The inquiries that are shaped and informed by English antiquarian approaches and the emergent discipline of British ethnology outline its presence but also contain and contextualize it in very specific ways.

Oral and Vulgar Tradition

The Irish fieldworkers utilized the contemporary repertoire of social research techniques and terminology. Some of these are borrowed directly from the survey's definitional lexicon. Contemporary Irish vernacular culture is described as aboriginal or vulgar tradition and the language itself as the vulgar Irish language. O'Donovan describes himself as investigating tradition, collecting antiquarian lore, traditionary lore, loco-tradition, or simply traditional accounts. As far as oral tradition was concerned he could be described as a demurring folklorist. This may be evidence of a tension between the survey's drive to define Irish culture and the fieldworker's more specific understanding of his own role within it.

He inquires from Co Carlow whether he should continue to mention anything related to tradition and seems to have thought that it was secondary or even irrelevant to his role as etymologist and orthographer, 'as the progress of the Survey is now so very rapid would it not be better for me to procure the correct names of the places without delaying to collect and write the traditions connected with them? ... I should think it more prudent for me to keep pace with the Survey than to lag behind writing traditions of holy wells and castles.'[51]

Oral tradition, considered as a source of some vague original truth, had been a subject of debate amongst antiquaries since the seventeenth century at least. What he calls the old fables or old romantic stories of the old traditionalists were often a source of frustration to him. What the survey's inquiries sought to highlight may not have been commensurate with his own understanding of tradition. He was preoccupied with infusing evolutionary English antiquarian or ethnographic research with contemporary scientific standards and recasting them in an Irish idiom.

On occasions O'Donovan complies with the constraints imposed upon him and sometimes borrows racialist or colonialist jargon such as kaffir, Hottentot, aborigine or peasant. On other occasions, however, his ambiguous role-playing and proleptic language re-appropriated the colonizer's discourse. Along with his colleagues in the survey he pioneered a methodology that re-integrated local history and local information within the emergent disciplines of the nineteenth century. It could be asked whether he was strictly working for the survey or availing of an opportunity to test and practice his own evolving ideas about antiquarian research and ethnology. It sometimes comes as a pleasant surprise to him when the methods actually work:

> I fear I have remained too long here. It is however, the most fertile spot in all the county for legends and traditions. I have discovered traditions connected with other places here that have been forgotten at those places themselves and which throw great light upon their history. I have met in the Townland of Labby an old man of the gifted tribe of O'Brollaghans who knows more about the traditions of Derry than all the rest put together.[52]

It is difficult to be certain at times, particularly in his earlier work in the northern counties, whether he is being ironic or serious in his many references to the Irish as natives or aborigines. He recounts what he calls got-up traditions as belonging to 'a period when the Irish were as silly as men could possibly be',[53] and were 'never at short for such useless stories'.[54]

The rhetoric of tradition sometimes depersonalizes the informants and information is often given simply as 'tradition says'.

In relation to a round tower at Lorum, Co Carlow, Thomas O' Connor states that 'tradition means here the local information of those who heard there was such a building here, but never saw it'.[55] A degree of reflexivity and objectivity is discernable in the discussion of tradition.

On the other hand the understanding of tradition in the letters sometimes mirrors that of the memoir-writers. This is qualified by the variegated nature of the tensions within the survey that took on different hues. The dominant tension was between imperial science and Irish culture. Within this there were also tensions between the shibboleth tradition of evolutionary science and the ancestral past of concern to the Antiquarian and Topographical Department. There were other tensions between the fantastical notions of antiquaries like Charles Vallancey, Louisa C. Beaufort, Charles O'Connor, Edward Ledwich and William Betham and the more problematic tradition of the Topographical Department. The latter school of thought, also a professional practice in itself, would later evolve into the professional discipline of ethnology and folklore in the twentieth century.

Tradition, viewed as an abstract notion, is often rejected as peripheral to the immediate and demanding imperatives of the task on hand. Eugene O'Curry shared this view, and in Tulach O'Felme, Co Carlow, he observes that, 'there is a holy well enclosed by a stone wall, round which the deluded peasantry do penance on the eve of the fair days of the year'.[56] O'Donovan is often at pains to distinguish 'some more certain record – than oral and vulgar tradition'.[57] Petrie, who directed the Antiquarian or Topographical staff and encouraged them to be interested in vernacular discourse, gives the following account:

> There are many individuals who look with much contempt on traditions and will think it a degradation of the understanding to listen to them . . . an individual relates one with an air of honest conviction and with a complete indifference (arising from the useless nature of the thing itself) to the hopes of any gain. He will give some neighbour as his authority, who will say that he heard it from his father and that he supposes his father heard it from his grandfather. This chain, it is clear, could not have been produced except by truth. It will go back without end unless we stop it by the actual occurrence of what has founded the tradition. An Irish tradition is a vocal history. It may be corrupted but it is never unfounded, and it is very easy to clear from those corruptions by a little consideration and by reducing it to the simplest form when necessary.[58]

In the field correspondence the debate about the reliability of oral tradition, ongoing in nineteenth-century antiquarianism, is continued. While O'Donovan, Petrie and Rev. Dr Charles Graves were sceptical, other antiquaries such as Richard R. Brash, John Windele, William Hackett and Nicholas O' Kearney were proponents. The latter believed that customs and superstitions shed light on older mythological beliefs of prehistory.[59] This debate itself is evidence of an increased concern with methodological rigour and scientific objectivity. After all, popular antiquities did not yet have its present English language name of folklore after the conclusion of the survey.

The Old Antagonist

In the letters tradition is associated with the mountaineers who lived in 'a world of romance swayed over by the imagination'. While O'Donovan offers a strained defence of such narratives he classifies the narrators in a particular way, 'these stories were truth itself in their time, and they are not in any degree more ridiculous than others which are still in honore. Iupiter, Temple of a Cat in Egypt. Juggernaut, fairies, monastic legends, well worship, prophecies, witches, sorcerers, miracle working relics.'[60] While in counties Cavan and Leitrim he disregards contemporary vernacular discourse, he does give narrators the title of shanachie, however, derived from the Irish language *seanchaí*. This recognizes their status in the community as more or less encyclopaedic raconteurs of vernacular knowledge and discourse:

> I shall have some conversation with the Shanachies of Meath who reside on the southern banks of Lough Sheelion. The men with whom I have conversed on the Cavan side of that Lough have nearly forgotten all the tales of old, and regret that I was not out in the time of their grandfathers who could tell a thousand tales about miracles, saints, chieftains and plunderers, which their posterity, being in the progress of being anglicized, think unworthy of treasuring up in the cabinets of their memory. One story about Dan O'Connell is now worth all.[61]

Popular narratives with their more obvious and explicit political implications remain beyond the controlled purview of the survey. It can be said with certainty that O'Donovan viewed oral tradition as problematic. In his correspondence he is particularly concerned with establishing etymological clarity and accuracy and was wary of the creativity and diversity of oral narratives:

> I have argued with a Dr. of Salamanca about Oral Tradition: he produced logic and ingenious subtlety; I produced facts that had fallen under my own eye; he adduced the accuracy of the Antediluvian pedigrees and other traditions; I produced the inaccuracy of what Oral Tradition has preserved in Donegal in 1835, a more tangible epoch (era). He urged the perfection of human memory; I urged the imperfection of my own, and the acknowledged treachery of those men of genius and sound learning . . . from the fact that no six persons will (or can) separately relate the same story in the same words, order, or detail.[62]

His clearly defined role did not interest him in nineteenth-century popular literature *per se*, particularly if or when this is understood as archaic belief and superstition, but in the identification and Anglicization of the place names. Nonetheless he is compelled to engage with what he respectfully calls his 'old antagonist'.[63] He maintains a guarded objective interest in it, for him it is 'a blundering Booby who has a clouded memory and muddy brains'.[64] Classifying such traditions as temporary phenomena he writes that:

> Traditions of this kind had only mushroom existence, as they had no foundation in fact or early superstition; for some traditions are founded on distorted facts, some on natural phenomena ill understood, and others upon superstitious beliefs; the three latter kinds of traditions are curious and worth preserving, as throwing light upon the manner in which people unacquainted with science accounted for natural phenomena, but there is another class of traditions worth nothing, as serving to illustrate nothing.[65]

He is dismissive of what he calls wild or got-up traditions and disregards many areas that in his opinion have nothing of 'antiquarian, superstitious or fairy interest'.[66] His attitude can be at least partly accounted for by a wariness about questions of representation or categorization. Vernacular discourse is often tested and proven false particularly in regard to ancient ecclesiastical ruins. As is still the case in ethnographic and folkloristic discourse the present is often secreted while the past is presented. In contradistinction the oral tradition freely assigned antiquity to old ruins in support of local claims to Christian origins:

Old Rhymes and Rags of Legend 149

> It is very curious that the present oral tradition refers the erection of the Churches called Granges to the primitive age of the Irish Church. Thus the Church of Grange in the parish of Killererin is said to be the third oldest in Ireland and even this Church of Grange is referred to a period so far back as the fourth century, which is ridiculous. No part, however, of this Church remains so that the antiquarian can pronounce no opinion on the accuracy of the tradition from an existing monument, which is always the surest clue to the age and history of Churches. A certain professor of anatomy and physiology in Dublin does not believe in ghosts because he cannot dissect them . . . and I intend to follow his example as far as old Churches are concerned for I will not believe in their antiquity until I see their features.[67]

The popular ascription of antiquity to such ruins is treated by him with scepticism and disdain as erroneous and unverifiable. Like a surgeon who will not believe in ghosts unless he can dissect one, he subjects tradition to a rigorous inspection. The evolutionary category of the primitive in this discourse is ambiguous and has alternately positive or negative connotations. It could refer to patron saints, buildings ancient and contemporary or even to the people themselves. It is not only evolutionary at times but exclusive and exclusionary. In Co Kerry, he likens an ancient cell to a modern cabin that was alongside it:

> This cell is now used as a pigsty, a purpose for (to) which it is far better adapted than to that for which it was originally intended. It may be curious to remark, however, that the adjoining modern cabin is not much more exalted in the scale of civilized architecture, and it is very much to be doubted if the present inhabitants be one degree more enlightened than those by whom the cell was originally erected. They speak the same language and learning seems more on the decline than on the advance![68]

It is likely that traditions of saints and holy wells are mentioned because of their proximity to ancient remains or their attachment to names of places and not from any concerted effort to seek them out. Referring to one such tradition in Co Galway he notes that:

> They have no tradition of the Saint or hermit of this island nor any tradition whatever in connexion with it except that it admits of no burial within. There never was since the time of the Saint a human body interred within this Church but one, and to the astonishment of

the neighbouring peasants (who I must confess do not know as much about the laws of nature, or the nature of laws, as Mr. Blake of Rinvyle) the sacred clay, or stones rather, of the sanctuary eructed it from its bosom, and it was found the next morning at a place called *Sméaróid*, on the opposite side of the continent where children are now usually interred.[69]

Thomas O'Connor describes tradition as sad stuff that should 'when it raises its head against history, be at once crushed unless that it might be curious to preserve for a display of inconsistency'.[70] Patrick O'Keefe dismisses the parishes of Ballynadrimney, Caddamstown and Dunfierth in Co Kildare as 'places of no historic importance and . . . tradition has almost disappeared in that district'.[71] Where tradition is thought to exist it is often described as ridiculous and where it is thought to be absent the places are described as being of no interest. The dubious honour of being traditional is conferred upon particular areas while the ambiguous quality of being bereft of tradition is applied to others. This black and white, active or passive, map of tradition remains largely unchanged at present.

The people in the fuzzy zones exist as if suffering from collective amnesia, scratching their heads in wonderment in a mute landscape of ancient derelictions. In Co Kildare O'Donovan notes that 'the natives pretend to be able to point out the site of St Bridget's house, oak tree and fire house, but I fear that one cannot rely safely upon their traditions'.[72] Unusually for him he discredits the manuscript tradition and the narrators' powers of retention in Co Derry as he searches in vain for the verbal vestiges of a more ancient, more classical past:

> I have also searched these Glens for some of the old Tuireadhs or Eligies sung at the wakes of the old families but I could not find one, except one fragment of an elegy composed for a Manus O'Kane who had lived near Garvagh, but at what time no one could tell me. They cannot recite any of Ossian's poems except odd lines here and there. These they have not from a succession of oral traditions but from hearing old men, (Irish scholars now dead), read them out of MSS. Now decayed or lost. Well has Doctor Johnston argued against McPherson that no illiterate person can preserve a long epic poem by the mere power of memory.[73]

Decrepit and fragmented tradition owes its repertoire not to memory or contemporary cultural creativity but to an older more complete and more perfect literate tradition. Vernacular discourse is denied any autonomous existence or creativity and must be traced back to a book or manuscript. It is ascribed to the older Irish scholars, as opposed to contemporary romancers or shanachies, who could read. O'Donovan lists the contemporary repertoire as he found it in the Dungiven district in 1834:

> I made every enquiry in the vicinity of Dungiven about Aileach but I could discover no recollection of it; the people there know nothing of Irish Monarchs or Irish Kings, and it is curious that Fionn MacCool and his followers are better remembered than Cormac MacAirt or Niall of the Nine Hostages. This can be very easily accounted for: some time ago in these glens there were Irish Scholars whose principal amusement was to read over at their fire side on Winter evenings tales and poems concerning the exploits of Fin and his Fians; such as the tale of the elopement of Gráinne, the wife of Fin, with Dermot, one of his warriors (their beds pointed out in various places). The poem concerning 'the huge woman' daughter of the King of Greece. The chase by Fin on Slieve Gullion. The defeat of Manus, son of the King of Denmark, etc., etc . . . a few lines of these are as yet repeated in the parish, but I could not meet one able to recite the entire of any one poem, or to tell any one tale correctly. These tales and poems are yet preserved in MS, but they are better recited in the mountains of Waterford and Kerry than in O'Kane' country. [74]

It is as though he expects knowledge of a pristine ancient Irish society to be still current. What begins as a relatively naive question about ancient topography and kingship digresses into a discussion of contemporary storytelling. When compared to the older elite classical tradition this appears to be disjointed, incomplete and incorrect. The tradition of the Irish aborigines is presented as hopelessly misleading, half forgotten, primitive, natural and wild. Like the song about Bennada Glen it is recounted in 1834 to survey headquarters in the nature of a curiosity:

> This song is frequently sung in the Glen. It was composed by a man of the name James Feary, who is remembered by the old people to have been a great poem-maker of great satiric powers. In one of his poems he compares a man's forehead to that of an old pig, and says that his face is as soft and smooth as the Dreen or black-thorn. He had wild mountainous ideas, and his poetry pleases the inhabitants of the Glens of Ullin and Bennada better than the most refined

productions of modern poets. They can savour strong natural language, and risible pictures, but no affected or abstract phrases. The style of Milton, Johnston or even Moore would either offend them or they could form no idea of the meaning.[75]

While remaining an interested sceptic O'Donovan displays an acute critical awareness of popular culture. From time to time he gives brief outlines or descriptions of popular categories. From Maghera, Co Derry, he writes, 'I find that when those old romancers cannot account for the origin of a name or building, they ascribe them to the Danes or to Fin Mac Cool's militia. They divide history into three epochs: 1. Fin Mac Cool's time 2. The Time of the Danes 3. The Wars of Ireland.'[76] He singles out creative renewal in vernacular discourse in so far as it epitomizes the futility of it all for him. Many of his arguments with informants regarding vernacular etymologies begin with inquiries about ancient topography:

> I have had great argumentations with the Shanachies of Glenuller, and other places, about the meaning of Aireagal . . . they can invent stories in an instant to account for the names of old Churches. The story of Banagher, Boveva and Balteagh is repeated here also . . . the more learned Shanachies of the County however viz. those who have read Robinson Crusoe and think well on't insist that the word Errigal signifies a field of Battle, and would not scruple to tell a long story about a bloody conflict that took place there some 2,000 years ago between the O'Cahan and McCuillin or between Fin Mac Cool and Manus, the son of the King of Denmark . . . the more visionary and foolish the derivation of an Irish name is, the better it goes down with those who know nothing about it, viz. the whole human race, with the exception of nineteen or twenty. I should be much bolder to appear before the world, if the world had known a little more about the subject, but when I come in contact with persons who argue that Derry is a corruption of Deire, i.e. the end, i.e. the extremity of Ireland, and Lough Neagh, the Lake of Ulcers! What can I do?[77]

The cynicism about the proliferation of the characters of the Fenian tales in place names is qualified by his own belief, if it is true, that the hero Fionn Mac Cumhaill was a historical figure.[78] What Sjoestedt calls the mythological geography of the *Dinnshenchas*, a vernacular genre in which there is a metrical account of place names, posed a particularly difficult conundrum for O'Donovan. The capacity of the Fenian place names to be everywhere and nowhere and to blend into narrative, as Ó Coileáin puts it, escaped the etymological eye of the orthographer and the limelight of the surveyor alike.[79]

Tradition, far from being privileged in the letters, is often described as a hindrance to the work of the survey. In the context of nineteenth-century antiquarianism, or the emergence of professional ethnology in Ireland, O'Donovan's position is best sought after somewhere between the elite colonialist discourse of the official survey and the emergent academic disciplines of Irish history, archaeology, geography, literature, folklore and ethnology in both Irish and English. Rather than being an instrument of purely alien external imperial imperatives it is the result of a complex process of negotiation with and within the survey that mediated global ideas flowing through it. The fact that tradition is a problem or an object of inquiry is testimony to the adaptive or transformative roles of the fieldworkers as interlocutors between the elite culture of science and the emerging science of culture. O'Donovan in particular addresses himself specifically to this question:

> The more I look into the traditions preserved among the peasantry to account for the names of places the less I think of their title to historical credit; and upon strict examination it will be seen that like Ovid's Metamorphoses they have all been founded upon the real or fanciful signification of the names . . . I must lay it down as a kind of postulate, if not axiom, that respectable written authority is preferable to any oral tradition and that the authority of Colgan and Cormac's Glossary is preferable to that of Vallancey, who never read the former and never understood the latter. I need not remark what an imperfect mode of transmitting knowledge oral tradition necessarily must be in consequence of the weakness of the human mind and of its love for the wonderful and wild when it remains for a succession of ages in a state of ignorance and romance. This character of the mind is easily traced in our own much boasted of nineteenth century, for we find that the serious page of the historian and moralist will not please us but that we must have recourse to the Fairy Tale and the Novel to indulge this strong propensity of our nature. If this rage for romance and creations of the fancy prevails so much at present in crowded and enlightened cities how much more so it must prevail in the wild glens and mountains of a County far removed from Books and all philosophers excepting those who thunder from the pulpits, the eloquence of Christianity and the belief of Demons, 'the Witch of Endor and Pharaoh's Druids'. The next argument against oral tradition is that no two persons will tell the same story alike. Its parts are omitted, distorted, ornamented and

augmented according to the creative powers of the fancy of the narrator; and the fact is that he remembers a few prominent features in each story and fills up the vacancies according to his own idea of things. But when the Shanachy happens to be at all acquainted with *Robinson Crusoe* or *Gulliver's Travels* the whole story receives a new form and feature, and interpolations are inserted to no end.

In pioneering a more scientific methodology in Irish antiquarian and ethnographic work O'Donovan positioned himself between a precolonial civilized Ireland of relic, antiquity and taste and a postcolonial uncivilized Ireland of fancy and error. The survey work necessitated that he journey in both worlds and in doing this he generally overlooked the third possible world, the contemporary world, except to disprove it:

> There is another very extravagant notion among them that the human race are degenerating every generation. Fin Mac Cool and Goll Mac Morna conversed freely with each other while Goll sat on *Suidhe Goill* and Fin on *Suidhe Finn*, about nine miles asunder; and Finn cast a stone from Slieve Gallan to *Suidhe Finn* that God only knows how many men could lift now. They also remember the gigantic size of their grandfathers in comparison to the present pigmy race. I have reasoned with several upon the subject, and brought down all the arguments that I could remember . . . all would not do. A giant's grave on Carnanbare proves that the men of that age were of gigantic stature and as for brazen swords, and spears and arrow heads – they belonged to the wee folk (the fairies) not to men! . . . these extravagant notions can never be eradicated from the minds of the peasantry. The most conclusive arguments, the clearest demonstrations, would not convince them of the contrary. Their imaginations are wholly engrossed with them. Nor can they be removed until other principles of knowledge and other habits of thought are introduced and established among them. The peasantry were so in the days of Solomon (as he himself tells us) and they will remain so until that happy period (the millennium) arrives, when the serpent shall loose his sting, and man shall be reduced to his original state of innocence! And all shall be equally sagacious and intelligent.[81]

Such philosophical speculation on the peasantry clearly places them in a more primitive state than their observer. The living vernacular supernatural landscape in the minds of the people cannot be reconciled with the archaic landscape of the antiquary. The people are generalized and their words paraphrased by the domineering discourse of the survey. They are finally framed by the prophetic piece of millenarianism, of biblical anthropology, that forecasts eventual intellectual equality. Some legends narrated to O'Donovan are quoted by him as arguments that they were 'in themselves of no value, but they illustrate, in a striking manner, the credulous simplicity of the people among whom they originated, and cause a sincere enquirer after the truth to lay down this historical axiom, "nothing should be received as truths from the ancients, but what agrees with the common order of occurrences in this age"'.[82]

The idea of tradition that informs this perspective encompasses behavioural, verbal and artefactual aspects of the vernacular culture. In this there is an early outline of the emergent ethnographic imagination of a more three-dimensional sketch of human experience and expression. The inquisitiveness is variously inspired by imperial administration or individual curiosity and illustrated by various examples tempered by aversion, superiority or admiration. In the Ballynascreen area of Co Derry in 1834 O'Donovan relates the eating of bullock's blood, boiled and mixed with butter as 'the most abominable custom that ever prevailed among any people'.[83] This, he continues, was 'considered a great luxury by the mountaineers'; he admits, however, that his opinion may be the result of the 'association of ideas' or 'prejudices of my own fancy'. While he makes the point that many of the people lived in extreme circumstances they were, in his view, the wretched descendants of the bards stranded by a receding cultural tide. He searches still further back for an idyllic ancient Irish culture. Writing on some old manuscripts that had turned up he comments:

> The writing is beautiful and the spelling perfectly grammatical and correct. From this specimen you may form an idea of the taste of the inhabitants of Glenconkeine for preserving the writings of their ancestors. The truth however is, that there can be no taste where there is no comfort – no love of learning amid the squalor of poverty and no cleanliness where the smoke and soot have rendered the house and everything in it 'black as ebony' and where the pig and gander feed promiscuously and roam about the house just as they please.[84]

His gaze is not always entirely innocent. At times it seems to be the same commanding and condescending eye of the colonizer. Occasionally he trains his vision on the vernacular world of his own country within the parameters of colonialist discourse. These are often the self same parameters of elite English language cultural discourse. Empowered to look at places and people he is master of all he surveys. From landscapes, ruins, people and interiors, he names and translates places and people, his 'march of the intellect' doubles as a form of intelligence gathering. The colonizer may behave at times like a member of the elite or a member of the elite may behave at times like a colonizer. Class or religious allegiance may determine status for the native while racial or religious distinction may determine status for the colonizer. In the discourse of the survey they are indistinguishable at times.

The Tomb of Oblivion

The appearance of such accounts in the letters is not due to a purely objective interest in antiquities but to a will to dismiss them as examples of the simplicity of a putative tradition that must needs be superseded by reason and science in the English language. The predominant notion of tradition informing the survey is of something invariable, uncontested, homogeneous, inexact, static and implicitly believed in. A few counties are described as though they were coping with an epidemic of memory loss.

Co Longford did not interest O'Donovan in any way: 'I never was in a County less interesting than this, the people have forgotten everything about its ancient state'.[85] These are places in which tradition is described as being broken, patched and conflicting like southwest Co Meath where 'the Scotic language has totally disappeared . . . and the ancient traditions are entirely forgotten, which caused me to get done in that neighbourhood as soon as possible'.[86] Similarly Co Carlow was 'very barren in ancient remains and more so in traditions, so that we have nothing to guide us in identifying old places'.[87]

Like the folklore collectors one hundred years later, his role is expressed in terms of an eleventh-hour chronicler, to borrow Ó Duilearga's phrase, trying to catch up with what is described as a rapidly disappearing tradition.[88] Folklorists, as Frykman and Löfgren say, 'conducted what they thought was a last-minute rescue operation, recording the last vanishing remnants of endangered customs'.[89] In Co Roscommon O'Donovan laments, 'if I had been here even two years sooner I could have met a man who knew a great deal about Lough Key and its neighbourhood, but all the old Seanchies are now dead and with them the traditions are interred in the tomb'.⁹⁰ Writing in 1837 he estimates that 'fifty years hence, the Connaughtmen will not believe in giants, fairies, ghosts, blessed wells, stations or prophecies. This will, however, not make them one degree happier, nor more honest.'⁹¹ Co Westmeath is a cultural desert where the people have 'lost the Irish language and forgotten their old traditions (*non sequitur!*). I was never in any part of Ireland where the people know less this way.'⁹²

Accusing O'Donovan of becoming a dry topographer, Dr George Petrie encourages him to 'get as much of everything as you can, manners, customs, traditions, legends, songs, etc., The opportunities at present afforded may never occur again.'⁹³ O'Donovan's interested, if ambiguous, deconstruction of contemporary vernacular tradition sits uneasily with this instruction. Still in Co Westmeath he regrets that some informants recommended to him died, 'almost all the old men mentioned by Mr. Geraghty in this neighbourhood are dead. The two Fagans of Slieve Boy, Merriman of Fore and the most intelligent old men of Faughallstown and with them a vast deal of the traditions and romantic legends of the district lies buried in the tomb of oblivion'.⁹⁴ Similarly he recounts from Banagher 'I have been travelling all day through the south of the ancient kingdom of Me (Mídhe) where, I am sorry to say, the old language is just dead and the people remarkably ignorant of their old history and traditions'.⁹⁵

The ambivalent but productive posturing between the roles of colonial ethnographer on one hand and patriotic antiquary on the other is further reflected in his Co Westmeath deliberations on what is understood as the eminent demise of tradition:

> The traditions formerly connected with our localities are fading away more and more every year and the Irish language will be entirely lost in less than fifty years hence. If, therefore, we do not now record the glimmerings of tradition that exist [remain] and connect them with genuine history, none will have the power to do so hereafter and the topography and history of this noble but unfortunate island will forever remain in a state of confusion and darkness from which future investigators, tho' ever so industrious, can never clean them. Proceed we then to our work.⁹⁶

The consistent referent is remote prehistoric tradition such as might include ancient mythology, genealogy or obsolescent knowledge surrounding place names or 'remains of the ancient time'. His unflagging determination in questioning people about this archaic world causes him to be ridiculed in some places. Some ancient names had simply been forgotten or were creatively misunderstood in contemporary forms. He is laughed at in Ratoath when he asks if any one had ever heard it called *Ráth Aodha Mic Bruic*.[97] On another occasion he complains that 'the Irish Saints are altogether out of fashion, and I cannot bring their names to my assistance in topographical research without being laughed at'.[98]

Ancient territories also became a problem, 'it is very few will ever bother their heads about them, for I don't meet a single individual that cares a pin about the extent or history of these territories, and they laugh at me for such an old fashioned enquiry'.[99] After an old man laughed at him for trying to read an inscription on an old stone in a churchyard he comments that he was 'becoming a Pinkerton for scolding and ridiculing'.[100] The ironic pun is made upon the eighteenth-century Scottish antiquary and Goth enthusiast John Pinkerton.

It could also be argued that the many references backwards in time appear to be at least partly an effort to affirm an oppositional sense of Irish civilization, 'if Ireland was ever civilized (I mean comparatively civilized) it is while the monarchy stood, for from the first dissolution of it in the time of Brian, until the final conquest of Ireland in the reign of James I this kingdom was one scene of warfare, barbarity and bloodshed'.[101] The qualification of the term civilized is pointed if not guarded and, at a time when the nature of the progress of man and civilization is preoccupying English society, he is careful not to offend the sensibilities of the survey's superintendents. He continues, however, to match contemporary British civilization with an ancient Irish one.

The ancient seems to suggest what he calls pure tradition or pure legend as opposed to oral tradition or the contemporary vernacular culture. The latter is sometimes understood as evidence of barbarity or backwardness. In Co Mayo he protests, 'I should not wish to have it imposed as a task on me to write the manners and customs of the natives of Mayo for the last 150 years . . . I say if you educate the people they will make their betters be better; as they are now fast doing. Active, all searching, unbiased, unbribable law is what will make all classes good. Anything else is jargon.'[102] Many raconteurs test his patience by relating narratives to him that he simply does not want to hear. A man whose surname is given as Magaghran was one:

Magaghran wore out my patience telling me how Sheridan lost the Townland of Derrysherridan, but it is so like the story about Carthage and the Bull's hide that it is not worth preserving. Is the Legend about Lough Sheelion given in the Dinnsheanchus? If so let me have it that all the old rhymes and rags of legends, romance and history may be preserved and illustrated for posterity, who will altogether forget the old language of Ireland.[103]

The following passage from Co Roscommon addresses some of Petrie's inquiries and shows the contradictory nature of some of O'Donovan's ideas on tradition. At times he discredits oral tradition as fanciful in preference for literate tradition, whereas here he appears to discredit the literate in favour of 'pure oral tradition':

I have made every enquiry for traditions connected with the monuments he alludes to, and have found that there is not the faintest trace of a tradition in existence relative to a single fort, *rath*, *caisiol*, or *cathair* except that they were built by the Danes. Anything else the natives know about them has been acquired by reading. I could at once distinguish between a story preserved by pure oral tradition and one manufactured from 'an oulde history'. Several stories have been told me about St Patrick. Mave of Croghan etc., but I have learned upon enquiry that they were not preserved by oral tradition, but read out of Keating, and Lynch's Life of St Patrick. The only traditions connected with cromlechs now, and perhaps for the last thousand years, is that they are giant's graves, or the Beds of Dermot and Grania. Not a word more. I deny that there is a single word more preserved by pure oral tradition about them; but in this age when there are so many men that can read, it is not easy to discover whether a tradition be simply oral or fabricated from the oral and written accounts, and if you mistake one for the other, you will arrive at very wrong conclusions.[104]

He expected that tradition would be inexorably and faithfully connected to monuments and that it would provide him with a ready-made archaeological account of cromlechs. Since monuments were the furniture of the antiquaries' landscape those who knew nothing about them, or who knew the wrong things about them, had no traditions. Monuments must verify narratives and narratives must verify monuments. In the parish of Kilcooley, Co Roscommon, he wonders at a ruin: 'tradition makes it a Church of St Patrick, but vague tradition cannot be received when there is no monument, such as well, or stone, or even any regular legend to render it probable'.[105]

If it could not be corroborated in this way it should be checked in written tradition.

The Most Irish Parish

George Petrie's tradition is confined to those described in English antiquarian terms as old Milesians, seers, sages, druids, traditionalists, the peasantry, 'oulde' Irish and shanchies of the old aborigines who speak the ancient, vulgar, Scotic or Celtic language. Whether it is by design, accident or the editorial supervision of the survey's superintendents, colonialist discourse seeps into the letters and they often follow the same outlines and patterns of inquiry. O'Donovan often recounts his journeys much like a traveller, tourist or explorer entering unexplored regions with great expectancy of surprise and awe. The representation of the Irish landscape in the letters is 'cultural rather than personal; it is aesthetic, artistic, often literally picturesque'.[106] From Co Donegal he writes in 1835:

> On Friday we travelled through the Parish of Clonmany and ascended the Hill of Beinnin. Clonmany is the most Irish Parish I have yet visited; the men, only, who go to markets and fairs speak a little English, the women and children speak Irish only . . . I never heard Irish better spoken, nor experienced more natural civility and innocence than in that very secluded and wild Parish.[107]

The qualities of being natural, wild, secluded and innocent while speaking an ancient language are reserved for the Irish. The parish of Cloghaneely inspires similar thoughts, 'Cloghaneely alone is equally Irish throughout to be itself in all its parts – for they are all Cloghaneelymen from Muckish to Bloody Foreland and from Geeadore to Assfeenan. These latter all pray in Irish, and receive instructions from their shepards (pastors) through the medium of that Celtic dialect, which flourishes there yet in wild luxuriance.'[108]

A series of romantic vignettes portray a simple religious people distanced from civilization and viewed as 'extensions of the landscape, as the wilderness in human form'. Moving on to Ballybofea in the same county O'Donovan is moved to write:

> Yesterday being a beautiful day for travelling, we directed our course westwards along the southern bank of the black rolling Finn, and after a journey of five miles we found ourselves in the romantic Gleann Finne in the very heart of a purely Irish country. We entered

a Chapel Yard and soon found ourselves surrounded by a crowd of the old longheaded natives of Glen Fin – the remnant of the men of Moy-Iha.[110]

The sense of discovery is palpable in many accounts in which he describes almost stumbling across the natives. They surround him in curiosity and amazement while he casts them as the last survivors of a lost and ancient tribe. In Dunglow he expresses wonder at the sublime nature of his surroundings, 'I had never thought there was any part of the Sacred Isle so extensively and desolately wild or so thinly inhabited as the region through which we have wandered since I wrote last. It is sublimely barren, and at night poetically gloomy and horrible.'[111] He depicts a vast empty frontier set back in time and space and awaiting discovery, improvement and civilization. From Kilcar, Co Donegal, he writes of the people of Glencolumbkille in 1835:

> What their forefathers thought, believed, said and did, a thousand years ago, they think, believe, say and do at present. They are primitive beings who have but few points of contact with the civilized world. They hate, as indeed they should, the travelling preacher, and cling to the notions of their fathers with dignified independence. Social immobility seems to me the dominant trait in the character of these people, who live in what might be called the extreme brink of the world, far from the civilization of Cities, and the lectures of the philosopher.[112]

The wilder and more primitive he perceives an area to be, the more he expects to add to the 'already considerable collection of ancient topography and history'.[113] In Ballinvicar in Dunquin, or Dún Chaoin today, Co Kerry, he writes of the local respect for an old grave as primitive simplicity, 'it bears no name at present, but the people, who are very Irish and retain all their primitive simplicity in this remote parish, would not touch a stone of it'.[114] With increases in the relative degree of Irishness in an area there is a commensurate increase in the use of categories such as primitiveness, remoteness or superstition in the descriptions. The inhabitants of the island of Innishmurray, Co Sligo, are described in more detail in 1836:

> There are 102 inhabitants reckoning men, women and children, on this island; they are of indolent habits and are supported by illicit distillation. The inhabitants of the neighbouring shores supply them

> with barley, potatoes and other articles of food for the stomach and the still, for which the islanders give them in return a defined quantity of *Uisce Beatha*. There is but a small portion of the island cultivated, the people being so very indolent and depending altogether for support upon the produce of the still and the success of a few fishermen. The Revenue Police, however, sometimes circumvent them and reduce them to a woeful state by seizing upon large quantities of the mountain dew. Yet, notwithstanding the dangers, they will continue to distil and invent new methods of evasion, rather than be driven to annoy the surface of St Molaise's island or their own lazy limbs by using the loy or the spade. I saw yesterday a group of them stretched in the sun with as little care as if this world were a paradise; as if Adam had never been ordered to earn his bread with (by) the sweat of his brow.[115]

Whenever the survey encounters potentially oppositional, political or popular aspects of local culture the ethnographic observations of antiquity quickly become moral lessons. In this account the indolent, lazy islanders whose pioneering role in the local economy, beyond the taxation system of the imperial government, is not seen as endeavour but as idleness. It is not their relative statistical or ethnographic state of improvement that concerns the writer but the amoral behaviour in acting 'as if this world were a paradise'.

In contrast to the descriptions of the wild romantic landscapes of the mountainous coastal counties, coterminous with today's *Gaeltacht* or Irish-speaking areas, are the plains of Co Meath, 'the Meathians are a dull *tóinleasc* people with very bad memories, big heads and regular Gothic faces, the natural consequence of the stupid rich flat on which they live. A Connaught herd in the mountain of Bin Bulbin or a native, even of Innishmurray, has ten times their life, animation and intelligence.'[116] Still in Meath he feels suffocated:

> With the closeness of the trees and the strong odours of meadows, so that I would rather see a part of it elevated into a Quilkagh or a Slieve Snaght than be oppressed with a dull level covered with meadows and trees. I think a level country is too stupid to live in, and that the inhabitants will become stupid! I would rather take one view of Bin Bulbin, the gap of Barnesmore or the Deep Valley of Glen Togher than enjoy the scenery of the Plains of Tailteann for years; and still I can see that a native of Meath would feel himself miserable at the Rosses or in Glen Gavlen! You see rich meadow and luxuriant fields of potatoes, wheat and oats in every direction, and still the people are starving – 'starving' they say 'in the midst of plenty'.[117]

Unlike the idealized 'peasants' mentioned above the stable point of comparison, 'the ancient Irish Gaédhil nobility', also feature:

> Dean Swift would send one to the Coal Quay in Dublin to see the descendants of the ancient Irish nobility, but I would wish to view them on their own foídeens *Fóidín's* (remnants of estates) in Connaught where they figure in all their pristine splendour of original misconduct, vanity and worthlessness. To this many honourable exceptions must however be made, but still traced to their true causes, viz. education and connection with respectable society.[118]

This nobility have fallen to become 'Irish rakes and fools' who lack education and respectability. Their forms and features distinguished the ancient families in Co Donegal:

> The O'Donnells are corpulent and heavy, with manly faces and aquiline noses. The O'Boyles are ruddy and stout – pictures of health when well fed. The mad Devits are tall and slender with reagh visages. The O'Dogherties are stout and chieftain like, stiff, stubborn, unbending, much degenerated in their peasant state, but have all good faces. The MacSwynes are spirited and tall, but of pale or reagh colour. Among them all the O'Boyles and O'Dogherties are by far the finest human animals. I do not believe that the men of Tirconnell are at all so much improved as those of the south offspring of intermarriage of the ancient Irish with the Normans and Cromwellians.[119]

In nineteenth-century evolutionary and ethnological terms intimate connections are made between race and place. The gaze is physiognomic as well as picturesque and the eye is drawn to faces and noses as much as it is to mountains and lakes. The character of the people often matches the characteristics of the landscape and manifests the cultural, moral or physical improvement thought to result from settlement or plantation. Co Fermanagh is described as 'very nearly as wild now in the nineteenth century as it was in the sixteenth, during the rage of Chieftain Fights and bad laws. Lough Melvin is one of the most beautiful lakes in the world and still its banks are desolate and dreary, not one gentleman's residence on either side of it.'[120] The land is depicted as empty and dreary and in want of civilization. The gentleman's residence adds the finishing touch to bare nature and complies with the elite idea of the picturesque.

Lofty Peaks, Desert Isles Forlorne

Although O'Donovan protests that such descriptions are 'more the business of the poetical tourist than of an antiquarian orthographer' he succumbed on occasion to reveries on Irish sublimity. He describes Killarney's 'lofty peaks, desert isles forlorne, and musical echoes' in terms of what Schama calls a Celtic picturesque or idyll.[121] He wrote a poem for an informant named Brian Man Mullen, 'his is the spot of romance and wild song, where superstition holds her powerful sway, where the wild gloom impresses the mind'.[122] Saint Kevin's Shrine in Wicklow is 'horribly beautiful! And truly romantic, but not sublime!'[123] In Co Westmeath he portrays the scenery while he is rowed across Lough Ree with Thomas O'Connor, 'yesterday being a glorious day, O'Connor and I hired a boat at Athlone and we were rowed down Lough Ree with considerable rapidity. The scenery is magical, but tame and tranquil, Slieve Baan being the only object which adds a little sublimity to the scenery. Lough Ree is thick set with very beautiful islands "sparsely scattered" in it, here a cluster, there a solitary island.'[124]

The employment of the idea of tradition in the descriptions of Irishness is overwhelmingly rhetorical, figurative and imagistic. It is seldom based on any substantial detail. The result is often similar when it is. The following extract from a letter from Co Donegal includes some detail:

> The inhabitants of these glens and mountains are fair specimens of what the Irish were in times of yore. They have no idea of comfort; the smoky cabin of the cottier is perhaps not much less comfortable than the slated house of the grocer or the leather-cutter; the wet potatoes that grow in the holme or bog serves them for food, and if they can procure buttermilk for kitchen (as they call it) it is deemed a luxury; everything else (eggs, butter, oats, pigs, sheep, etc.,) is sold to make the rent or to buy tobacco.[125]

The smoky cabin and wet potato, so often complained of by the fieldworkers, are also signs of Irishness. O'Curry writing from Castlecomer, Co Kilkenny, gives some ethnographic substance to his descriptions:

> I date this from Ireland, and I am certain, with great propriety, as well as with great satisfaction to myself, for I do feel that I am now on Irish ground, not so when at the east side of the Barrow, where they have nothing but bad English, white frieze, good potatoes, and the

girls all wearing shoes, stockings and bonnets; but here, glory to them, they all speak fine Irish and English, wear blue frieze, drink whiskey, dance and fight; the girls all going barefooted and bareheaded; In short, here is every good manner and custom that ought to distinguish a decent country town in Ireland.[126]

Perhaps based on remembrance or familiarity, O'Curry enjoys the luxury of showing both sympathy and respect for the people he describes. In 1834 O'Donovan compares the devotion of the people of Bennada Glen to their houses in Co Derry with those indigenous peoples he classifies in colonialist terms as Laplanders and Hotentots:

The Bennada Glen is wildly beautiful, and produces numerous tribes of herbs and shrubs; but its houses are wretched (bad chimneys – worse floors!) and its damsels are neither stately nor melodious. But as his own little hut is more agreeable to the Laplander than would a spacious Palace, and his Kraal to the Hotentot, so were the smoky cabins of the Bennada Glen, and its brown faced barefooted girls more agreeable . . . than those of the most cultivated spot in Ireland. It is pleasing to consider how the human mind can comply with circumstances.[127]

Little has been written about the public reaction to the doubtlessly conspicuous and carefully observed manoeuvres of the survey staff. In one area it was interpreted as a kind of Armageddon that would see the British military bombarding the countryside. This fearful image, like other local registrations of the work of the survey, was based on legitimate and logical reasoning if not experience. Other interpretations saw it as a military recruitment drive or an attempt to suppress agrarian unrest. Equipment was sometimes interfered with and the sappers bore the brunt of people's anger with their landlord on at least one occasion.

The fieldworkers were mistaken for tithe collectors or suspected of involvement with government plans and conspiracies of one kind or another. Misinformation was often given as were blank refusals to co-operate for various reasons from the political to the religious. Science became the new magic and opposition was folklorized as superstition. There is an additional sense in which the mythology of imperial science is matched by the mythological geography of the country that resisted mapping.

Within the survey itself there were surprising elisions on almost every level imaginable. Relations were tense between the survey and the public, the survey and London, the survey and the Topographical Department, the Topographical Department and the public, the Topographical Department and Griffith's Boundary Survey. There are many similarities between the discourse of the memoirs and that of the letters. This is best illustrated by the use of the definitional category of tradition itself and the customs and habits discourse that is reserved for the representation of the Irish.

Masquerading as ethnographic data, tradition is often a euphemism for primitiveness, the precise opposite of the survey's vision of English civilization. While O'Donovan resisted the willy-nilly folklorization of the entirety of Irish culture there is a marked degree of ambivalence in his parroting of totalizing terms like aborigine, primitive, kaffir or Hottentot. Stigmatized by what is constructed as a lawless popular culture of fairs, patterns, wakes, holy days, festivals, cock-fighting, hurling, dancing, fighting and absurd, ridiculous, worthless and uninteresting narratives, it is explicitly the Irish who are traditional. They are at worst a primitive people who inhabit primitive places, at best the motley and bedraggled progeny of bards, chieftains and petty monarchs reduced to smoky cabins. Like their traditions they are described as effaced copies of some pure original that is inevitably and inexorably lost to modernity.

The selection and inclusion of examples of vernacular culture, highlighted in schematized inquiries, is a purposeful display of the supposed state of social and cultural degeneration to be compared to a postulated genuine moral and philosophical history of progress. It is a recurrent trope within cultural theory that the self-consciously modernizing discourse constructs its opposite as moribund and disjointed tradition. In the survey this is comprised of the broken traditions of an archaic primordial Ireland quarried from manuscripts, inquisitions, monuments and the questionable archive of the living Irish language and oral tradition. In Fabian's words such representations are 'predicated on the Other's absence' where 'the Other is displayed, and therefore contained, as an object of representation; the Other's voice, demands, teachings are usually absent'.[128] The present is absented, an empty space, bare and barren like the landscape; in evolutionary terms a *tabula rasa* or a blank page begging re-inscription in the language of education, improvement, civilization and science: English.

NOTES

1. Henry Glassie (ed.), *Irish Folktales* (London, 1985), p.36.
2. Angélique Day and Patrick McWilliams (eds), *Ordnance Survey Memoirs of Ireland* (Belfast, 1990–98), 40 volumes, Volume 15, p.17. Hereafter abbreviated OSM.
3. Marshall Sahlins, *Islands of History* (Chicago, 1985), p.144.
4. Ibid., p.149.
5. Scríbhneoirí Ban Ros Muc, *Idir Mná* (Conamara, 1995), p.108.
6. Gillian Smith, '"An Eye on the Survey": Perceptions of the Ordnance Survey in Ireland', *History Ireland* (Summer 2001), p.40.
7. OSM, 3, p.44.
8. OSM, 8, p.87.
9. Benedict Kiely, *Poor Scholar: A Study of William Carleton* (Dublin, 1972), p.9.
10. Gloria G. Raheja, 'The Ajaib-Gher and the Gun Zam-Zammah; Colonial Ethnography and the Elusive Politics of "Tradition" in the Literature of the Survey of India', *South Asia Research*, 19, 1(1999), p.32.
11. OSM, 31, p.57. Thomas Richards, *The Imperial Archive: Knowledge and the Fantasy of Empire* (London, 1993), p.19
12. Rev. Michael O'Flanagan (ed.) *Ordnance Survey Letters 1834–1841*, 43 volumes typeset and bound in Boole Library, University College Cork, (Bray, 1927–35), Londonderry, p.21. These were written in the course of their fieldwork for the survey by John O'Donovan, Eugene O'Curry, George Petrie, Patrick O'Keefe and Thomas O'Connor. See Archival Sources for further information. Hereafter abbreviated OSL.
13. OSL, Louth, p.10.
14. OSL, Louth, p.110.
15. OSL, Meath, p.84.
16. John H. Andrews, *Paper Landscape: The Ordnance Survey in Nineteenth-century Ireland* (Oxford, 1975), p.92.
17. Ibid.
18. Ibid., pp.43–4.
19. Damien Murray, *Romanticism, Nationalism and Irish Antiquarian Societies, 1840–80* (Maynooth, 2000), p.41.
20. OSL, Fermanagh, p.51.
21. OSL, Londonderry, p.147; OSL, Westmeath, Vol.1, p.87.
22. OSL, Roscommon, Vol.1, p.56.
23. OSL, Sligo, p.37.
24. OSL, Kilkenny, Vol.1, p.8.
25. OSL, Cavan/Leitrim, p.106.
26. OSM, 2, p.11.
27. OSM, 2, p.26.
28. OSL, Londonderry, p.143.
29. OSL, Cavan/Leitrim, p.133.
30. OSM, 2, p.62.
31. OSM, 2, p.63.
32. OSM, 3, p.107.
33. OSM, 4, p.71.
34. OSM, 4, p.130.
35. OSM, 14, p.77.
36. OSM, 4, p.131.
37. OSM, 5, p.52.
38. OSM, 38, p.17.

39. OSM, 6, p.18.
40. OSM, 38, p.33.
41. OSM, 6, p.18. Written by J. Boyle, J. Stokes and T. Fagan.
42. OSM, 13, p.50.
43. OSM, 19, p.61.
44. OSM, 13, p.65.
45. OSM, 30, p.31.
46. OSL, Mayo, Vol.2, p.87.
47. Jennifer Schacker, *National Dreams: The Remaking of Fairy Tales in Nineteenth-century England* (Philadelphia, 2003), p.8.
48. Originally suggesting a hill or mound this appears in the modern Irish language in many combinations in vernacular discourse that speaks of the ancestral supernatural world such as *bean sí, poc sí, sí gaoithe*, that are now translated as fairy woman, fairy stroke or fairy wind. The translation of *sí* as fairy, although it is difficult to avoid, has quaint Victorian English connotations that are misleading. O'Donovan's rebuttal of the emerging nineteenth-century sense mirrors contemporary concern within folkloristics in Ireland to go beyond the simplistic colonial notion of superstition to the more psychological, social and cultural import of the term.
49. OSM, 22, p.15.
50. OSM, 19, p.104.
51. OSL, Cavan/Leitrim, p.3.
52. OSL, Londonderry, p.125.
53. OSL, Galway, Vol.1, p.159.
54. OSL, Galway, Vol.1, p.124.
55. OSL, Carlow, p.81.
56. OSL, Carlow, p.140.
57. OSL, Cavan/Leitrim, p.7.
58. OSM, 15, p.30.
59. Murray, *Romanticism, Nationalism*, p.21.
60. OSL, Roscommon, Vol.2, p.80.
61. OSL, Cavan/Leitrim, p.47.
62. OSL, Donegal, p.143.
63. OSL, Donegal, p.146.
64. OSL, Donegal, p.146.
65. OSL, Westmeath, Vol.1, p.58.
66. OSL, Galway, Vol.1, p.120.
67. OSL, Galway, Vol.1, p.234.
68. OSL, Kerry, p.40.
69. OSL, Galway, Vol.3, p.14.
70. OSL, Kildare, Vol.1, p.20.
71. OSL, Kildare, Vol.1, p.39.
72. OSL, Kildare, Vol.1, p.78.
73. OSL, Londonderry, p.24.
74. OSL, Londonderry, p.26.
75. OSL, Londonderry, p.33.
76. OSL, Londonderry, p.35.
77. OSL, Londonderry, pp.39–40.
78. Murray, *Romanticism, Nationalism*, p.25.
79. Seán Ó Coileáin, 'Place and Placename in Fianaigheacht', *Studia Hibernica*, 27 (1993), pp.55–7.
80. OSL, Londonderry, p.64.
81. OSL, Londonderry, p.98.

82. OSL, Longford, p.4.
83. OSL, Londonderry, p.113.
84. OSL, Londonderry, p.124.
85. OSL, Longford, p.53.
86. OSL, Meath, p.82.
87. OSL, Queen's County (Laois), Vol.1, p.24.
88. Séamus Ó Duilearga, 'Notes on the Oral Tradition of Thomond', *The Journal of the Royal Society of Antiquarians*, 95 (1965), p.147.
89. Jonas Frykman and Orvar Löfgren, *Force of Habit: Exploring Everyday Culture* (Lund, 1996), p.6.
90. OSL, Roscommon, Vol.1, p.113.
91. OSL, Roscommon, Vol.1, p.110.
92. OSL, Westmeath, p.74.
93. OSL, Roscommon, Vol.2, p.127.
94. OSL, Westmeath, Vol.1, p.104.
95. OSL, King's County (Offaly), Vol.1, p.58.
96. OSL, Westmeath, Vol.2, p.75.
97. OSL, Meath, p.113.
98. OSL, Roscommon, Vol.1, p.68.
99. Ibid.
100. OSL, Roscommon, Vol.2, p.23.
101. OSL, Londonderry, p.100.
102. OSL, Mayo, Vol.2, p.70.
103. OSL, Meath, p.34.
104. OSL, Roscommon, Vol.1, p.120.
105. OSL, Roscommon, Vol.2, p.36.
106. Johannes Fabian, 'Remembering the Other: Knowledge and Recognition in the Exploration of Central Africa', *Critical Inquiry*, 26 (Autumn 1999), p.55.
107. OSL, Donegal, p.10.
108. OSL, Donegal, p.81.
109. David Spurr, *The Rhetoric of Empire: Colonial Discourse in Journalism, Travel Writing and Imperial Administration* (London, 1993), p.165.
110. OSL, Donegal, p.95.
111. OSL, Donegal, p.99.
112. OSL, Donegal, p.124.
113. OSL, Mayo, Vol.1, p.15.
114. OSL, Kerry, p.40.
115. OSL, Sligo, p.10.
116. OSL, Meath, p.29.
117. OSL, Meath, p.9.
118. OSL, Roscommon, Vol.2, p.82.
119. OSL, Donegal, p.110.
120. OSL, Fermanagh, p.45.
121. OSL, Kerry, p.114. Simon Schama, *Landscape and Memory* (London, 1995), p.469.
122. OSL, Londonderry, p.58.
123. OSL, Wicklow, p.71.
124. OSL, Westmeath, Vol.1, pp.3–6.
125. OSL, Donegal, p.132.
126. OSL, Kilkenny, Vol.1, p.1.
127. OSL, Londonderry, p.33.
128. Johannes Fabian, 'Presence and Representation: The Other and Anthropological Writing', *Critical Inquiry*, 16 (1990), p.771.

CHAPTER SIX

Traversing the Country: Fieldwork, Mimicry and Method

Contemporary accounts of the survey often read something like this: from 1834 to 1841 John O'Donovan 'travelled the length and breadth of Ireland, sending back detailed accounts of the language, folklore, topography, manuscripts, and place names of the localities he visited'.[1] This exaggerates the function of the letters and the memoirs and the role and work of the surveyors. It has more to do with the ideological preoccupations of the twentieth century than with those of the nineteenth. The survey has some detail about language use; it was interested in English as much as, or more than, Irish. Tradition, or folklore, is more a category of the survey than a described phenomenon. It actually is Irish culture rather than an aspect of it or appendage to it. Manuscripts were simply a source supplementing or even superseding local information. There were many other Irish scholars outside the survey engaged in the writing and reproduction of manuscripts. Séamus Ó Duilearga, head of the Irish Folklore Commission, noticed this forty years ago when he wrote that 'the OS letters provide the historian and the antiquary with information on the field-monuments, but of the tradition of the unlettered Irish countryman there is scanty evidence in these much quoted documents'.[2] For some there may be something attractive, romantic, heroic or affirmative enough in the survey, in the idea of the Empire coming to the rescue at the last minute, to perpetuate it.

The main preoccupation of the letters, the copious descriptions of what could be called, in Varagnac's word, archeocivilization, are seldom mentioned. The idea of an interdisciplinary alchemy out of which modern Irish learning flashed readymade is partly mythical. The staff of the Antiquarian or Topographical Department were, 'the instigators of the scientific study of Irish antiquities' but the view of the surveyor as an intrepid pioneer, hacking his passage through the forest primeval, overlooks the actual process of knowledge accumulation and registration in favour of a heroic portrait of an explorer in the impenetrable jungle of pre-Famine Irish tradition.[3] The contribution of O'Donovan, O'Curry, O'Keefe and O'Connor is much more specific and more exacting than that.

The construction of the ethnographic occasion in the field is central to any assessment of their work. The ethnographic inskilment of the fieldworkers illuminates the processes that produced the correspondence routinely sent back to headquarters in Dublin. Their various roles as investigators, inquirers, etymologists, orthographers, antiquaries or topographers were crucial to the domestic success of the survey. It is this particular aspect of their work that secures their place in the story of folklore and ethnology in Ireland. In pioneering 'new modes of investigation' to inquire into the 'condition of man', O'Donovan confuted the speculations of a generation of English antiquaries in Ireland. Both he and his colleagues should be regarded as the first professional Irish ethnographers. In highlighting the anglocentric and even eccentric tendencies of their predecessors they 'formed a new model' within the confines of the survey. It was not bestowed upon them by the survey but something they cabbaged from it.

The correspondence details a resourceful engagement with vernacular culture in terms of methodology, the use of informants and sources. Competing with tourists, fishermen, revenue police, festive celebrations, fairs, markets and patterns, the fieldworkers came to a good working accommodation with the everyday life of the people. The popular cycle of petty sessions, holy days, Sundays and festivals was exploited to this end. On these days people had time to talk and the fieldworkers had time to send for, pick or have informants pointed out as they often put it. Local people often guided them or introduced them to others who could help. Letters of introduction were used to make contact with members of the elite. A careful methodology was worked out that almost proved too burdensome for the work of eliciting the pronunciation of place names and producing an Anglicized orthography. This sometimes led to the rejection of the ethnographic challenge or the hasty retreat from worlds that were only too familiar.

The role of O'Donovan is particularly interesting in this regard. His early employment as a cataloguer, editor, transcriber and, more significantly, as a translator of old manuscripts, was an ideal preparation for his work with the survey. It is arguable whether his name would be as well known today if it were not for this work. The names of many Irish scholars who were also part of the imperial civil service of translators employed by English or Anglo-Irish antiquaries receded into anonymity. It is also possible that, if he had enjoyed the similar freedom and status as English antiquaries for example, he might have been able to pursue his interests in a more professional manner outside of the survey.

The letters are detailed ocular surveys of ancient ecclesiastical derelictions such as towers, churches, monasteries, castles, forts and abbeys. They frequently mention holy wells and the burial places of unbaptized infants but these happen to be located close to early christian ruins in many cases. The landscape portrayed as a result is a Gothic wasteland. There is often an acute sense of exploration and discovery engendered by the elation of translating places anew into the pages of English history. At times O'Donovan speaks of discovering saints that are household names for local people. It is as though their translation into English is what finally unearths them and brings them into the light of a new day.

He interrogated informants in a tireless, detailed and determined fashion in search of what he understood to be ancient tradition. Surnames and even nicknames occasionally feature as do songs, legends and popular beliefs, although to a far lesser extent than is generally imagined. These are infrequent and brief appendages that occur in the course of archaeological observation and onomastic investigation. In constructing ethnographic occasions in the field it was simply impossible to avoid encounters with the vernacular culture. Although it is specified in the survey's directives it features most prominently as a backdrop to the toponymic investigations.

Enquirer after Fragments of the Olden Time

Before any question was asked, before any pen was put to paper, before any map was engraved, Irish culture was conceived of as traditional; from its own perspective the survey was simply documenting this fact. In tandem with the mapping of boundaries the survey mapped an ancient Ireland populated by alternately respectable or rudimentary primitives. The letters are responsive to overarching directives whose function was to identify, describe and define Ireland. They are tailored to the needs of a civilizing project of Anglicization and assimilation. As Spurr says, 'writing journals, scientific papers, maps, or documents of colonial administration – all of this writing illuminates the darkness with the light of the writer's own countenance, while at the same time it sets up, once and for all, the logical oppositions between writing and blankness, knowledge and doubt, culture and nature, civilization and savagery'.[4] Irishness was one of the critical and abiding ideas with which it worked. In the thesaurus of imperial rhetoric terms like primitive, traditional, vulgar, ancient or aboriginal are all euphemisms for non-Englishness. The exact nature of this Irishness, however, is contested and contradictory as the survey progresses. O'Donovan's initial enervation wanes as he makes his way further south and finds himself close to home.

The grand imperial rhetoric is diluted by the homely and familiar topography of his birthplace. When he finally arrives the depersonalized peasants or aborigines of colonialist discourse are boiled down to the personal pronouns I or we. In Co Kilkenny he finds himself back where he started and this homecoming is an apt and even moving metaphor for all of his work with the survey. The story of the survey is not always coherent: if it is not the story of Irish folklore salvaged then neither is it the story of victorious British colonialism.

The received view of the survey, in particular the view of O'Donovan's role in it, is complicated by the tensions between the official survey and the unofficial role of the fieldworkers. It is further coloured by the networks, relations and methodology employed at its headquarters. Add to this the personalities involved and their degree of complicity or originality in implementing official directives. He refers to his work as topographical observations, the investigation of tradition or the restitution of decayed intelligence. He describes himself as an enquirer after fragments of the olden time, an honest or practical investigator of ancient topography, an Irish Etymologist, a historical topographer and an antiquarian orthographer. Although officially employed as an orthographer and etymologist the different strands of investigation, observation, inquiry, topography, history, antiquarianism and etymology typify the disciplinary flux of the early nineteenth century. Emerging from what Ó Giolláin calls the literature of confutation of previous centuries, these investigations evince a shift towards enlightened inquiry where 'condemnation gives way to a more disinterested scientific curiosity'.[5]

Ó Danachair did not include O'Donovan or his colleagues in his overview of proto-ethnologists in Ireland while he counted Charles

Vallancey and Thomas Crofton Croker as being foremost amongst them.[6] O'Donovan's status in this regard mirrors that of a group of subaltern scholars that remained, both literally and metaphorically, in the shadow of the canon. It is as though behind every genteel English antiquary or ethnographer there is a native understudy. In an anglophone historiography the names of scholars like Patrick Lynch (Edward Bunting), Charles O'Connor (Charles Vallancey), Theophilus O'Flanagan and Murice O'Gorman (Charlotte Brooke), with their very names translated into English, are set adrift from the mainstream. Many of the pioneers belonged in identity-seeking elite English-speaking groups if they were not colonial officials or military officers. They were pioneers of English language antiquarianism in Ireland but, as Ashcroft *et al.* suggest:

> The fictionality of the narratives of such processes of 'discovery', which the process of mapping objectifies, is emphasized by the role of the native guide in such explorations, who leads the explorer to the interior. The prior knowledge of the land that this dramatizes, and which cannot be wholly silenced in the written accounts of these explorations, is ignored, and literally silenced by the act of mapping, since the indigenous people have no voice or even presence that can be heard in the new discourse of scientific measurement and written texts that cartography implies.[7]

For increasingly marginal(ized) indigenous researchers translation, and afterwards silence, was a significant factor in the formative period of professional Irish folklore and ethnology from the middle of the eighteenth to the beginning of the twentieth century. This period was marked by the unequal interplay of vernacular and scientific knowledge, methodology and epistemology across languages and cultures. In embracing what were global, as much as British, flows of information, theories and methods the fieldwork of these early Irish ethnographers contributed to, and symbolized, this disciplinary synthesis. A key aspect of this was the sometimes cogent and still relevant critique of their predecessors.

The Vallancey School of Antiquarianism

In the absence of formal academic structures or disciplinary development in nineteenth-century Ireland, the theory and method of antiquarianism or ethnography was advanced within the quasi-collegiate department of the survey, albeit a surrogate one. The broad outline of antiquarian research in Ireland and the prophetic view of its future is a significant part of the outwork of the survey. This cannot be understood without first recognizing the declamatory and colonialist nature of the discourse of both the survey and contemporary English antiquarianism. The Irish scholars are severely critical of what O'Donovan calls the visionary class of investigators. O'Curry calls these collectively the Vallancey School of Antiquarianism. Carefully enumerating the Irish scholars previous to both the survey and the English antiquaries O'Curry writes:

> There is a batch of Antiquarians for you, to whom if we give credit, will prove that the ould Irish people were the most godly creatures in the world, for wherever you turn your face to, you will, according to these sagacious writers, find lashings and leavings of hills, mountains, rivers, rocks, caves, cairns, etc., dedicated to their Gods, their Great Gods or their . . .? It is strange how the Vallancey School, without knowing anything of the Irish language either natively or by its affinity with other languages . . . could know the meaning of Irish words better than the Cuan O'Leochans, the Eochy O'Flinns, the MacFirbises, the O'Clearys, the O'Dugans, etc.[8]

O'Donovan is critical of what he calls the wild assumptions of Protestant unionist Sir Samuel Ferguson, the author of *Congal: a Poem in Five Books* (1872), regarding his positioning of Killeshin in Co Carlow:

> Our friend Mr Ferguson in his Hibernian Knights Entertainments published in the Dublin University Magazine, places Killeshin in O'Nolan's county, in which it never was as I shall hereafter demonstrate when treating of the territories in the Queen's Co. and Co. of Carlow. There is no end to the wild assumptions and fooleries of modern Irish historians, legend writers and topographers.[9]

He may be referring to Vallancey and Crofton Croker in his last comment as well as to those he later alludes to as 'ignorant Irish writers as Moore, Crofton Croker and others'.[10] Regarding what he sometimes collectively and dismissively calls hagiology or fairyology he comments in Co Mayo that, 'it is a pity that that little fairy elf, Crofton Croker, is not better acquainted with our ancient MSS., to give us the old history of the fairies of Ireland, for he seems to be well acquainted with their modern habits'.[11] Here he rebuts what might be called decontextualized fairy discourse, the commoditization of selective elements of popular belief that served to classify

Irish culture itself for English consumers of folklore. If the Irish are superstitious, gullible or naïve the argument for civilizing them is strengthened. By 'old history' he implies that there is no scientific or ethnological interest in the origins of the ancestral supernatural world. Fairy discourse continues to be a Pandora's box of fantasy and fiction. The Dublin Society and the Statistical Surveys are also critiqued:

> This account of Killeshin is written in a very unsatisfactory, unhistorianlike manner, which characterizes all the Statistical writers for the Dublin Society. The remark that the inscription on the doorway of Killeshin is in the Saxon character is strange indeed, and it would appear to me that by it he wished his readers to understand that the Church was built and the inscription built by Saxon people. *Vae victis!* But he will be laughed at when the antiquities of Ireland shall be a little better known.[12]

In the late eighteenth century the perception in Britain of protestant settlers as fully Irish troubled antiquaries like Edward Ledwich and William Beauford who had tried to integrate themselves into the 'mythology of the colonized'.[13] Sir Charles Coote's *Statistical Survey of the Queen's County* (1801) is particularly repugnant to O'Donovan:

> Such rigmarole! Such a conglomeration of nonsense! Such a disgrace to history and the human intellect ... all this vile stuff was fabricated by Beauford, a Schoolmaster at Athy, who was one of the *triumviri* and who imposed several fabricated inscriptions and historical pieces on Ledwich and others. It is time that such trash should be recognised and rejected by the honest investigators of Irish topography and history.[14]

The *Statistical Survey of the County of Galway* by Dutton, a land surveyor and extensive farmer, is also criticized by O'Donovan. Drawing satirically on the popular nineteenth-century belief in phrenology or the interpretation of personal traits from the contours of the skull he writes:

> I have received Dutton's *Statistical Survey of the County of Galway* and I looked over it, but find it will be of no use to me, as he gives no authorities. He is a regular helter-skelter Irish writer who has not the organ of order very prominent in his pericranium. He knows nothing about Irish history or antiquities, and has made no research

whatever in that way in this County. I shall send it back again as it is not worth carrying.

Later he comments that Dutton 'knows more about potatoes than antiquities which leaves him a rich man'.[16] Dutton died in 1837 while O'Donovan's work was ongoing and one year later he writes from Kilkenny that Dutton 'was a very useful man in Ireland, being of Malthusian principles and a great improver of bogs and, dreading a superabundance of population in Ireland, even though he was fond of potatoes . . . he applied himself to making books instead of land improvers'.[17] He was conscious of the pioneering aspect of his work within the survey as is evident in the following passage that considers all previous approaches whether Irish or English:

> We in Irish Antiquarian Research, like Lord Bacon in Natural Philosophy, shall have to strike out new modes of investigation. We must pursue the inductive system, that is, first amass a large collection of facts, and then draw such conclusions as these facts will warrant. Truth will never be illustrated in any other way. Hitherto we have had two parties of Irish writers of diametrically opposite feelings and notions. The one were Milesian to the backbone, learned, vain and pompous, but altogether wanting the acumen of intellect and that intimate acquaintance with facts which qualifies one for an investigation of ancient history. These may be compared to the old Seraphic Doctors of the Peripatetic philosophy, so beautifully ridiculed by Molière. The other were English or Scotch, learned, sharp (acute), severe and prejudiced. It was often their employment to turn everything connected with Ireland into ridicule in order that the conquered Milesians might become lessened, not only in the view of their conquerors and other nations, but also in their own estimation. In this they have most admirably succeeded. These latter writers were also unqualified for enquiring into the real state of ancient society in Ireland; many of them have wilfully falsified originals; others have distorted them, and others have been so blinded by prejudices as to draw the wrong conclusions.[18]

For O'Donovan these writers of the eighteenth century were 'a set of ignorant and dishonest scribblers without one manly or vigorous idea in their heads. Vallancey, Beauford, Ledwich, Roger O'Connor etc., were all either fools or rogues who were by no means fit (qualified) to demonstrate the truth of ancient or modern history'. Honesty, truth and order, in short objectivity, were not qualities he associated with these antiquaries.

Dancing Masters as Well as Farmers

It is precisely the drive to strike out new modes of investigation and engage critically with the evolving disciplines that makes a strong argument for the consideration of O'Donovan and his colleagues' as path breakers of Irish ethnography. One of the results of their methodological and critical deliberations was the wariness of the rhetorical category of tradition that colonialist discourse depended upon. This was the shibboleth of many of the English Celticist antiquarians. It was central to this critique that the methods and theories of these antiquaries were not sustainable in ethical, objective or scientific terms. By equating Ireland with fairyland and casting the Irish as deluded fairy believers they classified Irishness as primitive and childlike. By rejecting this kind of racialist image O'Donovan and his colleagues were innovative in theorizing the emergent branches of Irish research. This innovation encompassed oral evidence, manuscript evidence, the Irish language, the rejection of obscurantism, field research, ethics and the local. This pattern emerges from their overall achievement and reflects an ongoing concern throughout the course of their fieldwork. Occasionally this is expressed explicitly in ethnographic terms:

> What is the reason that man always wishes to know how his ancestors lived? That he burns with curiosity to find out how they built their houses, thatched their cabins, cut each other's throats, and peeled the potatoes? Why would I feel inclined, if I had the money, to set out for Alexandria to request the Magician Maugraby to shew me the Dagda reflected in a drop of ink? And why would another laugh at me to scorn? (Such and so different are the tastes of men, formed by circumstances and original temperament. That is all we are told by philosophers on the subject.) These are questions which will never be satisfactorily answered but the people will be always found to be ready to refer them to mental phenomenon in such a manner as to enlighten some and confuse others. Antiquarian research is of no importance to mankind in general, but still such is the variety of human pursuits resulting from modern civilization, that there will be always a number of people found willing to devote their whole time and study to the investigation of the condition of man from the period at which he was turned out of his garden for eating an apple . . . until he built a steam packet and formed a theory which accounts for the appearance of the Aurora Borealis. Whether such people will ultimately benefit mankind or not is a question on which Dugald Stuart has not thrown much light, but whether they will or will not,

no one but a man of refined (dwindled) civilization will ever set a value upon their labours; for he looks upon every branch of knowledge as connected immediately, or remotely with general science, and knows that he cannot form (draw) a perfect portrait of man until he is able to get a view of him in every state of civilization and for that reason he will set due value upon any kind of research which will afford him a glimpse at man in any condition or under any circumstances ... The Utilitarian, however, will always look upon the cultivation of the potato as the *summum bonum* of life and laugh at any philosophy except that which teaches him how to make two blades of corn grow where only one had grown before. (Half that opinion is also mine!) But for the present conformation of society it is found necessary (*recte*. pleasing) to have dancing masters as well as farmers, and Antiquarian and Historical as well as Horticultural and Agricultural Societies.[20]

This passage illustrates the spirit of disinterested, scientific ethnographic inquiry and the encyclopaedic nature of knowledge accumulation in the nineteenth century that is at the heart of the emergent academic disciplines. Dugald Stewart's Edinburgh lectures in moral philosophy were well known at the start of the nineteenth century. It was Stewart who coined the term conjectural history in 1793 and he directed his efforts both at the study of the human mind and political economy.[21] One of his lectures speculated on the origins of the American Indians.

The split between utilitarian political economy, the thoughts of Thomas Malthus, and the objective contemplation of civilized life is apparent in O'Donovan's thoughts. They reflect contemporary British as well as continental currents of thought about 'the physical nature and differentiation of man, which raised the problem of its universality; ideas about the nature of social order, which defined the specific content of civilization; and ideas about the methods appropriate to the study of human life and history'.[22] O'Donovan wished to introduce objective contemplation to the questions of method, content and theory of Irish civilization.

James Cowley Prichard, whose name is synonymous with the emergence of ethnology in Britain, was a student of Stewart's and a contemporary of John O'Donovan. The Irish scholars were familiar with the works of Prichard and philologist Professor Adolphe Pictet, who had called the survey a work *fort precieux*. O'Connor praised Pictet and Prichard for their efforts to 'elucidate the etymological history of Celtic languages'.[23] They were also familiar with the work of Franz Bopp whose 1816 publication of *The Conjugation System of the Sanskrit Language* marks the beginning of comparative philology.[24] The interest in ethnological topics is evident in the blend of natural science, moral history or moral philosophy and philology pursued by O'Donovan and his colleagues. It could be argued that it is O'Donovan who brings contemporary ethnological thought to Ireland in the first half of the nineteenth century. The anguish, challenge, creativity and curiosity of ethnographic encounters in the field were productive of conditions that were seminal in terms of the discipline's emergence.

Traversing the Country

O'Donovan's multiple roles and titles could be summarized more by the term epistemologist, understood as a surveyor of knowledge, than etymologist. He was as concerned with the origin, derivation and provenance of knowledge in general as he was with the origin, derivation and provenance of place names. He vernacularized or localized international currents of thought through his involvement with the survey and reassessed the disciplines of the day. Although he says bluntly in Co Roscommon, 'I don't look upon the letters I write as any part of my business',[25] he frequently describes his role as that of a pioneer, 'all I boast of is that I have made a beginning and pointed out to my followers (trotters) the sources from which information can be drawn, when the Records of England and Ireland are published'.[26] In Co Roscommon he protests 'all I am doing, ambition to do, or can do, during my hurried rambles through Ireland, is to connect history with topography, and thus give it a particular or local instead of general interest'.[27] In doing this he showed 'each lovely spot where Dagda trod (*Magh Luirg an Dagda*) Meva spoke or which Patrick cursed! And if Irish history ever happens to be fairly written, the inhabitants of each locality will see plainly by events which took place at their own doors how necessary it is to restrain the human race by strict laws.'[28]

Here his words resonate in the homely if uncharted world of narrative of which Nicolaisen says 'any attempt to pinpoint any of these places by means of the conventional network of latitudes and longitudes, or by any other kind of accurate grid, is bound to end in failure'.[29] Ó Coileáin argues that the name *Magh Luirg an Dagda*, is poetically projected into the mythological world, is de-onymized and ceases to be a place name proper as it spills over into the evocative topography of vernacular narrative.[30] To pinpoint such names was a task that could be likened to minding the local mice at every crossroads in Ireland.

At the end of his work in Co Roscommon he reflects, 'I have now done with the territories in this County, and have, to my great satisfaction secured information which if neglected for ten years more (i.e. longer) could never be recovered, and I may style my researches in that way, "Restitution of Decayed Intelligence"'.[31] In practice the letters digress from the direct and pressing duties of his employment and clearly foresee a future return to Irish antiquarian or ethnological research. That it would be a return is explicitly intended and he saw himself, as O'Curry did, as mediating or re-appropriating older indigenous practices even as he mediated Anglicization, 'if the ancient localities be not pointed out during the course of this Survey, i.e., before the Irish become totally Anglicized, future antiquarians will exclaim against us as a set of stupid idlers, who had the means, but not the industry, to preserve these things for posterity'.[32] The imperial archive would become a national archive in time. In Co Kildare he writes that he is 'amassing a vast quantity of national records and I should be very sorry that any one should hereafter have to say that they are full of mistakes, that they were "copied in a hurry and never compared"'.[33]

The fieldworkers understood their role as operating within the imposed structure of the survey for national ends. Although he refers in Co Roscommon to 'the great national undertaking – the Ordnance Survey of Ireland', O'Donovan pointedly qualifies the degree of indebtedness to the survey with the caveat, 'I wish to point out distinctly how much had been done before this Survey'.[34] On discussing with O'Donovan the almost clandestine efforts of the fieldworkers within the survey James Hardiman, a historian and commissioner of records, expressed scepticism, 'my friend Mr Hardiman told me when I was at Galway that we would never do it "but that it is of very little consequence whether we do or not as the up springing vigour of this nation will ere long complete it, even should the Ordnance Surveyors drop it now". I perfectly agree with him but it is too bad if the Surveyors shall have credit for forming the model only, after all the anticipation of the public.'[35] The underlying concern or goal of this epistemological espionage is expressed as 'doing it' or forming the model for Irish scholarship. That this was sometimes haphazard is clear from accounts of the everyday working conditions of the Irish scholars in the field.

Quills, Sealing Wax and Pens

Captain J.E. Portlock's instructions reveal the methodology that was employed. Each fieldworker had three days' work carefully outlined in advance. They were to proceed to the ground to carry out inquiries returning later with their notes and journals on separate days. On these days their observations were documented and they were expected to account for their time in detail:

> No one is to consider that because his services are principally directed to some particular object, he is on that account to neglect every other. On the contrary, all are labouring to promote the interests of a great work and each person should note all he sees and all he hears, whether in reference to his own immediate inquiries, to the social condition and habits of the people, to antiquities and traditionary recollections of all kinds or to natural history in all its branches, naming at the same time his authority for each statement.[36]

O'Donovan probably spent more time in the field than any other researcher in Ireland past or present. In spite of this little has been written to date on the everyday conditions of the fieldworkers. From time to time the fieldworkers complain of being inadequately supported from headquarters. There was a constant demand for stationery: in Co Galway O'Donovan writes, 'we have written ourselves out of paper again! And will have two quires bought in this town soon scribbled on. We also want quills, sealing wax and pencils.'[37] In Co Kilkenny O'Curry requests, 'paper, pens and sealing wax, having written out all the paper that you sent me to IarConnaught and 350 pages more and worn all my pens to the stumps. Mr Wakeman wishes to have some smooth paper on which he can draw with facility, and he says he cannot draw well on the square roughish paper which you sent us to Galway.'[38]

William Frederick Wakeman, who studied drawing with George Petrie, was employed to make sketches for the survey. In all, twelve volumes of sketches resulting from the survey drawings of Wakeman, Petrie and George Victor du Noyer were deposited in the Royal Irish Academy.[39] Frequent requests were made for the name books, indexes to them, maps, extracts of various kinds, books, vouchers, square paper, quills, pens and money. O'Donovan protests strongly from Co Sligo about the conditions under which he was working:

> It is the search for localities and striving to settle queries satisfactorily that occupy most time. If a person was not scrupulous in these things he could soon traverse the country. I find it a desperate (intolerable) onus to have to stick in old orthography for names in the books in these disagreeable country places, where one meets every annoyance, not having even an Index made to the Name Books, making the matter a thousand times worse. It is too bad to be forced to be making indexes amidst one's greatest hurry, delaying him from his proper business.[40]

From Co Meath he enquires about hiring a car, 'I am anxious to hear how I stand with respect to car hire; the expenses of this town are more than we can bear'. In Co Westmeath he asks for paper by return of post and complains about the careless treatment of parcels:

> I have retained the two books which you thought were lost. I was under the impression those which I sent from Ballymore Lough Sewdy were very well secured, but they pitch those things about very carelessly. Many of the parcels and letters from the Ord. Survey Office come to me opened (the wafers having given way to the rain) and the covers of parcels most generally burst open or torn.[41]

Parcels did not always arrive when they were needed most and he writes from Co Longford that:

> I was able to bring into Athlone only a few shillings of the five pounds you sent me. Think you then what a debt I shall have contracted before the pay arrives. If we could stop even for one fortnight in any one place we could manage better, but as we are situated at present we are very badly off indeed. We find it impossible to set up in public houses, nor do I see we should.[42]

From Co Roscommon he expresses concern that Thomas O'Connor 'will never have money enough to answer his purposes till the next pay day arrives, I wish he could buy me a port Mantua of good strong leather and bring me all the clothes I left in Dublin'.[43] Many letters show the accountability of the fieldworkers and the strict monitoring of survey resources. In the same entry he notes 'from the quantity of paper I have written on, it will plainly appear that I have been at work'.[44]

Notes from the Field

The letters come from a world on the brink of the photographic explosion precipitated by Louis Daguerre's dioramas of the 1820s.[45] Ireland is an established tourist destination increasingly designated as picturesque or sublime where the fieldworkers often find themselves in the centre of the marketplace having to pay for services like any other visitor. The people were already adept at exploiting the commercial potential of the modern aesthetic imaginative movement that English words and images created. Apart from the human interest of many of the letters the enactment of the theoretical and methodological framework that lay behind the fieldwork is also of interest here. This framework was not specific to either Ireland or Britain and was created from contemporary ideas about statistical, antiquarian and ethnological inquiry in which the British colonial regime was experienced from the impoverished urban areas of Britain to their more distant colonies. An examination of how this was put into practice further contextualizes the question of the knowledge that emanates from the survey and the manner in which it was produced.

Fieldwork is an uncertain process in any era and there is much evidence of this in the correspondence of the survey. Writing from Co Clare O'Donovan recounts some of his personal experiences, 'you talk of Clare are (air) in your last letter, but what do you think of wet turf, sleeping in bogs, damp beds, potatoes like turnips, half-baked bread, adulterated tea, "no meat" broken pains (i.e., of glass) and paying 2/6 per diem for an office to write in?'[46] In rare asides such as this lie some of the scarce portraits of nineteenth-century life. The lack of a suitable place to work in is a recurrent theme and from Co Donegal he observes that 'he who has a fixed office can do this more easily than we, who wander about from place to place like our Scythian ancestors, and who have often to study in a cold room annoyed by drunken Revenue police, and loudly speaking countrymen'.[47]

In contrast to the vast empty expanse and unpopulated picturesque landscape of the maps, the letters reveal a populous land that offered little in the way of peace to the fieldworkers. From the west of Ireland O'Donovan relates, 'I have travelled a good deal of *Conmaicne Mara* and met several features which I shall describe as soon as I can procure a quiet place to study in. Here I can write nothing to my satisfaction in consequence of the constant influx of visitors to Connemara who keep this little hotel constantly filled and noisy.'[48] In Co Meath he mingles with those attending the local fair,

'I set out from Kells for Athboy and upon arriving there I found the inhabitants preparing for a fair, and the Inn-Keeper would not let me have the smallest room in his house for less than five shillings per day!'[49] From Dunglow, Co Donegal he reports that he is 'wearied to death's door with the noise and interruptions of drunken fishermen in the head-inn of Dunglow'.[50] On a similar note O'Connor describes the Inn in Tyrrellspass, Co Westmeath:

> In all the Inns, all of which are called Sun Inns bearing each the Sign of the Sun (I wonder why) on a swinging sign over the door, and hotels I was forced to stop in, a person will be very sure to be roused out of his sleep with the daylight, by the noise of the female proprietors who are regularly out in the yard attached to these houses, cursing employed persons and servants for not doing everything to satisfy their wishes. And the great curse is 'God Blast you, why don't (or did you not) you'.[51]

Far from being the object of inquiry, vernacular culture is often a nuisance. It is nowhere and everywhere at the same time and even threatens to overwhelm the fieldworkers and suppress their purely antiquarian interests. Signing off from Co Fermanagh O'Donovan excuses himself to Myles J. O'Reilly with the cryptic comment, 'excuse hurry and Holly-eve night's disturbance in a wild country village'.[52]

As is to be expected the Irish weather is reported as a constant irritant and much of the fieldwork involved hours of walking, stripping off clothing to cross fields, being blocked by floods, facing snowstorms or bulls and standing for hours talking with local people on the roadside. From Co Louth O'Connor complains:

> In consequence of sudden transitions from heat to cold, getting wet sometimes and sometimes standing when heated for hours on a roadside getting names when we could not take men from their work together with bad accommodation in Clogher and no better in Dunleer, I (T. O'C.) am deplorably bad with cold from which if I do not get somewhat recovered before Monday we propose to go to Ardee rather than to Castlebellingham, as we are sure of meeting with a more comfortable place, which in some measure may relieve my cold.[53]

O'Donovan gives detailed descriptions of the conditions under which the work was carried out and the personal hardship endured:

> I got my head in to the Head Inn where I dried my clothes without taking them off and drank some beer and *aqua mortis*. I slept in a

very damp bed, the sheets of which I took off and slept between the blankets which absorbed all the oxygen or *vis vitae* from my body. I awoke in the morning at 6.30 in a fever; hoarse, tired, sneezing and in very good humour. Got up and took breakfast determined that neither fever, rain nor storm should prevent me from going on with my business and walked through the Parish of *Lis Molaise* imploring St Molaise to stop the rain, but in consequence of the weakness of my faith he suffered a shower to fall before I reached his old Church. However, I was not to be frightened, for though the clouds were murky, lurid and saturated with rain and the roads six inches deep I made my way to Killimor (six miles) and after being puzzled there about the old Church and some obsolete names of townlands I proceeded to Abbey Gormigan; and thence westwards to Loughrea where I arrived wet to the centre of my soul and benumbed with cold after having walked twenty miles through the slobbery roads and flooded bogs of *Síol Anmchadha*. My blood is now saturated with water which makes me believe that this fair world is a hell, and that I would be happier as a wild bushman, or a bear living in a forest, than the sort of rambler which I have latterly become.[54]

In many instances the work goes ahead apace even through illness. O'Donovan emerges as the spokesperson for the others, pleading their case with Larcom at headquarters. He writes on behalf of O'Connor a number of times, 'I tremble for O'Connor's health and request that you will allow him on his return to Dublin to change his hours of attendance, viz., from ten to five instead of from nine to four. A skilful physician should be consulted on the state of his lungs, which seems to me much worse than he himself is willing to acknowledge.'[55] From Co Laois (formerly Queen's County) he makes a similar appeal on behalf of Patrick O'Keefe:

> O'Keefe is getting into a very delicate state of health and will never be able to get on. He is not able to travel through the country without cars and he gets cold on cars. He is now unable to do anything and I think it is unjust to keep him out any longer, whatever will happen to me or Conor. I think it is my duty to let you and his father know the state he is in, as he would not complain himself nor write. He can do more for the Survey in Dublin than here and if he remains here another month, his health will be inevitably sapped and undermined. All now falls with a vengeance on O'Connor and me.[56]

He is continually aware of his own ill health and often dismissive of it, 'my health is beginning once more to fail, and I fear that another winter will kill me – no matter'.[57] In Co Galway he expresses a similar concern:

> I am sorry to say that in consequence of having slept in damp beds in wee country places and of having been constantly wet I have got a return of the old pain in my neck and shoulder, which has broken down my savage energy a good deal. It may continue to annoy me for seven or eight days during which I cannot possibly venture out. I have, however, a good deal to write and I trust that before I have finished writing the pain shall have disappeared; if not, God help me.[58]

From time to time a despondent note is struck in the letters. His comments from Co Meath on the question of employing a local schoolmaster reveal the sense of self sacrifice, 'this Peter would be of great use to us but as he has a large family, I should be very sorry that he would leave his house, school and potato fields for our uncertain speculations'.[59] This is clearly illustrated by his reflections on life and death from Galway where he expresses a sense of despair:

> I am just sick of Churchyards, skulls and mouldering walls of Churches, and the misery of it is that I have not reaped the proper benefit from them as the great philosopher James Harvey has done. I have visited Churches in order to ascertain their age by examining their characteristic features and measured skulls to compare their animal and intellectual bumps, not to draw any moral lessons from them, but for this I will be sorry hereafter, when cold old age comes upon me.[60]

He takes stock in Co Offaly (formerly King's County) in 1838 of his work with the survey and anticipates moving on to Co Mayo:

> Mayo, I suppose, will be next? Awful! My health is very much down, and I make no doubt but another winter's campaign would put an end to me, but I don't wish, as I have gone so far, to be killed, till I shall have examined all the oulde places of Ireland and the stories connected with them. I have now traversed, since the 8th of May last, the whole country extending from Lough O'Gara to Carlow, from Lough Sheelin to the Devil's Bit, which is a vast district, but I have injured my nerve by writing too much and sitting up too late.[61]

He is conscious of the potential usefulness of the work not just for the survey maps but also for those whom he imagines would follow in his footsteps. He returns to this point again and again when the exertions are taking their toll on his health. The potentially overwhelming scale and scope of the work is mentally and physically demanding. From Co Mayo he writes 'I have been writing names of holes and corners since five o'clock, and I can now scarcely hold the pen'.[62] O'Keefe recounts what he calls 'the walking part of the work' from Co Sligo. Here he is improvising and working on his own initiative while he awaits O'Donovan's arrival:

> When I came to Boyle, not knowing how long I should remain in the country, I thought it better to collect as much as I could, everyday expecting Mr O'Donovan. In doing so I did not adopt a course the most comfortable to myself, for I was not able to reach Boyle each evening after my days work, and so was obliged to remain the greater part of my time on the road-side, getting scarcely anything to eat but potatoes. I have now finished the walking part of the work, and will write out my remarks as soon as possible.[63]

O'Donovan gives an incisive account of his own arrival on the Aran Islands guided by three local fishermen. His footsteps on this now ethnographically epic journey were followed a century later by numerous writers, anthropologists and folklorists. The local knowledge acquired has the benefits of gaining access to the inn as well as to information for which he duly pays:

> We climbed up the big stones of which the little quay of Kilronan is built, and finding ourselves on the solid rocks of Aran, we proceeded by the guidance of two Claddagh men and one native Aranite to the Inn of Aranmore in which being chilly and fatigued, we were very anxious to get our heads in, and to lay down our heads to sleep for a few hours. I was astonished as I trod upon the smooth and level limestone pavement of the village and many wild associations of ideas flitted across my brain about creation, earthquakes and other incomprehensibilities, before we reached the Head Inn. On arriving at the house our sailors rapped at the door several times, but no answer was made, which made me believe that they brought us to the wrong house, at which the youngest of them pushed in the door! Immediately after this I observed a glimmer of light and heard a voice inviting us in. The man of the house was drunk, and having been very unruly the evening before in quarrelling with his wife and all that came in his way, he was after getting a beating from the priest, who deemed it his duty to beat him into something like rationality . . . I gave each of the sailors two glasses of little still whiskey and then after paying the Nauclerus 18 shillings, I took one glass of mountain dew punch and went to bed.[64]

In spite of the demands placed upon him in routinely describing antiquities, and the more pressing work of Anglicizing place names, the painstaking detailing of the vagaries, conditions and unpredictability of the fieldwork process is one of the most consistent features of his letters. His determination to reach St Kevin's Shrine at Glendalough, Co Wicklow, highlights his preoccupation with early ecclesiastical sites. Although sinking to his knees in snow he perseveres doggedly:

> One of my shoes gave way and I was afraid that I should be obliged to walk barefooted. We moved on, dipped into the mountain, and when we had travelled about four miles we met a curious old man of the name Tom Byrne who came along with us. We were now within five miles of the Glen but a misty rain, truly annoying, dashed constantly in our faces until we arrived at Saint Kevin's Shrine. Horribly beautiful! And truly romantic, but not sublime! Fortunately for us there is now a good, but most unreasonably expensive kind of hotel in the Glen, and when I entered I procured a pair of woollen stockings and knee breeches, and went at once to look at the Churches, which, gave me a deal of satisfaction. (I looked like a madman!) We got a very bad dinner and went to bed at half past twelve. I could not sleep but thinking of what we had to do and dreading a heavy fall of snow which might detain us in the mountain. At one o' clock a most tremendous hurricane commenced which rocked the house beneath us as if it were a ship! Awfully sublime! But I was much in dread that the roof would be blown off the house.[65]

He continues to describe the storm in detail finally regretting that he had 'not the paper to tell the rest'. The correspondence is more than a pragmatic channel of communication between the fieldworkers and headquarters. It functions also as a process of keeping fieldnotes or a personal diary. It provides an outlet for the expression of personal thoughts, feelings and frustrations.

Writing of the possible employment of the schoolmaster Peter Kenny the role of the fieldworker as a professional stranger is well expressed by O'Donovan, 'I would advise himself to do anything else in the world before he accepts of a situation which will oblige him to stop one night in Boyle, three in Sligo, two in Ballysadare, one in Glengavlen, a perfect stranger everywhere, and every one wishing to run away with his money'.[66] Outlining what he calls a wandering situation, he says that it is suited only to a single man, an old bachelor like Leyden who 'is touched with the madness called *amour patriae*'.[67] Michael Leyden replaced O'Donovan on the staff of the survey for a time. In spite of personal and professional difficulties a methodology was developed to construct the ethnographic occasion and access local information and informants. It was not just the received method of nineteenth-century evolutionary science but one adapted to suit local conditions.

A Little Parliament of Old Men

The fieldworkers took advantage of occasions in the vernacular calendar. O'Donovan says that they did 'more on a Sunday than any other day because the old sages of the Country are always idle on that day, and willing to talk about *Crom Cruach* or any other subject'.[68] Fair days were fruitful, as he records in Monaghan town, 'I have got on famously today, there being a fair in the town and many of the old aborigines having been sent to me to the office'.[69] From Co Donegal he anticipates a local fair saying 'I will have an opportunity of seeing the most longheaded farmers in the country'.[70] Foreshadowing the participant observation of modern anthropologists and ethnologists he attends a pattern in Co Roscommon:

> Yesterday I went to the pattern of Scormore, a place formerly celebrated for pilgrimages, but now for drinking whiskey and fighting ... there is a collection of round stones on a hill not far from the well to the north-east, and in one of these stones is shewn a hollow formed by the knee of St Patrick while he prayed there. I knelt on one of these round stones yesterday while some old Shanachies were pronouncing the names of the townlands in Kilmacumshy for me.[71]

In the style of the classical colonial anthropologist informants were sent to him in the office and it was sometimes the officers of the Royal Engineers who supervised this. The prospect for participants of being conveyed for interview by red-coated soldiers must have been daunting. In Co Fermanagh Major Bloomfield 'sent for the most learned Shanachies in the neighbourhood and told them I was going to make them Barons of Tooráá'.[72] In the same county O'Donovan records that a local man, a 'clever and enlightened Milesian', Con O'Neill 'collects all the seers in the Parish for me'.[73]

It was possibly the same man who sent into the mountain for the following informants: 'I have spent a long time in the County but I have not been a minute idle, day or night, and I have been under considerable expense in entertaining old Shanachies for whom I send to the mountains'.[74] He sometimes organized group interviews. In Co Roscommon he describes playing host to 'a little parliament of old men' whom he assembled 'about the Islands in Lough Ree'.[75]

Expenses are mentioned in accounts of encounters with informants. O'Donovan bought drinks for the fishermen who carried him across to the Aran Islands.[76] In Co Donegal he gave what he calls a glass of the native to his informant, 'for this I generally get to knock talk out of the fishermen'.[77] Boats, cars, guides, servants, boys or gigs are frequently hired. Dr D.H. Kelly sent both a gig and a servant to him while he was in Co Roscommon.[78] In Co Sligo Lieutenant Chaytor sent a labourer to guide him, 'after some time we procured a boat and four very wild men to ferry us across, and indeed it was not without difficulty, for, when they observed that I was determined upon going across, they were determined to take as much money out of me as they could, there being no other boat within three miles of the place'.[79] In Co Mayo he employs a guide to direct him up Nephin mountain:

> I employed a guide who knew the easiest way of ascent, which is on the southwest side, for it is inaccessible on the side looking towards Crossmolina, and for some distance around on every other part till one comes to where I was directed by my guide, who, having no shoes, but wearing a pair of trawheens (*troithíní*) made his way as speedily as a little goat would, up the side of the mountain.[80]

In Co Dublin O'Curry paid ten pence to a boy for collecting names for him for two days. The encounters reported as being most successful are often the result of a careful method that proceeded through letters of introduction and the use of local guides. In Co Derry O'Donovan relates:

> I made old Maelseaghlin McNamee (commonly called the Provost Lord John Eldon) travel with me into the Parish of Lissan to day. In our passage over the mountains we got wet in the clouds, which lay brooding upon them all day. We called upon the most intelligent old natives on the way and obtained a good deal of old traditionary lore, which yet lingers along the sides of Slieve Callann and in the deep and romantic Glens of Brackagh, Corick, Ballybriest, Mobwee and Claggan.[82]

In Co Louth, where O'Keefe and O'Connor worked together, O'Connor expected to do well as he was well acquainted there, as he put it, and knew whom to call upon. When they got there, however, they experienced some difficulty and were directed instead to the local Cess (tax) Collector.[83] This may have been a result of a popular perception of the work of the survey being related to taxation. The fieldworkers remained several days in towns to meet shanachies. O'Donovan stayed on in Maghera while O'Keefe and O'Connor remained for an extended period in Castlebellingham.[84] It must remain conjecture whether or not they were availing a little longer of comfortable lodgings, a luxury they were frequently denied.

In the northern counties the use of letters of introduction that took advantage of local networks was common. In Co Derry O'Donovan writes how he contacted Mr Bords to get a note from him requesting Mr James to give him a letter of introduction to Lord Roden.[85] Dr D.H. Kelly and Rev. Mr Irwin gave the following letters of introduction to him, 'which will open every opportunity of procuring information through the whole country': Rev. William Blundell, Mr Thomas Brooks and Owen Young Esq., of Castlereagh; Thomas Conry, Esq., Strokestown, Rev. Peter Brown, Knockcroghery, the Honble. and Rev. Archdeacon of Ardagh, Ballinasloe, the O'Connor Don, Clonalis, Castlereagh, Edmund Kelly Esq., Ballymurry, Morgan Crofton Esq., Boyle, the Dean of Elphin, Fitzstephen French Esq., Frenchpark, Rev. Joseph Seymour, Ballaghadereen, son of the celebrated Vicar of Kilronan, O.D.I. Grace Esq., Elphin, Rev. George Knox, Dunamon Castle, Doctor Kelly, Ballinasloe, Edward O'Connor, Esq., Bellangare and Jerrard E. Strickland Esq., Loughglynn.[86] The list of clergy, doctors and gentlemen reflects the composition of the elite in nineteenth-century Ireland.

The fieldwork brought O'Donovan through all levels of society and in concluding his letters from Co Derry, still at the start of his fieldwork in 1834, he reflects poetically upon this:

> I shall now bid farewell forever to *Oireacht Uí Chatháin* and *Gleann Choncadhain*, hoping that I have deserved the blessing of Eugenius, Luroch, Cadan, Finlagan and Muireach O'Henry and the goodwill of the living Saints whither Mass-men, Bible-men or Millianarians, as also of the Senchers, Doatards, and Irishians – of the Inn Keepers and shoe-boys – all of whom declared that I am one of the old stock. I have been long enough drinking of the pure waters of St Luroch's Well – dining with Parish Priests, and taking tea with old Mrs. Falls, the descendant of Brian Carrach O'Neill, and talking to her about old times. I shall now proceed to Lough Erne to hear the romantic stories of *Daimhinis and Oilen Badhbha*.[87]

In Co Roscommon he complains, with the usual touch of irony, that he is unable to write much and expresses his wish to leave as soon as he can 'as in the house of an Irish Prince of large fortune, too much time is passed (wasted) at dinners etc., which is so contrary to my habits of working all day and night, that you would soon have to set me down as an idler'.[88] Quiet days were spent, or so he would have it understood, going over the ground with local people. In Co Roscommon he writes, 'the best plan is to walk from one extremity of a parish to the other and inquire as I go along from the longest headed old Milesians or Anglo-Normans who speak the Scotic language'.[89] While in Co Offaly in 1837 he mentions working on Christmas Day with no thought of his own situation, the reason he gives for finally stopping is to allow his informant to enjoy the day.[90]

This is the Man for You

The practice of employing local people was well established. This was made easier by virtue of their nationality along with their use of the Irish language, although from time to time, O'Donovan was mistaken for a tithe proctor or preacher. The developing skill in interviewing informants is clearly illustrated in the following passage written by O'Connor from Co Westmeath; the interviewee is named as Jack Daly:

> On yesterday morning I got a car and brought Jack with me to Castletown and Newtown Parishes, where I picked out the most intelligent men I could get pointed out to me and got their pronunciation of the names, as well as they could give, which was elucidated very much by Jack's native Irish pronunciation. It served on this occasion as purgatorial of that of the unIrish inhabitants and could not be mistaken, in as much as it sent forth a tail of brilliancy as a comet does amongst the naturally dimmer lighted stars. I should not have employed the car, for I prefer walking to places that can be gone to within a moderate portion of time, and I find I can collect more information respecting the things enquired after when I walk among the people and fall in with them in discourses, than if I had fifty conveyances about me. They would not shew much weakness of mind as to tell the same stories to a person stepping off a car, as to one that is footing his way and is apparently not too hot upon his enquiry. If one does not effect an insinuation by apparent indifference, he can never succeed in eliciting the opinions of which the people are convinced, and must get his mind heaped with stories

got up by cunning persons to give present satisfaction or get it reconciled to the greatest absurdities by the ignorance of those who are ready to grant that everything enquired after exists. Knowing all these circumstances, I took a method of correcting Jack, if he were at all inclined to lead me into error, to do which I found afterwards, his very honest and candid mind could not allow him. And he acknowledged his ignorance of everything he did not know.[91]

Here there is evidence of an intuition and aptitude that is based on experience in the field. O'Connor is conscious of the impression travelling in a car could make upon the people and he prefers to establish rapport first by, as he puts it, falling in with people in discourses. He is also conscious of not leading people with questions but holding an apparently casual conversation and avoiding the pitfalls of an overly direct or anxious approach that could result in intentionally misleading answers. O'Keefe espoused a similar approach in Co Kildare, 'I received this information concerning this castle without putting the question to my informant, which species of information is always to be most valued'.[92]

O'Keefe and O'Connor both make insightful comments on the methodology, such as in Co Louth where they acknowledge the dispersed nature of knowledge in communities. Unlike many of the memoir-writers they recognize the internal diversity of the vernacular universe 'that we could collect all at one time is impossible, for we may meet one man, knowing one thing, and another may have a different story according to the attention paid to such matters'.[93] By asking several people questions chosen to eventually throw light on the place name required they achieved a good sample and an unselfconscious pronunciation, 'this we know by having asked several something relating to the Parish and having heard the pronunciation, they not being aware that our attention was directed solely to the accentuation of Bogoidigh. Thus we essay to steal wisely, which is all that a person without invention can do.'[94]

O'Donovan is constantly conscious of the necessity for a method, 'you must put your question in a cautious way or they will say that you are perfectly right'.[95] O'Connor relates from Co Sligo that there are three things necessary for what he calls a serious enquirer, 'I know three things are necessary for a serious enquirer; firstly, a method to elicit some information which can be acquired only by knowing the manners of the people, secondly to happen on persons calculated to give him information, thirdly, time and place i.e. good dry lodging without the noise of drinking people and smoke to digest, compare and write out what he collects'.[96]

While the first goal was to locate informants, the method must be based on a good knowledge of the people and their culture. Field-notes must be digested, compared and written out. This did not always succeed, however, and in Co Offaly O'Connor recalls a bad experience:

> The people there are not willing to communicate any information, from what cause I know not. I was introduced today to one of the oldest men in the Parish of Aghacon. He is a perfect oddity and if he were out in the time of the cynic philosophers, might be sure to graduate in their college with an expression of face indicating duplicity of mind, he made fabricated answers to every question put to him relative to the original names of places in the Parish and neighbourhood. When I told him that what he related was trifling and evidently made up by some persons who were in the habit of framing stories and giving them to swallow to others who had no means of discovering the untruth of them, he was afterwards wholly intent on giving negative answers and giving no information whatever about what was enquired after. Returning him thanks with words, though not at all feeling thankful, I came off. I wished that fate might not furnish me with another such character, and going on I found that the amount of information to be had in that part of the country is but very little.[97]

Here O'Connor tells the informant that what he is saying is nonsense and still expects him to continue after rapport has clearly broken down. Unfortunately the numerous references to the supposedly got up stories that irritated the fieldworkers so much are not elaborated upon. These are probably local or vernacular etymologies for place names, and as these can sometimes be humorous it may have been the case that the fieldworkers felt as though informants were being evasive or facile.

In some instances it appears that the informants are, if not refusing their co-operation, trying to outwit the fieldworkers. From Co Westmeath O'Donovan describes a sense of isolation, 'Mr Geraghty has not answered my letter and I am here a complete stranger; and as I look so like a swaddling preacher, the people, thinking that I want information to turn it hereafter against themselves, are shy in answering my queries'.[98] On this occasion he fails to create the conditions necessary for his work and his status as an outsider or even unwelcome intruder is not overcome.

Writing from Great Charles Street in Dublin to O'Connor who is in the field in Co Sligo, O'Donovan offers a good insight into the types of questions being asked 'what sort of fort is it? . . . Is there a name for the carn? Do you find a townland or tract or district at the fort of Bin Bulbin called *Ceathramhadh na Madadh* or as it may be Connaciter analogically Anglicized Carrow-na-maddoo? What is the local or legendry interpretation of *Bínn Bulbain?*'[99] He goes on to give further questions and continues to advise O'Connor about doing fieldwork in general:

> What do the Connacians understand by the word *scáird, dimín*? *Scárdán* and *Scor* in Skurmore? In ascertaining this you must be careful to distinguish between a local, native meaning, and one derived from printed books and Dictionaries; and you will observe that a country man will often tell you that he does not know the meaning of a word when in reality he uses it in common conversation, it being a fact that he does not know what meaning means! You must therefore, put the question in his own language, that is to say, in his own mode (style) of expressing his ideas.[100]

In this correspondence from 1836 it is clear that he is directing the work from TeePetrie on Great Charles Street. It is the local meaning that he is concerned with and this, he argues convincingly, can only be elicited by knowing the people and their language. Also in Co Sligo O'Connor takes O'Keefe travelling with him for one or two days 'in order that he might form his ear for the reception of the Connaught accentuation of Irish names before he goes to traverse any part of the county by himself. I deemed it advisable to introduce him wherever he could get a taste of the *Machaire Ou blas* which is the purest in the Province.'[101]

The few lengthy descriptions of encounters with informants are particularly insightful and illuminating. O'Donovan transcribed an interview with an informant in Co Derry saying

> on this very wild road I met a Tyrone farmer of the name Ó hEochadha. He saluted me in Irish and the following conversation ensued between us:
> O' H. What brings you to this wild country?
> O' D. The love of wild scenery.
> O' H. I have travelled the greater part of the three kingdoms, and I never met a tract of country so desolately wild or one that presents so black and savage an aspect.

O' D. I suppose there has not been a house here since the creation.
O' H. With the exception of the huts of the herds; that slate house on yon holme is not there many years, as its neatness and newness sufficiently shews; the proprietor seems to be a herd, who has cultivated the holmes of yon mountain stream, and sprung up to be a farmer. Do you know the name of this place?
O' D. Loghermore (Great rushes).
O' H. A very unfit name! If I were naming it I would call it Freaghmore or great heathy place, for you observe that no rushes appear except in the glens and holmes of the stream; however, it might have been anciently more rushy, for you may observe that the inhabitants of the lower country are gradually cutting the turf from the surface. I have observed in different places that when all the turf is removed heath generally springs up, and I am convinced that many of those slopes which now produce heath had originally produced rushes.
O' D. Do you suppose that any of this extensive district was ever cultivated? It is now waste as far as the eye can see in every direction.
O' H. It never was; for in ancient times the people were not numerous, and it was not worth while (*níor bh'fhiú*) to cultivate this apparently barren tract, while they had such land as that around Limavaddy naturally fertile. Yet I am convinced that all these slopes would produce most luxuriant crops of potatoes and oats. If I got a tract of it to cultivate, I would rather face it than go clear the woods of America. If these glens and mountainsides were properly cultivated, this wild tract would form a beautiful country instead of terrifying us as it does at present with its savage aspect. There is one advantage attending it at present; if the few inhabitants have retained any of the old spunk, they have no one to fight with.[102]

In a conversation that appears to be a translation from Irish, with a typical touch of the burlesque, it is O'Donovan who is questioned first. Carefully concealing his actual role and blending into the contemporary social landscape he answers that he is a tourist or traveller. The role reversal continues as the farmer tells O'Donovan about his own travels. O'Donovan's interjections are casual, brief and gently inquisitive allowing the farmer to speak at his leisure about the area in question. It is the farmer who asks O'Donovan if he knows the name of the place, exactly the information that is required, and he continues to give a knowledgeable and detailed description of the area. As the interview progresses the farmer's replies become more open and he alludes to emigration to America and the conditions he believes prevail there. A similar passage occurs in the letters from Co Westmeath where O'Donovan and O'Connor were working together on Carbry's Island, this time, one of only four or five such encounters. Here, the informant is female:

> I asked her how long she was living on the island and she said 40 years. Did you hear any name on the Church? No. Did you ever hear any name on the island but Hare Island? No. Why was it called Hare Island? From the number of hares and rabbits that used to be on it, but there is not one on it now. When did it cease to be a burial place? About 100 years ago, as the old people say. Is there any holy well on the island? There is. And she walked to the place and shewed it. It lies near the shore and is now nearly choked up with briars and rotten branches. Has this well any name? No. How do you know then that it is a holy well? When I came to live here about 40 years ago, I saw rags tied on the bushes which then grew over it, and the old woman who had care of the island before me told me not to use the water of it for washing or boiling potatoes, that it was a blessed well and that it might not be proper to use it. Is that old woman still living? She is. She lives in Cuarsan just opposite my finger on the other side of the water. Her name is Rose Killen. Do you think she knows the old names of the island, the Church and the well? It is very likely she does, because she, and I believe, her father before her, was born on the island, and she knows Irish, which I do not. Did you ever hear that there was any old stone with old letters and (or) crosses on it? Indeed then there was and I saw many gentlemen striving to read it, but no one ever did or could.[103]

O'Donovan's interrogative style is apparent with the woman, Mrs Duffy, giving short answers. The attention that the old stone on the island attracted, along with countless others mentioned in the letters, speaks of the heavy traffic of antiquaries in nineteenth-century Ireland. In Co Mayo in 1838, now after many years of experience in the field, he shows that he is a master of role-play:

> I am tired of travelling all day through *Partrigia de lacu*, which is full of ruins of Anglo Norman buildings. I visited the Abbey of Ballintober, which is a beautiful specimen of early Gothic architecture, and of the civilization of Charles the Redhanded O'Connor. I met within it an old man of the name Hennelly who keeps a little school in the *Sanctum Sanctorum* of the friars, and who knows a great deal of the traditions of the place. I addressed the old man with all the enthusiasm of a knight errant, and it is not improbable that I impressed him with the idea that I was after affecting my escape out of one of the Lunatic Ass-isle-ums of the province. 'Mr. Hennelly' sez

I, 'you are a most distinguished professor of literature holding your college in the *Sanctum Sanctorum* of Ballintober. Sir! Trinity College is but a pigsty in style or sublimity to your college. The sublimity of the place and the associations connected with it are calculated to raise the mind to Heaven, and to cause that divine intoxication of the soul for which the poets of all ages have taken of the waters of the pure-air-ian spring. Sir, the echo responds to your words and seems delighted with the symphony of your sentences.' 'You compliment me, Sir', says he, "the building is no doubt a fine one, but I am but a humble poor man, and have no claim to the praise which you have been kind enough to lavish upon me.' 'Sir', says I, 'did you ever hear any account of the founder of this abbey?' 'Yes, Sir, I did. It was founded by Charles the Redhanded or Cathal Croibhdhearg O'Connor.' OSL, Mayo, Vol.2., p.21

Such playfulness or mimicry is a hallmark of O'Donovan's letters. In the above passage this is emphasized by the fact that he was himself no stranger to the circumstances of his interviewee. The addressee could be an imaginary third party, the survey's superintendent or indeed future readers of the letters. The playful treatment of many aspects of the correspondence was frowned upon by the survey and reflects duplicity in his role. The irreverence, irony and humour contrast with the officious, serious and formal nature of the survey. The result is sometimes an ambivalent open-ended style. Writing in Co Donegal he makes a characteristic and pointed pun, '*turas* in Irish is the same as Tour (ass!) in English'.[104]

The phonetic transcription or intextualization of the voices or dialect of informants often presents them as caricatures akin to the stage Irishman. His intentions in doing this can only be speculated at but he was critical of the increasingly popular writings of authors like William Carleton and Thomas Crofton Croker. The cryptic and cunning nature of his writing, illustrated by frequent puns and shifting styles, could also reflect an attempt to interrupt the authority and fluency of the survey's discourse. In treating what he calls a laughable legend told by a simple people concerning crosses in the Church of Castlekeeran in Co Meath, which are thrown into the Blackwater by St Kieran, he gives the narrator's conclusion, 'and with this "my dear" he flings it into the Blackwater, where it is yet visible. Aye in tráath!' In Co Offaly an informant tells him that an island in Co Laois was called the Living Island, 'here was made formerly a pilgrimage, considered (they say) the greatest pilgrimage in Ireland, aye, in the world, Sir!'

Are These the Firbolgs of Baghna?

The final and most telling example comes from the Parish of Kilglass in Co Roscommon. O'Donovan is in a jaunting car, called a tax car, with Mr Kelly of Strokestown. Interestingly they take a particular route to evade two local families named Hanly and Lee. He records that he actually feared for his safety:

> It was the market day of Stroketown and I was struck with the prevalence of black hair in the district, 'hair black as the raven's win' . . . are these the Firbolgs of Baghna who have assumed Gadelian names. It is very probable they are; for a similar process is now going on with regard to the names of the Gaels being assimilated to those of their conquerors; as in Slieve Baan, Quilly to Cox! MacUiseoige to Lark! MacShane to Johnson, Breen to Brune! O'Braochain to Brougham! So that in seven hundred years hence, if the world should exist so long, an inquirer into the different races of mankind, observing the difference of physiognomy between the Johnsons of Ireland and those of England, will ask in Ireland as I do now about the *bolg-thuath Baghna*, are these Irish families who assimilated their names to those of their conquerors.
>
> We passed on in a N.E. direction thro' the Parish and when we had arrived at the foot of the hill of Mullaghmacormick, we met on the road a respectable looking old man of the name MacShane or Johnson, whom as soon as Kelly knew (kened) at a distance, he said to me, 'this is the man for you! He knows more about Kilglass than any man now living'. He told Kelly not to take the horse any further, that the road was too steep and rough, but to put him into his own stable and that he would shew me the Parish from the summit of Mullaghmacormick. This was done, and old MacShane or Mr. Johnson shewed me the townlands and pronounced the names of all the features in north Kinel-Dofa very satisfactorily (he knows every single foot of the Parish) . . .
>
> MacShane then walked with us to *Cara na dTuath* Bridge . . . to shew me the islands in Lough Boderg. All this he did cheerfully and most willingly, but then he turned to me saying: 'I would not go with you one foot or tell you one word, were it not for Mr. Kelly's sake'. He shewed me the house of O'Hanly of Lavagh now in a most wretched condition, tho' the O'Hanlys should be very well off, having three hundred and sixty acres of excellent land free of rent; but the curse of St Patrick is still over them and it is said that they obtained this little property by murdering a man of the name Igoe or MacIgoe. Tradition in Kilglass observes that notwithstanding the sanctity of St Barry, O'Hanly, his brother, was cruel and barbarous, and made very little scruple of hanging a man for nothing, and there

is an island in Lough Lagan called *Príosún an Dubhaltaigh*, where Dwalto O'Hanly used to confine people till they were starved to death . . . they are great rakes and sublime rascals! There is not an assize (assise) or court held in Roscommon or Strokestown or Carrick-on-Shannon, to which they are not summoned for debt or some piece of misconduct. They are now members of the aristocratic religion of the State, 'because' as MacShane remarks, 'they were too wicked to belong to the true Church'. But he told me, that if he thought I had any notion of writing down one word of what he told me, 'one word he would not tell me'. The O'Hanlys brought out the tongs and the pot-hangers one day to murder MacShane because he called a friend of theirs a turncoat. One of them (Malby) was drinking whiskey in a Sheebeen house on the side of the road when we were passing by, and when he saw Mr. Kelly, he hid, because he was not sufficiently well dressed to show himself. Having got so much information from MacShane, I offered to give him a glass of grog (a word which I never wrote before) in the Sheebeen house, but he said that he could not drink any whiskey since he was drunk one night when he broke a man's collar bone with a loaded whip (for which he has still to stand his trial). 'But I can take some at home', says he, 'and we'll mix it with milk'. And so he did and had some bottles of his own, of which he drank very freely, and swore by St Barry's Crozier and by other far more sacred things that we too should drink, and he shewed us a long pitchfork, a blunderbuss and two bulldogs to convince us that it was dangerous to say a word against him. I would have given three shillings at the time to get out of his clutches, but finding that I was in the 'County of Kilglass' I thought it better to let him get on his own way. 'Only for it is Friday,' says he, 'I'd kill some of these geese for you (I have thirty-five of them) and I'd shew you that I could drink whiskey for twenty four hours without getting drunk or sick'. I made every effort to keep the conversation upon St Barry and his Crozier, and MacShane told me the following story about St Barry . . . 'Jenkins advised people to drink when they are thirsty, but the way it is with me I generally drink when I am not thirsty at all, to make myself thirsty, and still I'll live till my wife will be tired of me'. 'But Sir,' says I, 'we'll leave the drinking alone now and talk about St Barry'. 'Well Sir, whatever you like' . . . MacShane gave us a good dinner, and came with us as far as Kilglass Town, the capitol of the County of that name, where he should get more drink. Here he got beer which was as sour as cider and I resumed the conversation about Barry and the families of Slieve Baan . . . we got rid of him at last, and Kelly has not stopped damning him ever since. On our return we met the Firbolgs returning home from the market, staggering; men and women drunk!'[107]

O'Donovan sets out on a market day accompanied by a local man. This gives him the advantage of knowing not only where and who to go to but also where and who to avoid. With this guidance he goes further into the vernacular milieu of nineteenth-century Ireland than he did on any other occasion that he describes. It is an area with a vibrant local identity described as 'the County of Kilglass'. It is a market day and many people are socializing. There is a sense that the encounter might be too close for comfort.

At the outset O'Donovan's approach is like that of a physical anthropologist observing a primordial culture using ideas about physiognomy. This culture is lurking behind the superficial surface of an Anglicized nomenclature; they are the mythological *Fir Bolg* with English names. His fears for his safety and his antiquarian mindset highlight two elements that could potentially marginalize him and deny him access to the contemporary world. MacShane, an Irish speaker, was identified by Kelly as suitable for O'Donovan's purposes. MacShane shows him the area and pronounces the place names for him. Suddenly he makes it clear to O'Donovan that he would not co-operate in the slightest way if it were not for Mr Kelly.

Later, after speaking at length about a feared local family, MacShane lets O'Donovan know again that he would not tell him anything if he was going to write it down. This family had threatened MacShane and he obviously considered either the information or the encounter sensitive. This does not dissuade O'Donovan and he continues to make a detailed record of it although it is not relevant to the place name research or antiquities. In their travels they encounter one member of the family drinking in a roadside inn or sheebeen but he conceals himself from them for some reason. O'Donovan speculates, probably incorrectly, that it was due to his poor clothing. It is likely that MacShane refused his offer of a drink in the sheebeen because of this man's presence.

The tensioned background and the nervous nature of the contact with this informant, set against a backdrop of a busy market day, left O'Donovan's position precarious and uncertain. Very much at the mercy of the informant, O'Donovan records that he would have given three shillings to get out of the situation. This is unusual in so far as MacShane clearly had an encyclopaedic knowledge and information was flowing from him. O'Donovan also felt intimidated by his familiar carefree manner. MacShane was soon to be tried with breaking another man's collar bone with a loaded whip. O'Donovan finds it difficult to keep control of the conversation because

MacShane wants to talk about other things besides antiquities and saints and does not fit the profile of the credulous aborigine or old Milesian too readily. Having spent the day in his company O'Donovan says 'we got rid of him at last' and finally comments on the drunken Firbolgs staggering home from the market.

It is precisely the distance maintained by the colonialist trope of the superstitious, guileless peasants and an implicitly-believed-in folklore that is subverted in this encounter. The empowered position of the surveyor is undermined, if not threatened. The survey's somewhat superficial repertoire of barely remembered or forgotten antiquities recedes into the background and the present overshadows the romantic view of picturesque landscape. Here there is also an added element of participant observation that reveals the unpredictability of categories in action, of actual interaction and participation in the life of the informant and, for O'Donovan, the all too real social and cultural context of vernacular discourse. McShane does not conform to the image of the passive addressee and disrupts the fluency of the fieldwork process. Passivity and gullibility are replaced by agency, knowledge and action.

The Mimic Man

The term aborigine may have referred to a tribal name from pre-Roman times but is more usually derived from the Latin *ab origine* or original inhabitant. It is not in itself a derogatory term but it could be argued that it became one. It was used in English to refer to heterogeneous indigenous peoples encountered in the colonies.[108] In colonialist discourse its use suggested a primordial state of civilization, a pre-modern or pre-colonial one close to the original state of the human species. It has been used variously since the late six-teenth century to describe the indigenous inhabitants of places encountered by European colonizers, explorers, adventurers or seamen. As a totalizing or racist signifier it often designated the indigenous peoples of colonized territories throughout the British Empire. Like the word native it was employed in nineteenth-century British accounts in particular as a quasi-racist expression of inferiority.[109]

Lieutenant G.A. Bennett's comment in Co Down about 'the more aboriginal and less civilized' inhabitants underlines this usage.[110] O'Donovan uses it frequently in his work on the memoirs and letters in reference to Irish speaking people or the Irish in general. He also uses terms like bushman, kaffir and Hottentot. His colleagues, O'Keefe, O'Curry and O'Connor, appear not to have used it at all. Ó Duilearga notes what he calls the contempt of O'Donovan and O'Connor 'for the rich tradition then so readily available to them'.[111] On the other hand he notes O'Curry's inclusion in his letters of 'a number of folk-tales, such as the *Glas Ghaibhleann* story, a version of the Swan Maiden or Melusine type, notes on the sunken land of *Cill Stuifín* off Mothar, of which Golden Domes appear once in seven years, and in his valuable notes to Petrie's *Ancient Music of Ireland* (1855) he includes much of interest to the folklorist'. Ó Duilearga's own work in Clare a century after the survey belies the imperial image of a country virtually bereft of tradition or culture.[113]

O'Donovan's options were anything but limited given that he was more than familiar with the diverse range of prior vernacular names and denominations for the local inhabitants of the various localities where he worked. If he was not directed to use it then he may simply have borrowed or adopted it from the lexicon of the survey in which it was standard. Other technical or quasitechnical terms like unexplored regions, settling, laying down place names, proceeding to the ground or traversing the country were borrowed from survey jargon. This is one of the factors in his correspondence that suggests a degree of complexity or ambivalence in his role within the mission of the survey.

In spite of Ó Duilearga's reservations there is a sense in which O'Donovan personifies Bhabha's mimic man. The mimic man is the effect of a flawed mimesis 'in which to be Anglicized is *emphatically not* to be English'.[114] For Bhabha this is not just an aping of the colonizer but a reflexive imitation or an ironized definition itself that refuses to be defined. O'Donovan's anomalous position marginalizes 'the monumentality of history' in mocking its power to be a model. The letters reveal what could be called an eccentricity in the context of the survey that imitates colonialist discourse one minute and mimics it the next. O'Donovan, whom Larcom said dressed like a peasant and whom the people often took to be a preacher, acts the aborigine, is the aborigine and is not the aborigine all at the same time.

Bhabha calls this the partial gaze 'where the observer becomes the observed' and the gaze of the observer is refracted, disturbed or distorted, 'it is a form of colonialist discourse that is uttered inter dicta: a discourse at the crossroads of what is known and permissible and that which though known must be kept concealed; a discourse uttered between the lines and as such both against the rules and within them'.[115] His comments about bogs from Co Donegal are typical, 'nature never intended them for tillage, no more than she has Lapland for oranges or the skull of the Caffarian for philosophy, or the Irish skull for steadiness (oh! impiety). I may be wrong, for time changes all things – men into dust, caterpillars to gaudy butterflies, and Barbarian Britons to Bacons (Beacons) and why not Negroes to philosophers and Irishmen to Scotsmen?'[116]

This mimicry de-authorizes the normative knowledge of the colonizer, it disturbs the fluency of representation, jars and breaks what Raheja calls 'the illusion of consent'.[117] Elsewhere Bhabha calls this 'the articulation of the ambivalent space where colonialism creates hybridity rather than repression and total cultural closure, the effects of colonial disavowal are reversed and that other 'denied' knowledges enter upon the dominant discourse and estrange the basis of its authority – its rules of recognition'.[118] It becomes a counter-authority. It escapes the demand that is central to colonial power 'that *the space it occupies be unbounded*, its reality *coincident* with the emergence of an imperialist narrative and history, its discourse *nondialogic*, its *enunciation unitary*, unmarked by the trace of difference'.[119] O'Donovan's Irishness shatters the survey's racialist identikit, it is liquid, dialogic and multiform leaving traces of itself everywhere. Charged by Larcom with effacing official documents O'Donovan was facing unofficial ones.

In Search of Aborigines

The progress of the fieldwork was determined by the success or failure to locate informants. O'Donovan writes that, 'the length of my stay in any town, depends upon the success I meet with in finding out old sages'.[119] With the exception of four or five, the informants were overwhelmingly old males, many of them octogenarians and most of them so called Irish-speaking aborigines. O'Donovan generally translated their names into the English language when he recorded them. O'Curry, O'Keefe and O'Connor differ somewhat from him in their correspondence. O'Curry openly sympathized with aspects of vernacular culture and criticized the British Empire on one occasion. He generally gave his informant's names in the Irish language and did not use the term aborigine. Most of them, however, depersonalized informants on occasions and distanced themselves somewhat by using less obviously racialist terms like peasant or peasantry.

The work is increasingly conceptualized in ethnographic terms as it develops wider cultural interests. Much time was spent searching for aborigines, as O'Donovan puts it in Co Derry, 'I have travelled through the Parishes of Clondermot, Faughanvale, Aghanloo and Finlagan in search of "Aborigines" but found none except one in Clondermot. The inhabitants of that level and rich district extending from Derry to Limavaddy and for miles around the latter town have not the slightest idea of the significations of the names of places.'[121] He makes numerous references to conversing 'with some of the most intelligent aborigines' who are then unexpectedly dismissed as 'knowing nothing'.[122] In Co Monaghan in 1835 he calls 'upon several of the aborigines, but found very little ancient traditions among them:'

> The most intelligent old fellow I came across is named MacKahy. He was formerly a Cess Collector but is now living in the most lamentable state of misery. He is as naked as Adam was before he clothed himself with the leaves of the fig tree! When I entered his cabin he had not a single article of dress upon him – nothing but an old blanket thrown across his shoulders. He sat upon a little stool and stooped with his face, hands and knees over a small fire which habitually roasted him.[123]

In Co Donegal he explains that he cannot finally send off the name books for Conwall and Donaghmore until he has consulted 'the aborigines at the head of Glensoolie for the names of Conwall'.[124] In what he calls the unexplored region of Co Fermanagh, near Lough Melvin, he writes in 1834, 'I got the Irish names of the islands and promontories in the lake from several of the old aborigines I met on the road'.[125] In Co Derry it appears that he is referring to Irish speakers when he uses the term aborigine, 'I intend to travel through Magilligan. Few of the "aborigines" or Irish speaking people are to be found in it; but I am told that a few have retained the language in consequence of their communication with the opposite side – Inishowen'.[126]

A further distinction is made between those who could speak Irish very well but did not understand it.[127] In a style reminiscent of the memoirs he gives an account of the principal aboriginal families in Lurg, Co Fermanagh, amongst them the Muldoons, 'no longer chiefs, nor higher than the rank of farmers, but they are said to be very decent respectable men, fond of justice and able to fight'.[128] Many informants are archaized as 'descendants of the old chief' who were 'the most numerous of the aborigines' of every district.[129] The informant is often styled the oldest and longest headed man who can speak the old language. J. Stokes' replies to the queries of G. Downes in Co Derry in 1835 quote a local informant or shanachie as saying 'I sat where the hearth fire of my fathers had burnt for six generations before me!'[130]

The frequent references to the Irish language as old or ancient aestheticize and esotericize what was, and still is, a contemporary language of everyday communication. The following is a typical description of the nomadism and methods of the fieldworkers:

> Last Friday I travelled from Mohill to Drumsna, from thence to Leitrim, and from Leitrim northwards to Drumshanbo, where I slept on Friday night and met the Priest and also the oldest and by far the most learned and intelligent of the natives, Fergal Moran, who stopped with me the greater part of the following day. On yesterday I travelled southwards through the barony of Leitrim and called upon several longheaded natives (who are nearly all called Moran) and arrived in Mohill about an hour after sunset.[131]

Lieutenant W.E. Delves Broughton describes two of his own informants, Cauhall Sharkey and Daniel Gallagher in the Parish of Tullaghobegley, Co Donegal, in the memoir for that county:

> Said Sharkey is a man of sincerity and truth, of morality, piety and Catholicity. He is also a man of undaunted courage and mostly sees those supernatural beings in the daytime. The last interview he had with them was in or about 16 months ago. They appear to him in the human and ordinary size. He tells that he knows or is acquainted with several of them, i.e. such of them as he knew in this life or before they died, although they appear very pale and discoloured and in mutinizing and fighting appearance. He tells that they have their regular chieftains and officers, and sometimes the Connaught part offers violence and abuse to him but always relieved by the Ulster division. Nor would he always escape their violence with safety, were it not for the interference of a young man by the name O'Neil from Connaught, who was killed in a battle fought in the parish of Conwall near Letterkenny, in a place called in Irish Brissue-na-scaribheshoilse ... I have to observe that there is another man, a resident of the town or village of Thoar, by the name of Daniel Gallagher, whom I have heard telling that he was coming home after night with 2 cwt of meal

in a sack on his back, he met with a crowd of the above species. They played the violin very melodiously and travelled slowly along with him. They all appeared to be in men's dress with the exception of 1 woman. They asked him to dance to their music. He consented, as he dare not refuse, and danced with the woman. They lifted the sack on him again, the woman always desiring him not to fear. They played their music and left him safe at his own house, and went off without any harm.[132]

This portrait of informants recounting personal experiences of interaction with the ancestral supernatural world frames them as hopelessly a-historic bearers of pseudo-knowledge. The traces of vernacular knowledge and the cognitive map onto which it is intimately grafted is burlesqued. What is styled as the informants' interviews with the fairies mirrors the British officer's social survey interview of the informant himself. Inverted and skewed in the lens of the observer Sharkey is ironically depicted as a surveyor of the supernatural, the illogical consequence of the lack of true science. The tangible imprint of pre-colonial culture is folklorized in a new evolutionary rhetoric of reason. It is relegated to the realms of the unreal, the removed and the insubstantial.

O'Donovan, on the other hand, describes old Merly from Glen Fin in the same county sympathetically, 'I marked as by far the most intelligent and skilled in ancient lore, an old man of the name Merly (Ó Mearlaoich) whose forehead and features spoke good health, good nature and intelligence. He is intimately acquainted with the situation and meanings of the names of the Townlands, and the repertory of the legends, stories and prophecies of Glen Fin'.[133] Merly also gave an account of the battle of Scarve Sollus mentioned by Sharkey above. In Co Derry he describes a mason called Francis O'Brollaghan giving his full pedigree and title in vernacular terms:

I have met in the Townland of Labby an old man of the gifted tribe of the O'Brollaghans who knows more about the traditions of Derry than all the rest put together. He is a mason by trade and has travelled much thro' the district; he has such taste for these things that he comes in to me whenever he can command an hour. His name is Francis, mac Francis, mhic Shemus Oge, mhic Shemus More, mhic Teige, mhic Rory, and he says he will never write his name Bradley again, but will preserve the spelling used by the Braher Bane and all his illustrious ancestors[134]

Fishermen provided O'Donovan with a plethora of coastal place names that were not engraved on the maps or recorded in the name books. Place names like *Poll Cam, Leac na Sagart, Tobar na Sagart, Aill an Chaor, Colbha na Téide, Ceann Toll and Poll Tír Aodh* in the Bay of Liscannor were surplus to requirements.[135] In many coastal areas he sends for some fishermen, as he puts it, and plies them with drink to knock talk out of them. In Ross Goill, Co Donegal, the fishermen offered their help and their services to him:

> On arriving we enquired for the most intelligent old fishermen who were acquainted with the bays and points on the coast of Ross Goill, and as we soon discovered one, who was the more anxious to give us every information, as he had in the first instance learned from us that we might perhaps employ him to convey us across the wide mouth of Sheep Haven to Horn Head. His name was Donagh O' h-Oireachtaigh or O'Herraghty; the colour of his face indicated his having spent much of his time on the sea, while his attention to us spoke his natural civility, and his art to get at earning a few pence.[136]

In the same county he becomes infatuated with thoughts of progress and evolution. His mind is too preoccupied evaluating the progress of man, understood in evolutionary ethnographic terms, to hear the fisherman, whom he casts in a primitive world far behind his own, although he was standing beside him. So near and yet so far.

> I stood for some minutes on the wall of the Cashel to enjoy the bold scenery around me, and to ponder on the mutability of human affairs; for strange thoughts crowded into my excited imagination, as I considered the progress of man from the period that he built the rude habitation of stone without cement, constructed the slide car to be drawn by the tails of horses, and formed the slender currach of twigs and the hides of animals, until he formed the alabaster Palace, with windows of stained glass, and curtains of embroidered silk, drove the rapid car on the land, and ploughed the ocean by the unconquered arm of steel . . . these thoughts took such a forcible possession of my mind that I could not well hear what the fisherman was telling about the names of the rocks.[137]

Terence or Torlogh Carran, Toirr McCarthainn in Irish, was noted as a 'very intelligent old man . . . deeply versed in traditionary and legendary lore'. Although he gives him many place names in Co Fermanagh O'Donovan does not want to be distracted by recording the discourse surrounding the names.[138] Patrick Maguire of

Letterbaily in the same county was the most intelligent he met.[139] He describes 74-year-old Harry McGuiggan in a complementary way as an Irish scholar who told him a vast deal of wild stories.[140] In Co Meath two farm labourers, George Hand or Seoirse Mac a'Láimhe and Thomas McCoy or Tomás a'Céide, 'to whom Hand referred me as to an oracle' provided the Irish pronunciations of place names.[141]

In the townland of Moneyneeny in Co Derry an 88-year-old man called Murray, explaining the townland name as the Hill of Wonders began, 'to tell a string of stories about the wonders that are traditionally handed down as having happened in it such as two hands seen without a body hewing down a tree with an axe; heads of pigs seen feeding without bodies etc. etc'.[142] Maurice Fox of Curristown, Co Meath was recommended to him because he was in the habit of reading old Irish Manuscripts.[143] O'Curry records the following in 1839 from a female informant regarding a stone near Tullow in Co Carlow called *Cloch an Phoill*:

> It was the practice with the peasantry to pass ill-thriven infants through the aperture in order to improve their constitution. Great numbers formerly indulged in this superstitious folly . . . my informant, on this occasion, was a woman who had herself passed one of her infants through the aperture of this singular stone. She informed me that some of the country people talked of having it cut up for gateposts, but a superstitious feeling prevented them.[144]

In many areas O'Donovan is refused information or meets with reluctance or suspicion. He explains this as resulting from the shyness and fear of those who did not understand. In Co Derry for example, 'the people, I find, are very "shy" in giving their names because they are afraid that they might be wanted for the 'service of war' or some other plan of the Government. The more intelligent however fear nothing because they know that no radical change will take place in their condition without the consent of the Parliament.'[145]

Racial tensions sometimes determine the limits of the work but it is also the case that such distinctions were not clear-cut. He is reluctant to approach Scottish settlers or to stay too long in Coleraine.[146] In Co Donegal, however, a wealthy Protestant farmer called Graham who spoke Irish well gave him the pronunciations of place names.[147] In Macosquin and Aghadowney in neighbouring Co Derry fieldwork proved unfruitful due to the caution of what he calls black Presbyterians. Interestingly he goes into the mountains near Garvagh instead and discovers the 'lowly residence of the oldest branch of the celebrated sept of the O'Mullans'.[148] In Co Louth people refuse information and O'Donovan tries to persuade others that his work is not related to Tithe affairs.[149]

One informant named Sarsfield told him and his colleagues 'that if we were persons who were in any one way connected with Tithe affairs, he himself would not tell us even one half a word concerning anything which would be asked of him by us, but on my explaining to him what information we wanted and for what purpose, he then enquired our names and said my name was a good one, but he said O'Keefe was a foreigner; he consented to answer our questions with pleasure'.[150] The last comment is repeated in Co Mayo in a way that suggests O'Donovan may have used some explanation of his own pedigree to persuade informants of his allegiances. Talking to Edmund O'Flynn he notes 'as soon as O'Flynn had learned that I was one of the oulde stock he commenced to give me a most curious account of his own family and of himself'.[151] This man was hiding from the Sheriff who was to arrest him for a debt. The parish of Dromin in Co Louth held out the prospect of a regular welting for anyone suspected of being a tithe proctor.[152] O'Connor's justified expectation of getting on well in a county where he was well known was shattered by suspicion and the blunt refusal of cooperation.[153]

The theme of decay arises again and again. In Co Derry, 'the old traditions are nearly effaced from their memories. Any recollections that remain are so very faint that they are almost useless. We travelled up the face of Slieve Gallion where the people are more Irish than in any other part, but met no sound Shanachy. Their general saying is that all the old men who knew those things are dead.'[154] What exactly they are referring to in saying this is not made explicit but it may well be the derivations of ancient names of places or information relating to them. This was a topic that, apart from the necessity to investigate place names, preoccupied O'Donovan at all times. When his inquiries do not produce results the people are dismissed. In Monaster Evin in Co Kildare they were, 'entirely Anglicized and have lost all their ancient traditions' as well as being 'boorish and unobliging' or simply 'completely ignorant'.[155] Here the 'natives have no traditions among them which would throw any light upon ancient localities'.[156] Here the Irish language, like ancient tradition, is said to be 'ready to be interred in oblivion with the present generation'.[157]

O'Donovan's ambivalence vis-à-vis his role within the survey, or indeed his representation of his informants, could be added to the incoherencies mentioned earlier. His apparent willingness to borrow the totalizing terms of colonialist discourse, the alternate acceptance and rejection of the authority of the survey, the bending and testing of rules, may be explained by his anomalous position both within the survey and within the wider context of Irish British relations. Such conditions may have led to this discourse-at-the-crossroads that evades the definition of the colonizer and the native alike while also working within them. The letters reflect a combination of O'Donovan's personal complexity and the survey's official complexity.

The fieldwork illustrates the magnitude of the survey as an undertaking and the voluminous paperwork that resulted from it. The constant preoccupation with paper, quills, pencils, sealing wax, extracts, indexes, maps, books and vouchers highlights the role of the bureaucratic British Empire as the first information society and its careful administering and monitoring of this process. It offers insights into the routines of hiring gigs, cars, guides and boatmen or buying drinks for informants. Otherwise this presentation of the fieldwork underlines the contribution of the antiquarian researches to ethnography and its early relationship to nineteenth-century evolutionary knowledge producing techniques like cartogrtaphy. In Ireland this relationship was mediated, by necessity to some extent but also by the overriding imperial imperative, by translation.

This contribution can be understood on a number of levels. The most important must be the way in which the field was read and the resultant methodology that emerged from that reading. Set against the ever-present background of vernacular culture and the Irish language the idea of the local became central to the practice. Local events, local guides, local dialects, insider information or local meaning often determined the success or failure of the work. The Irish fieldworkers critically appraised previous research and scholarship and recast the question of Irish antiquarian research in a new and more objective idiom that included tradition, topography, etymology, ethnology and history.

The deliberations of the fieldworkers introduce ideas of onsite observation and increased objectivity that had emerged in natural history. Hostetler says that 'ethnography distinguishes itself from other, earlier descriptions of foreign peoples through its reliance on direct observation, rejection of non-verifiable information, concern with method, and the urge it reflects both to collect and systematically categorize information'.[158] O'Donovan and his colleagues began to imagine serious inquiry for its own sake and rejected the speculative dabbling of English antiquaries in fairyology or hagiology.

This innovation combined the latest technological and theoretical advances that the survey instituted with the thought of nineteenth-century natural scientists, moral philosophers, utilitarians or ethnographers like Dugald Stewart, Thomas Malthus and James Cowley Prichard. Although framed as before the survey, in archaic society, or after the survey, in the future, it differed from them in attempting to recombine such knowledge in order to advance Irish antiquarian or ethnographic research.

The correspondence provides evidence of the ethnographic inskilment of the fieldworkers themselves. Much of the deliberation and speculation in it attests to the development of an ethnographic methodology that included interviews, the practice of going over the ground and becoming familiar with local dialects. The writing of the fieldnotes themselves and the manner in which they were used as both personal and professional diaries to be digested, compared and written out, mirrors to a great extent the practice of the next generation of professional Irish folklorists and ethnologists.

The fieldwork was not a seamless process of folklore collection but a contested and negotiated one that contended with the contemporary vernacular milieu in order to look behind it to a more distant archaic society. The records of the experience of locating informants, and the sometimes ambivalent attitude revealed towards them, points to the complicated nature of the fieldworkers' roles as insiders or outsiders by turn both within the survey and in the field. The result was an alternate acceptance or rejection, complicity or contestation with the imperatives of the imperial archive and the conventions of colonialist discourse. The contradictory themes of discovery, rediscovery or recovery of native culture in the English language are further coloured by their later retranslation in neo-traditional terms within an emergent Irish academy. In terms of post-coloniality the survival of colonialist discourse continues to have significant if unexplored consequences.

The fact that the category of Irishness is configured within colonialist discourse is often overlooked. This oversight is unnoticed at times due to the strong cocktail of colonial ethnographic discourse allied with the invisible hand of translation that looks through, and not at, a pre-colonial culture rendered transparent that is 'somehow outside of time and history'.[159] The fieldworkers, like Cohn's anthropologist, followed in the wake of the impact caused by the agents of change, and then tried to recover what might have been, 'the anthropologist searches for the elders with the richest memories of days gone by, assiduously records their ethnographic texts, and then puts together . . . a picture of the lives of the natives . . . the peoples . . . like all God's children got shoes, got structure'.[160] That the structure was imagined, transposed or translated is emphasized by the pre-eminence of the English language in the Anglization of Irish place names.

NOTES

1. Robert Welch, *The Oxford Companion to Irish Literature* (Oxford, 1996), p.424.
2. Séamus Ó Duilearga, 'Notes on the Oral Tradition of Thomond', *The Journal of the Royal Society of Antiquarians*, 95 (1965), p.136.
3. Patricia Boyne, John O'Donovan (1806–1861): *A Biography* (Kilkenny, 1987), pp.xiv–xv.
4. David Spurr, *The Rhetoric of Empire: Colonial Discourse in Journalism, Travel Writing and Imperial Administration* (London, 1993), p.163.
5. Diarmuid Ó Giolláin, *Locating Irish Folklore: Tradition, Modernity, Identity* (Cork, 2000), p.43.
6. Caoimhín Ó Danachaír, 'The Progress of Irish Ethnology 1783–1982', *Ulster Folklife*, 29 (1983), pp.3–17.
7. Bill Ashcroft et al. (eds), *Key Concepts in Post-colonial Studies* (London, 1998), pp.32–3.
8. Rev. Michael O'Flanagan (ed.) *Ordnance Survey Letters 1834–1841*, 43 volumes typeset and bound in Boole Library, University College Cork, (Bray, 1927–35), Kilkenny, Vol.1, p.54. These were written in the course of their fieldwork for the survey by John O'Donovan, Eugene O'Curry, George Petrie, Patrick O'Keefe and Thomas O'Connor. See Archival Sources for further information. Hereafter abbreviated OSL.
9. OSL, Carlow, p.32.
10. OSL, Roscommon, Vol.2, p.96.
11. OSL, Mayo, Vol.2, p.87.
12. OSL, Carlow, p.32.
13. Clare O'Halloran, *Golden Ages and Barbarous Nations: Antiquarian Debate and Cultural Politics in Ireland c. 1750–1800* (Cork, 2004), p.64.
14. OSL, Carlow, p.64.
15. OSL, Galway, Vol.1, p.111.
16. OSL, Galway, Vol.1, p.113.
17. OSL, Queen's County (Laois), Vol.1, p.16.
18. OSL, Meath, pp.81–2.
19. OSL, Meath, p.76.
20. OSL, Mayo, Vol.2, p.70.
21. George W. Stocking Jr., *Victorian Anthropology* (New York, 1987), p.31.
22. Ibid., p.9.
23. Gillian M. Doherty, *The Irish Ordnance Survey: History, Culture and Memory* (Dublin, 2004), pp.25, 110.
24. Stocking, *Victorian Anthropology*, p.23.
25. OSL, Roscommon, Vol.1, p.114.

26. OSL, Mayo, Vol.2, p.96.
27. OSL, Roscommon, Vol.1, p.120.
28. OSL, Roscommon, Vol.2, p.6.
29. W.F.H. Nicolaisen, 'The Past as Place: Name, Stories, and the Remembered Self', *Folklore*, 102, i (1991), p.5.
30. Seán Ó Coileáin, 'Place and Placename in Fianaigheacht', *Studia Hibernica*, 27 (1993), p.56.
31. OSL, Roscommon, Vol.2, p.44.
32. OSL, Cavan/Leitrim, p.50.
33. OSL, Kildare, Vol.2, p.59.
34. OSL, Roscommon, Vol.2, p.95.
35. OSL, Galway, Vol.1, p.181.
36. Angélique Day and Patrick McWilliams (eds), *Ordnance Survey Memoirs of Ireland (Belfast, 1990–98)*, 40 volumes, Vol.36, p.117. Hereafter abbreviated OSM.
37. OSL, Galway, Vol.3, p.204.
38. OSL, Kilkenny, Vol.1, p.57.
39. Máire de Paor, 'Irish Antiquarian Artists', in Adele M. Dalsimer (ed.), *Visualizing Ireland* (London, 1993), p.129.
40. OSL, Sligo, p.92.
41. OSL, Westmeath, Vol.1, p.49.
42. OSL, Longford, p.62.
43. OSL, Roscommon, Vol.2, p.73.
44. OSL, Roscommon, Vol.2, p.73.
45. David Brett, *The Construction of Heritage* (Cork, 1996), p.79.
46. OSL, Clare, Vol.2, p.82.
47. OSL, Donegal, p.89.
48. OSL, Galway, Vol.3, p.5.
49. OSL, Meath, pp.66–7.
50. OSL, Donegal, p.107.
51. OSL, Westmeath, Vol.1, p.115.
52. OSL, Fermanagh, p.46.
53. OSL, Louth, p.75.
54. OSL, Galway, Vol.3, pp.7–8.
55. OSL, Louth, p.165.
56. OSL, Queen's County (Laois), Vol.1, p.34.
57. OSL, Roscommon, Vol.2, p.126.
58. OSL, Galway, Vol.1, p.181.
59. OSL, Meath, p.66.
60. OSL, Galway, Vol.1, p.220.
61. OSL, King's County (Offaly), Vol.2, p.79.
62. OSL, Mayo, Vol.1, p.234.
63. OSL, Sligo, p.65.
64. OSL, Galway, Vol.3, p.208.
65. OSL, Wicklow, p.71.
66. OSL, Meath, p.66.
67. OSL, Sligo, p.5.
68. OSL, Cavan/Leitrim, p.69.
69. OSL, Armagh/Monaghan, p.21.
70. OSL, Donegal, p.150.
71. OSL, Roscommon, Vol.1, p.129.
72. OSL, Fermanagh, p.50.
73. OSL, Fermanagh, p.72.

74. OSL, Fermanagh, p.77.
75. OSL, Roscommon, Vol.1, p.36.
76. OSL, Galway, Vol.3, p.208.
77. OSL, Donegal, p.30.
78. OSL, Roscommon, Vol.1, p.19.
79. OSL, Sligo, p.9.
80. OSL, Mayo, Vol.1, p.208.
81. OSL, Dublin, 14.
82. OSL, Londonderry, p.120.
83. OSL, Louth, p.114.
84. OSL, Londonderry, p.35; OSL, Louth, p.93.
85. OSL, Londonderry, p.1.
86. OSL, Roscommon, Vol.1, p.19.
87. OSL, Londonderry, p.149.
88. OSL, Roscommon, Vol.1, p.11.
89. OSL, Roscommon, Vol.1, p.23.
90. OSL, King's County (Offaly), Vol.1, p.21.
91. OSL, Westmeath, Vol.1, pp.114–15.
92. OSL, Kildare, Vol.1, p.65.
93. OSL, Louth, p.25.
94. OSL, Louth, p.57
95. OSL, Sligo, p.18.
96. OSL, Sligo, p.140.
97. OSL, King's County (Offaly), Vol.2, p.44.
98. OSL, Westmeath, Vol.1, p.87.
99. OSL, Sligo, p.28.
100. OSL, Sligo, p.29.
101. OSL, Sligo, p.24
102. OSL, Londonderry, pp.28–9.
103. OSL, Westmeath, Vol.1, pp.5–6.
104. OSL, Donegal, Vol.3, p.126.
105. OSL, Meath, p.8.
106. OSL, King's County (Offaly), Vol.1, p.59.
107. OSL, Roscommon, Vol.2, pp.57–60.
108. D.W.A. Baker, *The Civilised Surveyor: Thomas Mitchell and the Australian Aborigines* (Melbourne, 1997), p.xiii.
109. Bill Ashcroft *et al.* (eds), *Key Concepts in Post-colonial Studies* (London, 1998), p.4; K. Schaffer, 'Colonizing Gender in Colonial Australia: The Eliza Fraser Story', in Gregory Castle (ed.), *Postcolonialist Discourses: An Anthology* (Massachusetts, 2001), p.370.
110. OSM, 12, p.5.
111. Ó Duilearga, 'Notes on the Oral Tradition of Thomond', p.136.
112. Ibid..
113. He writes that he recorded nearly 500 tales and a large body of seanchas or encyclopaedic vernacular knowledge and discourse, between the years 1929 and 1935. See Ó Duilearga, 'Notes on the Oral Tradition of Thomond', p.136.
114. Homi K. Bhabha, *The Location of Culture* (London, 1994), p.87.
115. Ibid., p.89.
116. OSL, Donegal, p.105.
117. Gloria G. Raheja, 'Caste, Colonialism, and the Speech of the Colonized: Entextualization and Disciplinary Control in India', *American Ethnologist*, 23, 3 (1996), p.495.

118. Homi K. Bhabha, 'Signs Taken for Wonders: Questions of Ambivalence and Authority under a Tree outside Delhi, May 1817', in Henry Louis Gates Jr. (ed.), *'Race', Writing, and Difference* (Chicago, 1985), p.175.
119. Bhabha, 'Signs Taken for Wonders', p.176.
120. OSL, Armagh/Monaghan, p.38.
121. OSL, Londonderry, p.27.
122. OSL, Armagh/Monaghan, p.33.
123. OSL, Armagh/Monaghan, p.37.
124. OSL, Donegal, p.99.
125. OSL, Fermanagh, p.37.
126. OSL, Londonderry, p.32.
127. OSL, Cavan/Leitrim, p.9.
128. OSL, Fermanagh, p.58.
129. OSL, Londonderry, p.129.
130. OSM, 15, p.107.
131. OSL, Cavan/Leitrim, p.103.
132. OSM, 39, p.173.
133. OSL, Donegal, p.95.
134. OSL, Londonderry, p.125.
135. OSL, Clare, Vol.1, p.108.
136. OSL, Donegal, p.28.
137. OSL, Donegal, p.29.
138. OSL, Fermanagh, p.64.
139. OSL, Fermanagh, p.67.
140. OSL, Londonderry, p.69.
141. OSL, Meath, pp.103–4.
142. OSL, Londonderry, p.107.
143. OSL, Westmeath, Vol.1, p.81.
144. OSL, Carlow, p.123.
145. OSL, Londonderry, p.21.
146. OSL, Londonderry, p.37.
147. OSL, Donegal, p.28.
148. OSL, Londonderry, pp.35–6.
149. OSL, Louth, p.10.
150. OSL, Louth, p.2.
151. OSL, Roscommon, Vol.1, p.68.
152. OSL, Louth, p.65.
153. OSL, Louth, p.110.
154. OSL, Londonderry, pp.115–16.
155. OSL, Kildare, Vol.2, p.39; Vol.1, p.65.
156. OSL, Kildare, Vol.1, p.76.
157. OSL, Londonderry, p.129.
158. Laura Hostetler, *Qing Colonial Enterprise: Ethnography and Cartography in Early Modern China* (Chicago, 2001), p.82.
159. Bernard S. Cohn, *An Anthropologist Among the Historians* (New York, 1990), p.19.
160. Ibid., p.20.

CHAPTER SEVEN

Derry with Derrida: Translation, Anglicization and Culture

'My job is to translate the quaint, archaic tongue you people persist in speaking into the King's good English.'[1] This is how Owen Hugh O'Donnell, cast as John O'Donovan, explains his role to his own community in Friel's *Translations*. The relevance of translation to the survey has been contested since the play, one of the first created for the Field Day Theatre Company of Derry, was premiered at the Guild Hall in Derry in 1980. Place names posed problems for mapmakers because they were the only culturally specific aspect of the increasingly proficient and universal practice of nineteenth-century cartography. One of the essential qualities of the modern map is that it can be interpreted by anyone trained in reading the genre without recourse to the local language.[2] The place names were cultural fingerprints on the otherwise pristine paper landscape of science. More signs than signatures, they signalled previous use, authorship and authority.

Colby's 1825 *Instructions for the Interior Survey of Ireland* stipulated that, 'persons employed on the Survey are to endeavour to obtain the correct orthography of the names of places diligently consulting the best authorities within their reach'. The procedure to be followed was clear. The common spelling was given one column, the variants that appeared in written documents a second and the authority for the name a third.[3] A note on the situation and a brief accompanying description was also appended. The place names called for attention as soon as the survey commenced in 1825. To avoid confusion or legal ambiguity it was thought necessary to Anglicize the names of the townlands and later to standardize the spellings of their different elements. The survey itself was familiarly referred to at times as the townland survey.

In 1829 Captain Waters requested clarification from headquarters 'as to whether the original and descriptive Irish names or the ones generally received and spelt should be put on the plans'. The names generally received and spelt were the Anglicized forms that were adopted for the maps, some of which were inherited from previous colonialist mapmaking ventures like the Civil and Down surveys. Colby's reply was that 'printed records are generally to be preferred as authorities, but that every authority should be consulted and entered in the Name Books'.[4] Ó Maolfabhail points to this as the founding moment of toponymy, the study of place names, in Ireland. The authorities entered were more often than not the names of the elite, ministers, doctors, respectable farmers, gentry, magistrates, officers of the British army, teachers and heads of councils.

Amongst the numerous records quarried by O'Donovan in his researches were the *Annals of the Four Masters*, inquisitions, Grand Jury maps, as well as those of Norden, Beaufort and Mercator, Johnson's dictionary, Giraldus Cambrensis, statistical accounts, surveys, Calendars, *The Book of Lecan, The Book of Lismore, The Tripartite Life of St Patrick, The Book of Fenagh, The Book of Clann Firbis, The Annals of Kilronan, The Book of Kilkenny*, Pope Nicholas's Taxation, Gough's *Camden*, the 1641 Journal of Rebellion, Usher's *Primordia*, Campion's History of Ireland, English Law Deeds and *Leabhar na gCeart* (the book of rights). Writing from Monaghan in 1835 he questioned whether he should restore ancient names or use contemporary ones, 'you will find that I have given the correct name in every instance, but the question is, is it proper to restore the correct ancient name in defiance to established custom?'[5] Writing from Ballina in Co Mayo in 1838, O'Connor says that he was unaware of the system for Anglicizing the names:

> As the system adopted for anglicizing the names has never been made known to me, which was not necessary as Mr O'Donovan was always at hand to perform this part, if I have in the names anglicized by me deviated from the letters adopted, for instance in putting i for e as Bin for Ben, or vice versa, or o for u as poll for pull, or vice versa, or in a similar manner with respect to other letters, the indexes must correct them.[6]

The deference to the Irish language led to the identification of some place names that were transfixed in the colonial paper trail of Anglicization already. This Anglicization was a bequest of imperial bureaucracy itself as most names were commonplace in the Irish language. Writing on Telltown, Co Meath, O'Donovan outlines such Anglicizations that many have wrongly attributed to the survey:

> The names of Meath present a very strange aspect to the Irish Etymologist; they have almost all assumed an English appearance, and the most ridiculous transpositions have been made to anglicize them; town has been stuck as a tail to the greater part of them and the word *rath*, instead of being placed first as is the Irish custom, has been placed last, which frequently gives a name a very exotic look. Ex. gr. Maperath for *Rath a'Mhabaidh*, Calliaghstown for Ballynagalliagh. To comply with the general custom of sticking town as a tail to as many of those names as possible, the ancient name of Queen Taillteann, the daughter of Mamore, was changed to Telltown, as if it were to make it impossible to tell what town it anciently was the name of! Fortunately however, it happens that the Irish language is still spoken in the neighbourhood, which enables me to put it in record that the place which the English speaking people call Telltown is invariably called Tailteann by the Irish, which, joined with the traditions connected with the Rath and with its description by Colgan as near Donaghpatrick, perfectly identifies it with the Olympic Games of Looee.[7]

O'Donovan's method can be appreciated from the last comment. It adds historical support to oral testimony as if to copper fasten the discovery, although the name was, of course, familiar to local people. The documentary bias in the methodology is revealed in another comment from Co Meath, 'I hope the historical extracts for Meath will soon arrive for I am wearied with expectation, as I am passing over very curious places without knowing what to enquire about'.[8] It is ironic that the topographer, employed to enhance the survey's knowledge of place and the production of maps, is lost in a landscape that cannot be read without the direction of a document.

There were disagreements between the Antiquarian or Topographical Department and the British officers working for the survey. Captain Waters expressed their frustrations saying that the officers 'are not unfortunately Irish scholars' and that his own knowledge of the Irish language was inadequate.[9] O'Reilly received a letter in March 1830 stating that 'the directors of the Survey of Ireland, now making by the Board of Ordnance, are desirous to have the Irish names of places mentioned in the Survey put into proper orthography, and to have translations of those names into English with derivations etc'.[10] Proper here means standardized English orthography and the translations were added in the name books.

Although in favour of extending the memoir project to the whole of Ireland, Thomas Davis criticized the survey's approach to place names saying that 'whenever these maps are re-engraved, the Irish words will, we trust, be spelled in an Irish orthography, and not barbarously, as at present'.[11] As Ó Maolfabhail writes 'what Davis is saying, of course, is that the spelling is "barbarous" in the Greek sense (cognately *balbh* in Irish meaning "mute"), not conveying the meaning of the names, and rendering them what is technically termed *opaque*, while the correct Irish-language forms of the names would, ideally, give the name their meaning and render them *transparent*'.[12] The tension lies between competing conceptions of English correctness and Irish barbarity on the one hand, or Irish correctness and English barbarity on the other.

O'Reilly's work as orthographer was very brief. He began in April 1830 and died in August of the same year when he was 65 years old. Two months later O'Donovan took his place and expressed his role on a few occasions as that of a translator. Although he addressed questions to the boundary surveyors, his powers were limited. Significantly he did not have the authority to decide which spelling would be engraved on the maps. Ó Maolfabhail argues that both O'Reilly and O'Donovan were subject to the authority of the military staff of the survey. Larcom had sight of all questions and queries before signing them and forwarding them to the officer working in the field.[13] Between 1830 and 1832 Richard Griffith himself and Henry Buck, who was the District Boundary Surveyor in Co Antrim, decided the spellings. Following a vexed reply from Lieutenant Robe of the Royal Artillery to some of his queries, O'Donovan resigned in January 1833. Michael Leyden and Patrick O'Keefe took his place briefly but he returned following the intercession of Dr George Petrie. Larcom, possibly as the result of an argument, expressed reluctance to employ him again.

Derry with Derrida

The debate sparked off by Friel's dramatization of the survey is focused upon the place names. Andrews says that translation 'has recently become a metaphor symbolizing all the cultural mischief done by Englishmen in Ireland, though, oddly enough, outside the domain of this metaphor translation is everywhere accepted as a form of recognition and acknowledgement: we show our respect for a writer by translating him'.[14] He points out that 90 per cent of Ireland's place names are of Celtic origin, or in other words, are in the Irish language. Describing the process as one of dictation rather than translation, he says that many names rendered in English orthography like Money for *Muine* are mnemonic devices that standardize recurrent elements and that these dictations 'like translations . . . are gestures of respect. The translator respects the sense, the recipient of dictation respects the sound.'[15] There is some evidence that this was the case, O'Donovan writes in Co Derry:

> The pronunciation of the names of townlands has been taken down from the viva voce of the most intelligent of the Irish natives in every parish, by one intimately acquainted with the general and local pronunciation of the language, and with the orthography of the ancient and modern language of Ireland. This will reduce the orthography of names of Irish places to a standard which must finally place the authority of this Memoir in such matters beyond dispute. [Crossed out: in spite of the devil.][16]

He emphasized the importance of this approach on many occasions, 'I insist upon it that the interpreters of Irish names will do a vast deal of mischief unless they go over the lands and hear the viva voce pronunciation of the natives'.[17] It is not just the method employed in acquiring the names and their pronunciations that is relevant here, however, but rather the overall context in which the Anglicizations were decided and the strategies that were evident in deciding them. In addition the place names are only one aspect of a far more comprehensive ethnographic project.

Even if the status of *Translations* as a national classic did derive from some misguided manipulation of the root metaphor of translation, this neither explains nor negates why or how it rings true for an intuitive consensus regarding cultural process in nineteenth-century Ireland. It reflected a cultural interpolation and may have spoken as much to the present as it did to the past. The Irish language was commonplace in the nineteenth century and had not yet undergone the aesthetic revaluation of an Anglicized official nationalism. Furthermore the retranslation of John O'Donovan into Seán Ó Donnabháin, the fifth master of Irish language revivalism, was equally ideological, although no less legitimate for that.

This is expressed by Ó Laoide's 1905 *Post-Sheanchas* that gave their Irish language names to the post-offices. The same year Dublin Corporation erected bilingual street nameplates. Local authorities opted to use Irish language names such as *Dún Laoghaire* or *Port Laoise*, but were not given official recognition until the independent state established the Placenames Commission in 1946.[18] The issues of legitimacy, power and representation, precisely what the play is perceived to highlight, reflected growing theoretical concerns. Friel's achievement has been summarized as the application of theory to Irish culture, the mixture, in Longley's phrase, of Derry with Derrida.[19]

The Anglicization of the place names is a significant aspect in its own right but it can be viewed as a physical trace of a wider translative energy that fuelled the survey. Kiberd says, 'the attempt to write all the new names into a book represents the colonizer's benign assumption that to name a thing is to assert one's power over it and that the written tradition of the occupier will henceforth enjoy primacy over the oral memory of the natives. A map, in short, will have much the same relation to a landscape as the written word has to speech. Each is a form of translation.'[20] The place names were only one aspect of a more searching project that encompassed more than the maps themselves. It also produced a corpus of ethnographic and statistical prose, specifically the memoirs and the letters, that also relied upon translation.

The survey may be a window on the nineteenth century but the gaze is carefully controlled and supervised. Most of the information is contained by an evolutionary or colonialist philosophy that subscribed to a view of the providential order of humanity. In this order a supposedly superior civilization may justifiably subsume a supposedly inferior one.[21] Many of the recent theoretical concerns that have arisen within the triangle of knowledge-fields known as cartography, ethnography and translation problematize the procedures and conditions under which knowledge is created. The multidisciplinary tentacles of the survey touched upon topography, toponymy, physiognomy, ethnography, geology, antiquities, zoology and botany. It is worth reflecting finally on some of the more direct implications of translation since the survey could not have proceeded without it

The Dominion of Transparency

Translation Studies broadly defines translation as the rewriting of an original text in a different language; this can be creative, destructive, innovative or repressive.[22] It is creative or destructive depending upon the strategy of the translator or the purpose of the translation. The translator's role as an interpreter is central: 'translation is a process by which a chain of signifiers that constitutes the source-language text is replaced by a chain of signifiers in the target language which the translator provides on the strength of an interpretation'.[23] Al-Shabab's definition also makes interpretation central, 'translation is the interpretation of linguistic/verbal text in a language different from its own'.[24]

Many questions have been asked of translation. Why it is necessary? Is the original inadequate? Who translates what and for what reason? How is the text to be translated selected? What are the criteria for selecting it? What is the context? How do those of 'the receptor culture' know if the text is well represented?[25] Or, as Manus asks Owen in Friel's play, 'what's incorrect about the place names we have here?'[26] The means do not always justify the end; it is not just a matter of a correct or incorrect translation.

Translation has to do with authority and legitimacy and, ultimately, with power, which is precisely why it has been, and continues to be, the subject of so many acrimonious debates. Translation is not just a 'window opened on another world', or some such platitude. Rather, translation is a channel opened, often not without a certain reluctance, through which foreign influences can penetrate the native culture, challenge it, and even contribute to subverting it.[27]

Translation in the survey is more than simply a movement of linguistic signs across languages; it is the result of an extended ethnographic entanglement representative of, 'the entire system by which one culture comes to interpret, to represent, and finally to dominate another. It includes, in other words, the discourse of colonialism as produced in such forms as imaginative literature, journalism, travel writing, ethnographic description, historiography, political speeches, administrative documents, and statutes of law.'[28] It is not restricted to linguistic transfer alone but is a vehicle through which 'cultures (are made to) travel – transported or "borne across" to and recuperated by audiences' elsewhere.[29]

Translators have traditionally remained invisible behind fluent translations that seem not to be translations at all. Many may well ask what all the fuss is about, as it can almost appear as if nothing has happened. It can become a technique of exposition and cultural sanitization in the hands of imperialist administrations with a strategic interest in other cultures. As such it serves to smooth the rough and resistant texture of otherness. Fluent translations provide the reader 'with the narcissistic experience of recognizing his or her own culture in a cultural other, enacting an imperialism that extends the dominion of transparency with other ideological discourses over a different culture'.[30] The survey extended a dominion of transparency over Ireland in the triad of map, prose and place name.

It is generally conceded that 'the study and practice of translation is inevitably an exploration of power relationships . . . that reflect power structures within the wider cultural context'.[31] Following the trail of translation involves examining the production of knowledge and its transmission, relocation and reinterpretation in another culture. The approach to a culture, even one's own, unavoidably involves translation of one kind or another. It has been used as a metaphor for ethnography itself that casts the ethnographer as a translator of the symbolic nuances of alterity. In the broad theories of post-structuralism and post-modernism the idea of the irreproachable, objective, disinterested production of knowledge collapses into an array of subjective and interested individuals who conjure it out of personal, professional or political bias.

Knowledge of history, language, customs and manners, ostensibly disinterested and detached, can be abstracted into strategic processes of estrangement, familiarization, exoticization and naturalization for any of a number of reasons of a personal, political or cultural nature.[32] The hegemony of institutional scientific or colonialist discourse over the indigenous creates and maintains a dominant mode of representation.[33] The creation of a dominative exegetical normality encroaches upon the symbolic resources or repertoire of the subject culture often rendering it a colourful but ineffectual part of itself and regulating all representation of it. In a manner reminiscent of nineteenth-century evolutionary discourse the other's knowledge is marginalized and detached from the reigning objective reality in a realm of psycho-cultural suspension. It is conceived of as inarticulate, sub-objective, unreal or even unhealthy. It is conflated artificially with its own progenitor. It is articulated as a degenerate version of a singular and universal form of knowledge. The colonizer as translator is poised like a ventriloquist or puppeteer over the native whose natural disposition is to collapse and become mute and inanimate in his absence. It is these extra linguistic factors that have made translation an increasingly interdisciplinary concern. It has long been realised that to translate necessitates an ethnographical knowledge of people.[34]

The use of translation as a root metaphor positions the ethnographer as a mediator whose role is 'to go behind the baffling chaos of cultural artefacts, to discover order in the foreign, and to transfer implicit meanings from one discourse to another'.[35] More recent translation theory argues that the mediator is traceable through patterns of rhetoric, the figurative props of objectivity and the perceptual structures of authority. Each representation can, and should, be located within a particular discourse and historical context.

One of the contexts highlighted has been the overarching power structures that determine translations. Irish was not a minority or lesser-spoken language in the nineteenth century but a majority language of low status, power and prestige. On the other hand English was the language of power, domination and colonialist identity. The paradox of much translation activity in the colonial context is that the scholars and translators who were most to the fore in defending the intrinsic value of native Irish language and culture made a significant contribution to the strengthening of the English language in Ireland and to the marginalization of Irish in the public life of the country.[36] In pre-academic pre-independence Ireland, however, circumstances dictated that the role of the Irish scholar would be that of subservient translator.

As well as being an element of it, translation is an appropriate metaphor for nineteenth-century cultural process in general. That it also speaks of hybridity in vernacular cultural process is undeniable but it is important to situate it in this instance within an overarching state epistemology of governance. It is also a discursive tug of war implicated in the powerful politics of identity.[37] Translation from Irish to English was a strategic intercession that followed wider patterns of domination and mediation of cultural flow. The translated text 'indicates the relative submissiveness or superiority of the translator and the authority of the receptor culture vis-à-vis the source'.[38] In a context even approximating that of colonialism translation is the direct result of asymmetric power relations. It is an exercise of power in which 'the colonial Other is translated into terms of the imperial Self, with the net result of alienation for the colonized and a fiction of understanding for the colonizer'.[39]

In the survey, to Anglicize often meant to archaize as well. This issued in a form of imaginary cultural closure where the translation becomes the epitaph for the translated. The source language is conjured as a curious vestige, a heritage of esoteric hieroglyphics. The living language becomes the preserve of the philologist or archaeologist who deciphers its mysteries for a bemused reader. The reader may even sympathize with the apparently mesmerizing expertise of the translator in salvaging meaning from an apparently chaotic oral archive of pre-history. This contrasts greatly with the vernacular perception of the modernizer as in fact lagging behind. Many people are amused by O'Donovan's obsession with ancient inscriptions for example.

The elation at unearthing the names of saints or places is balanced by the fact that they were well known all the time. It was perhaps the people themselves who were modern while the survey bewitched their world with the antiquarian's wand and the surveyor's compass. The archaeological metaphor in nineteenth-century translation practice when 'the translator like the archaeologist rescues records from oblivion. The "site" of translation is the patient unearthing of the language and literature of ancient civilizations.'[40] In the survey the process was often an anachronistic archaeology of a living language and culture to be displayed in the museum of the map. The map takes on the character of an antique for the genteel Anglophile collector.

Translation can be a beginning as much as it can be an end. It facilitates the further appropriation of the language and culture in what could be called its afterlife. The dominant language realizes and objectifies itself while it constructs the other language as source within an alienating discourse of heritage, national culture and patrimony that is crucially consigned to the past. Colonialist discourse often outlives its authors and it has, as Cronin says, 'powerfully affected discourse on the Irish language to this very day. By positing translation as an act of retrieval, the implication is that the other language and culture is lost to the reader in its original form. They are lost both in the sense of coming from a very remote time and being condemned to the oblivion of obsolescence.'[41]

Loss, unavoidable and inevitable, and recovery, timely and heroic, are the stable defining characteristics of the survey in general and its later integration into national life. The tropes of this discourse are isomorphic and proliferate in the artistic, literary, historical, linguistic, ethnographic, archaeological, journalistic, psychological, social and political discourse of the present day.

Colonialist discourse is an umbrella term for an aggregate of patterned conceptualizations that are imagined to exert a coercive, strategic or dominative influence over indigenous knowledge and information. Translation, one of the most exercised limbs of the octopus of nineteenth-century evolutionary science in Ireland, is invariably into dominant languages. It constitutes a way of speaking or writing that reflects actual power transactions. Irish is translated into English, whether it is by surveyors, writers or antiquarians, because the Empire, whether British or Anglo-American, is leading the way. This may not occur through personal or officious agency alone, but through the unspoken and unperceived assumptions that make up the reigning ideas and exegetical rules that guide all interpretation. Colonialist discourse includes 'ways of talking, writing, painting, and communicating that permit ideas to pass from one discourse . . . to another in order to make possible the ends of colonial control'.[42]

In speaking of the violence of translation that resides in its very purpose and activity Venuti goes further than this. It wrenches the values, beliefs and expressions of the source language and replaces them with the values, beliefs and expressions of the target language, 'translation is the forcible replacement of the linguistic and cultural difference of the foreign text with a text that will be intelligible to the target-language reader . . . whatever difference the translation conveys is now imprinted by the target-language culture, assimilated to its positions of intelligibility, its canons and taboos, its codes and ideologies'. In self-consciously civilizing, assimilationist projects 'translation serves an imperialist appropriation of foreign cultures for domestic agendas, cultural, economic, political'.[43]

It is pressed into the construction of national identity for exogenous elites and plays an important role in geopolitical confrontations. Translation, in the service of a hegemonic language and culture allied to colonialist discourse, is highly ambivalent. It creates as it destroys, erases as it writes, praises as it condemns and ignores as it notices. It territorializes and domesticates the foreign space by imagining and objectifying a putative universalist superstructure in which all knowledge is organized. It also:

> Effaces its own mark of appropriation by transforming it into the response to a putative appeal on the part of the colonized land and people. This appeal may take the form of chaos that calls for restoration of order, of absence that calls for affirming presence, of natural abundance that awaits the creative hand of technology.

Colonialist discourse . . . appropriates territory, while it also appropriates the means by which such acts of appropriation are to be understood.[44]

It exerts a multidirectional influence linguistically, culturally, ideologically and politically. Any reflection on the ethnographic or cartographic impulse of the survey must therefore also account for the transcendent role of translation in it.

An Anthropologist of Providence

The triad of ethnography, translation and cartography illuminates the recesses where the work of delineating, translating and describing was done. In re-animating the processes that constructed the conjectural knowledge of the archive they complement each other. This knowledge appears stable, fixed and self-evident, it claims universality when it is particular and part illusory. It is a rite of rhetoric or a ritual of arraignment. The wonderful knowledge of the survey, a self-fulfilling prophesy, is a projection of the desire to know. It was not always fulfilled: as Richards says 'an Empire is partly a fiction. No nation can close its hand around the world; the reach of any nation's Empire always exceeds its final grasp.'[45] Although the survey was not ostensibly a charter for dispossession or resettlement, it formalized and legitimized the precedence of the English language and institutionalized the juridical and political prerogatives of Empire.

Griffith's Boundary Survey reorganized many territorial divisions. The translation of the townland names was done to accommodate the valuation of property for the purpose of taxation. It also served to familiarize the otherness of the landscape in a more fluent and streamlined English administration. This in itself was a form of colonization, displacing or offsetting the originary names in a symbolic half-world akin to what Nicolaisen calls a 'non-etymological landscape of symbols'.[46] This risks 'collapsing the cultures and histories of . . . peoples into the English histories' and silencing the absent voices of history while claiming to represent them.[47] Translation 'means precisely not to understand others who are (inhabitants) or to understand those others all to easily – as if there were no question of translation – solely in terms of one's own language, where those others become a usable fiction: the fiction of the Other'.[48] The construction of a chaotic or anarchic culture and language crying out for order and rescue occludes a complex and internally diverse universe of discourse.

In authorizing the names suggested by the manorial class the survey was adhering to the Old English meaning of townland as tún-land or land forming a tún or manor. In England it was used to refer to local districts of a large parish or to a particular manor. Its use in Ireland to designate territorial divisions as units of estate management or government carries with it the connotations of a township. Petty used the term in 1658 and it is estimated that there are up to seventy thousand townlands in Ireland. It was applied universally by the survey and took precedence over a plethora of prior vernacular terms that designated divisions of various sizes and caused much confusion for the surveyors.[49] Given the cultural and linguistic diversity that lies behind this pre-eminent English term it is useful to look over some of the vernacular appellations.

Its use is thought to have been based upon one of the main terms in the Irish language thought to be its equivalent. *Baile biataigh*, thought to mean the land of the victualler, was alternatively spelled ballybetagh in English orthography. This term comprises of two Irish words familiar to every Irish person, *baile* meaning home and *bia* meaning food. In any language in the world they suggest two fundamentals of human life. Five thousand of these names have the word *baile* in them and this homeliness touches upon its significance in daily life and social relations that has carried on into our own time.[50] The *baile* was a piece of land or place belonging to one family, group or individual and reflects social organization and settlement patterns in early Ireland. The eminent Irish language novelist Ó Cadhain emphasized the centrality of the *baile fearainn*, an alternate of ballybetagh, as 'the linchpin of patrimony, community, reciprocity . . . of life and death'.[51]

Vernacular denominations like *seisreach*, a ploughland or team of six plough horses, *baile* or *leathbhaile* (half of a *baile*), *triocha céad* (cantred or barony) or *leathtriocha* (half of a cantred), *cos* (cosh or foot), *gníomh* (gneeve), *fearann* (faran or land), *ceathrú* (carrow or quarter), *cartúr* (cartron), *baile bó* (pasture) and *cnagaire* (croggery) were superseded by the English 'townland' on the survey's maps. A *gníomh* was one twelfth of a *seisreach* and was itself made up of *ceithre cosa* while twelve *gníomh* made a *fearann*. Two ballybetagh was one thirtieth of a barony, a barony or cantred was thirty hundreds of land or thirty ballybetagh or three thousand, six hundred quarters or one thirtieth of a *cúige*. Alternatively twelve *seisreach* amounted to one ballybetagh. O'Donovan writes of these denominations in 1844 that:

> There are few, if any, townlands now so extensive as the ancient Irish ballybetaghs, thirty of which made a *triocha céad*, or 120 quarters, and that the denominations of land in modern times called townlands are generally quarters of the ancient Irish ballybetaghs. In many instances the ancient names of the ballybetaghs are lost, and the names of their subdivisions only are retained as townland names; but in some cases the name of the ballybetagh remains, although it is not applied to as large a tract of land as it was originally.[52]

Keogh notes that a *ceathrú* or quarter was the fourth part of a townland and a *gníomh* or gneeve was the sixth part of a quarter just as a *cartúr* or cartron was a fourth part of a quarter.[53] Robinson says of the four townlands of the Aran Islands for example that 'each are divided into carrows or quarters, each of which is divided into four cartrons, each of which contains four croggeries or fourths . . . a *cnagaire* is sixteen acres, the nominal holding that 'could feed a cow with her calf, a horse, some sheep for their wool and give sufficient potatoes to support one family'.[54] Estyn Evans adds the common land measure of cow's grass by which land was often let and sold, 'they judge of the dimensions of a holding by its being to the extent, as the case may be, of one, two or three "cows' grass". They have divided not only into the fourth part of a cow's grass, called a "foot", but into the eight part of a cow's grass, or half a foot, denominated a "cleet".'[55] Divisions were further complicated for the English by the use of the Irish acre, one and two-thirds the size of a statute acre, and the allowance for rock, forest or unfertile ground. The cantred or barony was of equal status with the *tuath*, derived from its earlier meaning of a population capable of supporting three thousand soldiers in an emergency, but twice the area. The *tuath*, today the modern Irish language word for the countryside in general, was anciently ruled by a king and second only in importance to the *cúige* that today suggests a province.

For the English colonizers of America the fence was the essential sign of civilization and cultivation that defined one individual's property as private vis-à-vis another's.[56] The lack of fences was not understood as a different organization of land but as a non-organization of it. The question is, 'can one translate the idea of place as *property* into an idea of place the terms of which the West has never granted legitimacy?'[57] Behind the translation or

Anglicization of the townland names there was a selective reorganization or stratification of spatial categories that relegated vernacular terms to the detritus of a new monoglossic history. This brought English culture, organization and improvement to what was already Irish culture, organization and improvement. It was achieved, however, by conceiving of the latter as anarchic non-organization, empty or devoid of English qualities.

Translation was central to the perceptual apparatus of the survey, 'from its beginning the imperialist mission is, in short, one of translation: the translation of the 'other' into the terms of the Empire'.[58] The Antiquarian or Topographical Department practiced translation routinely in eliciting place names and conversing with informants in the field. The memoirs and letters are the direct result of a purposeful process of translation at the coalface. The unintelligible, only to the anglophone surveyor it must be remembered, grammar, intonation and shape of the other language was smoothed over by Anglicization and translation.

Translation is just one of a number of sites that brings the colonial subject into being: 'conventionally, translation depends on the West-ern philosophical notions of reality, representation, and knowledge. Reality is seen as something unproblematic, "out there"; knowledge involves a representation of this reality; and representation provides direct, unmediated access to a transparent reality.'[59] The translative–interpretive function of the survey reinforced a dominative vision of the colonized culture as particular, peculiar, static and anachronous, left behind by a providential universal civilization fated to succeed it.

The inevitability of this was established by an evaluative examination of the local culture. The ancient or vulgar Irish language was handed over through translation to modern English civilization. It was not an incidental change but a historic reorganization of geopolitical boundaries that was presumed to be final. Translation in the survey provided the materials for a mosaic of antiquity incorporating landscape, place names, personal names, oral tradition, mythology and pre-history. By absenting the present this achieved a textual fiction, it constructed an antiquated Ireland, imprinted in Gothic script, a kind of cultural watermark on the ultimate map of English Ireland.

O'Donovan introduced the ideas of contemporary interdisciplinary research into an emergent Irish scholarly discourse of national heritage, education and improvement.[60] In his search for expert conservators and clues to a primordial fossilized culture in the archive of a living language, he was influenced by the strong cocktail of botany, geology, palaeontology, history, antiquities, ethnography, philology and comparative linguistics. He was a time traveller amongst the verbal and artefactual remains and ruins of another time. He was, as Olender says of Herder, 'an anthropologist of providence'.[61]

Cultural Operator

Until the 1980s translation was viewed as a revelation, a project of cultural understanding in which the translator played the role of ambassador. In this light the survey was understood as an accord or reconciliation between what was moribund and what was being born. This was reinforced by the idea that cartography, like ethnography or translation, was mimesis, that it actually represented a matter of fact reality that was unproblematic. The map was not viewed as a cultural artefact itself, a constructed image that resulted from a particular historical context and relationship that was shaped by a specific discourse.

Most of the traces of difference, often portrayed as the imprecision of a flawed tradition, were subsumed within an idiom of domination. British colonial ethnography was rooted to a significant extent in the Celticist Orientalist paradigm of the eighteenth and nineteenth centuries which translated the primitive or traditional not just into another language but by extension into modernity itself, 'the desire to translate is the desire to *construct* the primitive world, to *represent* it and to *speak on its behalf*. What the discourse of ethnography traditionally represses, however, is any awareness of the asymmetrical relations between colonizer and colonized that enabled the growth of the discipline and provided the context for translation.'[62] The reality posited was in the mind of the observer, as Fabian says, 'the Other is never simply given, never just found or encountered, but *made*'.[63]

The images constructed through translation formulated 'an identity of the source culture that is recognizable by the target culture as representative of the former – as "authentic" specimens of a world that is remote as well as inaccessible in terms of the target culture's self'.[64] Rather than simply representing other cultures ethnographers invent them, sometimes through translations that are 'always *producing* rather than merely reflecting or imitating an "original"'.⁶⁵ In other discourses, literary or vernacular, this can be a source of creative renewal, innovation and cultural replenishment in itself. On the level of an official interference in the symbolic universe of another culture it was not so much enlivening as it was embalming.

Niranjana's exposition of the writing of Charles Trevelyan shows how liberal nationalists, firstly accepting others' representation of them as primitive, attempt to imitate the civilization of the colonizer. Very much alive and well, such discourse is itself a derivative of colonialist discourse. Both contain internal heterogeneity and complexity while projecting homogeneous hegemonic versions of the colonized.⁶⁶ The dominant language exerts control over representations made within its discursive domain, it patterns and regulates thought and expression, 'when a writer from the colony seeks to enter the domain of the colonizer, he seems to have no option but to deploy the symbolic order of the English language which already has an existent repertory of discourse'.⁶⁷

The historic and contemporary repertoire of the vernacular culture, both oral and literary, was disassembled piecemeal as a tattered and partial knowledge. The implication was that its proper place was as an appendage to the whole, entire and healthy target culture and language. As an omniscient arbitor the empowered language refracts the other language and effectively puts it in its place. Hewson and Martin describe the role of the translator as that of a cultural operator in the space between two languages. This space is a liminal area whose boundaries are constantly changing. The competence of the translator is measured by his ability to analyse, compare, and convert cultural systems. The translator is not culturally neutral and translation is a 'culture-bound activity . . . carried out in the perspective of the Target Language and of the forthcoming translation'.⁶⁸ The interpretation, once it is adopted, converts the circulation of meaning in the language or culture to be translated; it becomes 'the ground of a potential *difference, distinction, or tension*'.⁶⁹

The text to be translated is in danger of becoming something other than itself. It accrues a series of differences, anomalies and problems through a kind of hyper-reading. While translation theory normally restricts itself to literary texts, the question of the place names and the accompanying ethnographic prose speak eloquently of the interplay of two cultural systems. Although translation is also a site of hybridization, creative understanding and misunderstanding alike, the place names are removed from the open proliferation of lexical meaning into a poetics and politics of identity that fixes, rarefies and regulates.

The Turn of the Tongue in English

It appears that O'Donovan's position was similar at times to Niranjana's nativist who, accepting the colonizer's view of his culture as primitive, sought at times to politely depict an oppositional Irish civilization set back in time. Translators usually belong to the target language culture and in this case it could be argued that O'Donovan's sympathies lay, if not philosophically then pragmatically, with English. He had been a translator of manuscripts, place names and personal names all of his professional life. He believed, wrongly as it turns out, that the Irish language would be dead within thirty or fifty years of the survey. In 1902 Flannery noted O'Donovan's disregard for the living language.[70] Walsh wrote in 1957 that O'Donovan 'was too fond of placating the eye of the English reader'.[71] On this point Andrews adds that 'he knew that the Survey's customers were overwhelmingly English-speaking and he was careful to seek forms that would not offend the orthographic instincts of the English reader'.[72] His role was to settle the place names in an Anglicized spelling that was easy on the eyes and ears of an English speaker.

He often refers to his own role in the survey's jargon as settling, deciding, laying down or reducing the place names to standard. In the light of later romantic reassessments it is important to remember that it was his role as orthographer and etymologist that took him into the field and that all of the correspondence results from this role. He was answerable to the survey's authority in relation to the place names although he often had an additional responsibility to Petrie regarding antiquities.

Some questions have been asked of O'Donovan's habit of distancing himself from his compatriots by drawing upon the lexicon of colonialist discourse. De hÓir points out that he continued to associate with protestant societies in a time of religious animosity and acrimony. Some of his relatives were protestant, however, like the Carolines of Wexford to whom he alludes in a letter from that county. He is writing about the burning of 195 protestants in Sculboge near Newbawn, Co Wexford, during the 1798 rebellion, 'among these victims (holocausts) was William Caroline (a near relative of the writer of these notices) who was so well liked by the rebels that they would spare him if he would condescend to "bless himself" but he would not and therefore they cast him into the midst of the conflagration. Unfortunate man!'[74] Although there is a degree of uncertainty about it, English appears to have been spoken in his home but there is no doubting that he soon became immersed in the Irish of his locality.[75] Indeed if he had been a monoglot Irish speaker he would have been of no use to the survey.

The ambiguity of his stance is emphasized by the fact that, by the time the survey had reached the south of Ireland, the peasants and aborigines of his official transactions had become we and I. Having finally arrived home he permitted himself a degree of recognition that was absent from the rest of his voluminous correspondence, 'we, the inhabitants of the Barony of Ida and Ofa in the Co of Kilkenny, always call this place in the Irish language, which we speak very well, by no other name than *Fiodh Ard* and that we are right can be made to appear from the Irish Annals'.[76] He says of Faithleg in Co Waterford, 'I have been acquainted with the name since I was a child'.[77] What Fabian calls remembrance finally surfaces and the alienating distance and superiority almost dissolves.[78]

O'Donovan was always careful to emphasize the wider cultural aspects of place names. He wrote to Larcom in 1834 that 'no person is fit or should be allowed to meddle with those names except one acquainted with the whole circle of Irish lore, and with the peculiarities of pronunciations that prevail in the different districts'.[79] This statement was not a nostalgic reflection on Irish tradition and folklore but a matter of fact one. He learned in Co Derry that his own lack of knowledge of the dialects of Irish misled him in several interpretations.[80]

> There is another peculiarity which has led me astray very much. They never admit the combination cn into their dialect, and hence their Crock instead of Knock. Not knowing this I conjectured Crockdooish was derived from *Cruach*, a stack or round hill as in Croaghpatrick; but Crookdooish in the North is the same as Knockdooish in the South, the former always using cr for the cn of the latter. The cn however, is the orthography always used in the MSS. both of the Northern and Southern *Gaidhil*. This peculiarity has so much disguised the names of some Townlands that it would be impossible to trace them to their origin without conversing with the natives. Ex. Gr. Culnagrew (in the Charter Culnagow) . . . first it must throw off its Anglicized dress and appear a mere Derry Irish word *Cúl na gCró*, and next to make it intelligible, its provincial R must be changed to

N thus *Cúl na gCnó*, i.e. the Back or Retired Place of the Nuts. This is its undoubted name confirmed by the Charter but the provincialism must be retained, as the natives of the County could not articulate the combination cn at all; and they contend that cr is a more refined form and more easily pronounced. This I allow, but cr . . . has never appeared in any book or MSS. of authority.[81]

This passage is interesting in the manner in which it unfolds O'Donovan's translation strategy. Some of the tensions of the translations are evident in the interpretative interplay between Irish and English, orality and textuality, standard and dialect and the ancient and the modern. The submissive metaphor of undressing the language to reveal 'a mere Derry Irish word' and make it intelligible underlines the powerful patriarchal perspective of the receptor language. The words are obviously intelligible in their own language. The book, often a colonial charter or manuscript, was the final authority. He sometimes views the spoken language as barbaric, advising that O'Curry 'would certainly mistake the Northern pronunciation, and would do so until he would become well acquainted with their peculiarities and barbarisms'.[82]

The language, much like tradition, is described as local or unfixed, 'and so must every language be as long as its preservation depends upon the memory of the peasantry'.[83] Notwithstanding his view that a good knowledge of Irish lore was a prerequisite for understanding Irish toponymy he often considered it unreliable or even ridiculous, 'the more I look into the traditions preserved among the peasantry to account for names of places the less I think of their title to historical credit; and upon strict examination it will be seen that like Ovid's Metamorphoses they have all been founded upon the real or fanciful signification of the names'.[84]

This reveals an interpretation of the source language as fanciful or ridiculous prior to translation or Anglicization. He was dismissive of many local or vernacular etymologies, later recognized as a genre of folklore in their own right. More surprisingly he dismisses many of the place names that are derived from the popular Fenian cycle or *Fiannaíocht*. His scholarly exactitude led him to argue the toss regarding the derivations of place names with his informants as he did with the shanachies of Glenuller.[85]

He often discounts the indigenous culture in a semi-ironic tone and distances himself from the people. Like the officers of the Royal Engineers his was also a surveying eye and in this case it may be necessary, as Spurr argues, 'to ironize the ironizer'.[86] In Co Derry he discredits the Fenian place names of those he calls old romancers:

> I asked several old Irish men if they ever heard the Irish name of the Sconce; one told me that it was called Sconsa na bhFian, i.e. the Sconce of the Fenians or Fingallions, and another that it was called *Dún Oisín*, i.e. Ossian's Fort. This however had not the slightest weight with me, because I find that when those old romancers cannot account for the origin of a name or a building, they ascribe them to the Danes or to Fin MacCool's militia.[87]

Some place names are considered old although still in use, as with the *Cluain na gCluas* noted by O'Connor in Co Carlow, 'the old form of this name was *Cluain na gCluas* (the name is still pronounced thus in the country)'.[88] In general the ancient names, those verifiable in historical records, preoccupied the Antiquarian or Topographical Department and O'Donovan is strangely surprised that 'the peasantry have names of their own' for places, 'it is curious to remark how the peasantry have names of their own for the Parishes, and it is very seldom they know their ecclesiastical names. Thus Clondavaddog they always call *Fánaid*.'[89]

It is an anomaly in his system of recasting an ancient Irish civilization within a modern English one that the old is often novelized and the new is often archaized. Responding to Lieutenant Deloss Broughton from Co Westmeath the interesting remark is made by Lyons that 'there are not any persons near me who understand the Irish language further than to speak it. I do not know one who writes or reads it in the Gaelic character.'[90] The depersonalized Irish speakers are described as alienated and marginalized from their own language. The dominative discourse empties them of significance and proceeds to describe them from its own perspective. An informant in Co Kildare pronounced the place names in four parishes for O'Connor, a considerable achievement in itself, and told him that Taghadoe was called *Tigh Tuaith* in Irish. He told O'Connor, however, that he could not 'make the name speak English'.[91] O'Donovan thought that many modern names should be ignored, 'I would advise you not to give on the Maps names of loughs and places called after farmers now living, or people gone many years since to America, because such names will give way to others in a very short time'.[92] He reiterates this opinion in Co Mayo:

My time is entirely taken up with the small names on the coast, and I can spare but very little time to write letters. Many of these names are arbitrary and known to very few only and most of them are modern. The same may be said of many of the internal names, which are frequently called after people still living. Many of them are not names at all, and in my opinion should not be given as they may be changed in a few years.[93]

As a counter modernity the living language and contemporary vernacular culture of the nineteenth century was bothersome. The modernizing imperial survey was not trying to represent a modern Irish culture or civilization on its own terms. Archaic names were privileged over the living names of the present. Robinson, while mapping place names in contemporary Connemara, noted this on numerous occasions. The same preoccupation with the medieval world echoes through several of his interpretations of common names that are still known almost two centuries later. One name for a well is given by the surveyors as Tobergollankillane in Anglicized orthography while the Irish is given as *Tobar Galláin Coilleáin* or the well of Collin's standing stone. The ancient and conspicuously archaeological connotation takes precedence over the everyday and well known *Tobar Cholm Cille* or Colm Cille's Well.[94] In the townland of Glinsce the surveyors recorded a lake's name as Lough Clugacommen or in Irish *Loch Clog an Choimín*, the lake of the bell of the commons, but an Irish speaker was able to tell Robinson that it was simply *Loch na Cloiche Caime* or the lake of the crooked stone.[95] The bell and the standing stone replace the well and the crooked stone. While he often denigrated local raconteurs or shanachies as untrustworthy conservators of ancient topography he also complained of the abundance of the modern, 'the names of creeks, rocks, holes, clefts, dumhachs, glens, mungs, cwees and alts on these coasts have nearly worn me out of patience'.[96]

In Donegal where a young child of six years of age gives the enlightened surveyor an impromptu Irish lesson, the constant casting of the Irish language as vulgar, old or ancient is thrown into stark relief:

> In *Fánaid* I met a rock called *Carraig na bhFaochóg*, and on asking for the English meaning of the word, I was told it meant the Rock of the Wilks. But when I said that I knew not what either *faochóg* or wilk meant, a little child aged six went to the strand and carrying with him the shell of one of them resought the house in which I was, and holding the shell between his fingers said, '*Sin faochóg!*'[97]

The point here is not so much that the survey renamed Ireland or invented new names, the process of colonial Anglicization began earlier with land confiscation and plantation. Still others were what O'Donovan calls vulgar peasant translations brought about by what he was told was the 'turn of the tongue in English'. Local landlords had a part in this process also, as one informant told him:

> Gentlemen know nothing about Irish names of townlands – that they wish to harden, shorten and make them look like English names – and as this work is going on since the reign of Oliver Cromwell it is now most difficult to come at the oult name, for even the farmers are now forgetting the names which their grandmothers used to call these lands and adopting the hardened and shortened names which they hear with their landlords.[98]

The problem facing the surveyors was firstly to Anglicize the spellings and secondly to standardize these Anglicized spellings. Ó Maolfabhail makes the point that the standardization of the names in the Irish language would have been uncomplicated if the meanings were clear. The survey's idea of what was settled or unsettled is interesting in itself. Far from being unsettled the Irish language had developed a standardized spelling, syntax, assonance and phonetics that it shared with Scottish Gaelic. Irish place names were not unsettled, they just looked unsettled to the English eye. The strategy used in Anglicizing the names was to 'reduce Irish names to forms that appear significant to an English eye' as O'Donovan noted in many earlier Anglicizations.[99]

In deciding the spellings of the place names the preferred authorities were documentary and the spellings of the earlier surveys, as well as those scattered here and there in the administrative paper trail of the colonizer, were privileged. Captain Waters had suggested this before the survey sent the Antiquarian or Topographical Department into the field: he asks 'cannot a list of names be had from the Down Survey, which would, no doubt, give the names nearer their original Irish meaning than we have them at present?'[100] A few years later O'Donovan concurred, 'I am of the opinion that we should quote the Charter of Londonderry and the Down Survey in our lists of the names of Townlands; as it would add weight to orthography'.[101] Some years later, writing from Co Roscommon he reiterates this, 'the only evidences now remaining to prove the extent of ancient territories are the legends of the saints, the Irish Annals and the early English Inquisitions; and the points on which these fail to afford satisfactory evidence, must remain for ever disputable'.¹⁰²

The work of retracing a primordial Ireland was not exactly urgent in his position as orthographer, and his deference to the prototypes of colonial survey as adding weight to his own work is interesting. In Co Derry he explains that the Irish termination *buí* was left as 'boy' in the toponym Mullaboy because it occurred in English records, 'the termination buidhe "yellow" which so frequently forms the last syllables of names of places in Ireland, is pronounced in the original language bwee, but as it has been so invariably Anglicized boy in the names of men and places in all the old English records relating to Ireland, it has been thought the best spelling to make general on the maps'.¹⁰³

O'Donovan frequently expresses his unhappiness with the work of Griffith and the officers in his service from whose dictation he worked at times. From Co Roscommon he writes:

> I am sick to death's door of Lochawns, and it pains me to the very soul to have to make these remarks, but what can I do when I cannot make the usual progress? Here I am stuck in the mud in the middle of Loughs, Turlaghs and Curraghs, the names of many of which are only known to a few old men in their immediate neighbourhood and I cannot give many of them utterance from the manner in which they are spelled.¹⁰⁴

The careless name taking of Lieutenants Boteler, Chaytor and Corporal Berry annoyed O'Connor.¹⁰⁵ It is clear that the tensions between the Irish fieldworkers and the British military sometimes precluded the establishment of an agreed methodology. It was not always simply a case of the Irish scholar having a free hand to restore the place names of a lost nation. Henry Buck, one of Griffith's boundary surveyors is heavily criticized, 'Mr Henry Buck, District Boundary Surveyor, and Mr Taylor, Agent to Sir A. Brocki, had furnished orthographies which (I suppose) they expect will be adopted on the Ordnance Map. Both are guided by ear, and have no right to set themselves up as authorities, and Buck has frequently adopted ridiculous spellings.' ¹⁰⁶

O'Donovan furnished the detail but the real authority lay beyond him. Complaining of the officer's dictation and the spellings arrived at in the offices of the Boundary Survey he wrote:

> I think if all the officers would imitate Mr Beatty that those Name Books could be prepared long before the plans are drawn. He has employed a man to walk into every Townland to ascertain the name of every feature to be marked on the map, by which means he has all the names and their exact situations before the plans are drawn. To employ men who have a smattering barbaric knowledge of the Irish language to guess at the names in an office in a country town is truly ridiculous, and must finally lead to error or be of no use whatever.[107]

In Co Derry he glosses the name Moneyhanegan while reflecting on the early years of the survey:

> The most analogical Anglicized spelling of the name of this townland is Moneyhanigan, and it should be so spelled for the future, though the orthographical department of the Ordnance Survey was not sufficiently matured to decide upon the proper spelling of it, at the time that the names were required for engraving. The names in fact were at this time furnished by persons who had little or no acquaintance with the analogy or significations of Irish names of places; and when persons were found who had some acquaintance with the Irish language, it has been discovered that they were altogether carried away by the mania of etymological speculations, and have not infrequently furnished not only wrong interpretations of terms but also wrong names.[108]

Griffith's boundary surveyors often sought the advice of landlords and ministers of the Church of England in establishing townland boundaries. Large townlands were subdivided by attaching the words North or South, Upper or Lower to them and establishing them as new townlands. Smaller units of land and the demesnes of the gentry were established as townlands in their own right. O'Donovan frequently advises Larcom that the officers should use more local authorities to avoid putting incorrect names on the maps. He notes the discrepancies while in the field:

> The people deny that many denominations set down as townlands on our maps are townlands, and we have many denominations that are considered as comprising several townlands. I find that the Down Survey comes nearer the townlands of each Parish than ours, and I am inclined to believe that Mr Griffith frequently divides parishes into townlands, more from his own fancy than from the authority of the people. But as we have nothing to do with this, I took no trouble to ascertain how far he did so, or did not.[109]

Many of the difficulties that the Boundary Surveyors encountered, and many of the liberties taken by them, doubtless lay in the plethora of vernacular denominations of land divisions and the dialectical diversity of them.

The Powers of English Consonants

The strategies employed in handling Irish place names included different stages and strands of Anglicization, translation and standardization. This had the effect of archaizing the originary names: even in dictating or transliterating them, it fixed them in English orthography in which they were lexically meaningless. This was carried out in the perspective of the English language. In translating related documents in the course of his fieldwork O'Donovan expresses concern that his translation 'being too literal, reads rough and un-English like'.[110] In Co Fermanagh he is doubtful about the English spelling or Anglicized spelling to be adopted.[111]

At times he defers to Larcom for approval, 'I should like to have your opinion respecting the termination *án* when long and accented; whether is aun or awn the more comfortable with the English orthography'.[112] In Co Donegal he is anxious to avoid the obvious analogies with English that resulted from many transliterations, sometimes with humorous results, 'I am not yet satisfied with the Anglicizing of *Gaoth Dóir* and *Gaoth Beara* – they are made *Gui* in the Inquisitions and Guy in the modern authorities, but I fear that both would be pronounced by an English Scholar like Guy, Earl of Warwick. Please do consider this. Gwee Guee?'[113]

He takes liberties with some names like *Bealach an Chaoláin* that he made Ballyheelan in place of Ballyhillan, 'I have spelled it Ballyheelan, as I see plainly that Heelan, not Chaoláin will become the pronunciation as soon as the present generation shall go home to their fathers, for they will be succeeded by a race who will not be able to pronounce ch guttural'.[114] It is apparent that he understood his own role as that of improver and educator. In Co Monaghan he recounts another argument that gives an insight into his approach, 'many persons have told me that the name of a townland as well as the name of a man, might be spelled any way, and I made answer that it might in a rude age and among a barbarous people but that we should now try and reduce them to standard as well as the words of a cultivated language'.[115]

The strategy of Anglicization ultimately involved making cultivated English-looking words of barbarous Irish-looking ones. In arriving at the spelling Toneduff (*tóin dubh*), which made a single word of a noun and an adjective in the Irish language, O'Donovan notes, 'Dhu is rejected as altogether foreign to English and Irish orthography. In pronouncing the names of places in Ireland when they are Anglicized, they must be pronounced according to the powers of English consonants. In the original Irish the consonants are either thick or liquid, and it is impracticable and even incorrect to attempt to represent their sounds by any combination of English ones.'[116] The result is often a desemanticized hybrid drawn from Irish vocabulary and compacted into English orthography. The fluid and thick consonants of Irish (unsettled, rude, barbarous, plain) must submit to the power of the English ones (settled, civilized and comely). The distortion of the pressure to Anglicize leaves the Irish words in a linguistic dead zone and creates a hallucinatory form of English.

The Cultural Process of Translation

Al-Shabab lists five stages of translation: editing the source text, interpreting the source text, interpreting the source text in a new language, formulating the translated text and editing this formulation.[117] He summarizes these as:

> Editing the source text, when it is required, focuses on the source text and is carried out in the light of the source language. The interpretation of the source text obviously concentrates on that text. The central stage in the process of translation is interpretation in a new language, which focuses on the source text, and which produces the translated text, which in turn results from the active involvement of the translator as a human agent. The translator links the elements needed for interpretation in a new language, a stage in which the target language systems and culture are used to achieve the translation. The formulation stage focuses on the translator and the translated text, and is carried out in the light of the systems and culture of the target language. Again, editing the translated text is accomplished in terms of the systems and textual and cultural norms of the target language.[118]

These processes, diagrammatically represented in the name books, are remarkably similar to those followed by the survey. Interestingly they also mirror vital stages in the ethnographic process from research to writing. If the word ethnographer is substituted for translator this becomes an insightful summary on the process of ethnography itself. Larcom, writing to Colby in Southampton in 1842, after some eighteen years of the survey, summarized the process leading to the name books as follows:

> The orthography of names of places is unsettled everywhere, but peculiarly so in Ireland and the same name appeared to be spelled in such various ways by different persons, in the country, and in documents of considerable authority that it was thought desirable to procure access to other documents, and to employ some person conversant with the Irish language to assist in choosing among such conflicting evidence . . . the rule adopted was after collecting as many modes of spelling as possible to send the Irish scholar to the ground with the abstracted information in his hand where from examination and enquiry the original Irish name was discovered. The books were then returned to this office and we adopted that one among the modern names most consistent with the ancient topography. [119]

It is interesting that Larcom uses similar language to O'Donovan in this description. The name books show the final editing of the source text and lists authorities; the old aborigines of the letters and memoirs are not listed as authorities. The interpretation of the source text was represented by a translation into English, an approved Anglicization and a list of similar variants.

The letters and memoirs, however, are also interpretations of the source text whether for a particular name, legend or ritual. This was a process that involved the application of both linguistic and cultural criteria in order to produce a text for a specific audience in a specific social and linguistic environment. In this case the Anglicization was produced on behalf of Her Majesty in the perspective of the English language for an English-speaking consumer. The aim was obviously to provide Anglicized place names of townlands. Like the ethnographic process this was, 'masked as a mere sequence but was in fact a ritual dramatization of spatial distance between the sites of observation and places of writing'. [120]

Interpreting the source text in a new language is defined as 'transformulating a linguistic/verbal text or part of it, after first interpreting it, to a language other than its own'; a new text is thus produced.[121] Here there is a transgression of linguistic boundaries, translation by any other name, the place names in any case are made

English. In a broader sense the other was still present but only residually and on the terms of the target language. The English speaker was brought home by a strategy that familiarized the foreign while the Irish speaker was alienated by a strategy that foreignized the familiar. For the latter an alien reading experience was staged.[122] The study of translation is also the study of interpretations, judgements and cultural shifts. Through translation concepts, texts, tropes and notions are appropriated. I have chosen three examples that further illustrate this process. It should be added that narratives such as these are not numerous in the archive of the survey and their inclusion is often intended to negate rather than advocate tradition as if to show what must, by moral necessity, be subjugated to evolutionary science and reason.

Aghalurcher

This name appears in the statistical memoir for the parishes of Co Fermanagh and was written by Lieutenant Greatorex of the Royal Engineers in 1835 while working in the field on behalf of the survey. It is likely that Lieutenant Greatorex could not understand the Irish language from which the Anglicized name was derived. Explaining it as 'the field of the second shot' Lieutenant Greatorex relates the following aetiological or explanatory narrative:

> The general pronunciation by the peasantry is Agh-a-lurrgh-er, with a strong guttural; it is also sometimes pronounced Aghadurcher, but very rarely. The explanation given by the peasantry... is very absurd, but deserves being recorded from the general belief it obtains among them. The story as related to me by several individuals is as follows. In this period when St Ronan (who was a disciple of St Patrick) flourished, the saint was requested by the inhabitants of this district for a site whereupon to build a church, and there being some dispute as to the most appropriate spot, he determined to leave it to the chance throw of a stone, and accordingly the first stone was thrown from the opposite side of the lake and fell in Innishcollen, which being an island and very much out of the way, was strongly objected to for the intended erection. St Ronan was therefore entreated to take a 'second cast' pronounced in Irish '*darra urcher*'. The second cast determined the situation of the church (now a heap of ruins) in the townland of Aghalurcher. The only variation in this superstitious tale that Finn MacCool was the thrower and not the saint, and the asserters of this fact point out triumphantly the stone in Innishcollen and ask how it came there if the story be untrue.[123]

The narrative speaks of the fated and magical selection of the area by the patron saint who is rendered all the more powerful by his connection to St Patrick. In the Irish language the name is a mnemonic that recalls multileveled meanings that resonate within local identity. It is an expression of popular cosmology that, mixing religion and narrative, draws upon a complex shared knowledge as well as providing knowledge of a particular cultural landscape. The rational scientific discourse of the modernizing and civilizing mapmaker supersedes the charisma and supernaturality of the vernacular narrative. The new map and the new language replace the old map and the old language, the patron saint is replaced by the surveyor. The narrative is quoted as an example of the absurdity of a peasant culture epitomized by an implicit belief in superstition. Prior to translation and Anglicization the name is interpreted as a ridiculous manifestation in itself that typifies the misguided machinations of Irish culture.

The phonetic transcription given as the pronunciation of the name and the Anglicized spelling has the appearance of a cartoon orthography for an exclamation of pain 'agh-a-lurrgh-er'. The narrative is finally discredited by the rhetorical suggestion of a deluded and credulous peasantry who consider the presence of a stone as evidence of a mythological moment. The vernacular map is wiped clean as the place name is Anglicized as Aghalurcher for the new English map. Story, belief and meaning in the vernacular discursive mnemonic are consigned to the past. A full English translation, however awkward, would carry forward the lexical meaning where the compressed Anglicization silences it. Contemporary vernacular agency or efficacy is neutralized in its new context, it becomes ahistorical, erroneous and anachronistic. Looking English and trying to sound English, the new form satisfies the administrator but starts to transmit semantic noise. It produces incoherent messages of pain or even hunting with crossbred dogs. If the translation of names in the survey is fiction then the names are fictions of translation.

Cloghagaddy

This is a townland in Galoon in Co Fermanagh and the story attached to the name was noted in the statistical memoirs by Lieutenant Durnford of the Royal Engineers and also in the letters from the same county by O'Donovan. O'Donovan gives the following narrative:

> In Galloon you will find a Townland called Cloghagaddy, meaning the Thief's Stone. This name is derived from a very remarkable stone in the Townland about the height of a man and terminated like a sugar loaf. The name is accounted for by a story about a thief who was stealing a sheep. He had the sheep tied on his back by a rope around his breast, and when he was passing this stone he leaned his burden against it but the sheep slipped over the stone, and the rope slipping up to the thief's neck, actually hanged him.[124]

This narrative refers to a feature on the local landscape that gave the townland its name, *Cloch an Ghadaí* or the thief's stone. The stone, described by O'Donovan in the idiom of the antiquary as very remarkable, is a natural feature in the environment serving to remind the community of the consequences of theft, of the fatal and fateful retribution meted out to the thief. Externalizing their values and beliefs it linked the lifestyle of the community with a physical and moral environment. It was one resting place in the flow of vernacular discursive life. Presumably it also contains some ethnographic detail of methods of carrying sheep. Whether the story recounts an actual occurrence or is an imaginative association based on the shape and location of the stone is immaterial. O'Donovan recounts the narrative in his research with a view towards Anglicizing the spelling and he compresses the encoded epitaph into the opaque English looking puzzle Cloghagaddy.

Erne-Head Lake

After the fashion of the officers of the Royal Engineers O'Donovan prefaces the following narrative by saying that 'these legends are in themselves of no value, but they illustrate, in a striking manner, the credulous simplicity of the people among whom they originated'. This in itself represents an interpretation of the source text:

> Tradition, as preserved by Farrell Linchy – who is now near one hundred years old – says that *Tobar Gamhna* was in the oulde times, a well of great sanctity; that a woman profaned it by washing dirty clothes in its pure waters – the greatest insult which, according to the Jews and Irish, could be offered to such a thing – that a calf which was underground sallied forth at the insult, and ran north west in a serpentine direction, and that he was followed by a river which, when it arrived at a deep valley formed itself into a lake now called *Loch Gamhna* or the Lake of the Calf (*Lactus Vetuli*) . . . it is probable that

his name will soon be forgotten and swallowed up in Erne-head Lake, a name which the Lord of the Soil intends to establish in spite of the calf.[125]

Once again this story is quoted as evidence of the simplicity and credulity of the people and not as a prised example of folklore rescued from oblivion. The absurdity, superstition, simplicity or credulity of the peasantry is actually the true referent. Its presence, like much of the information, is wholly contingent and framed by its placement in a discourse that is actually talking about something else. It is an example of an inferior culture in need of improvement and civilization. The century-old narrator Farrell Linchy, who may have been born around 1737, represents tradition. The story, the storyteller and the place name are all magically and fluently translated taking on a new meaning in the target language. The storyteller is depicted as a caricature, some of his speech being intextualized as 'in the oulde times'. The legend, explaining the origin of the lake as emitting from a holy well called *Tobar Gamhna* in Irish or the Well of the Calf, is full of mythological echoes and speaks of the magical sacred and moral landscape representing a living and lived-in ecology.

When the ordinary or profane world, a woman washing clothes, is brought to the sacred place the calf emerges from the earth in protest miraculously drawing its waters with it and forms *Loch Gamhna*. The intimately known and spiritual landscape is rendered subjective, pagan and superstitious. The local landlord, whose social and cultural privilege and supremacy was protected by the survey, called the lake Erne-head Lake. It is recast as a picturesque ornament signifying civilization and improvement in elite English language discourse. The calf, autochthonous spirit of the moral supernatural vernacular landscape, goes underground again. Like the woman washing her clothes in the sacred water, the survey becomes a cultural transaction.

The Anglicized place name to be engraved on the map was the humble if not heroic product of the survey. The place names cannot be considered in isolation, however, as a kind of harmless wordplay. They must be viewed in conjunction with the processes that produced the comprehensive ethnographic prose that constitutes the survey in a fuller sense. These examples show significant aspects of the interpretation of the source text in a new language, or a desemanticized interlanguage, actually produced through a primary process of translation. The Anglicized place names are a shorthand for a more ambitious, protracted and systematic translation process. While both correct and incorrect lexical meaning can be, and is, restored to semantically opaque place names the rhetorical and symbolic Anglicization is still a product of power. [126]

Here the translator's, ethnographer's or cartographer's privileged position is visible. He becomes a cultural operator actively shaping and constructing modes of representation, dictating and relocating knowledge, place and people. As a mediator managing strategic themes of loss and recovery, manoeuvring within the tropes of tradition and modernity, barbarity and civilization, stasis and evolution, Irish and English, the other knowledge is subjugated or made secondary.

In a ritual of assimilation, the Irish language is conjured as an appendage to English. It begins to take on a synthetic and syncretic similitude to the English language. Archaizing the present, discovering the well-known or displacing the modern, the fictions of fluency and transparency silence the semantic otherness. The language and tradition, mythology and poetics of the other are constructed as barbarous anachronisms. The language and landscape are described as vulgar, unfixed, unsettled, peculiar and curious and, like the people who speak it or live in it, they must be civilized, settled and laid down. The enigma of the other is solved, on paper at least.

NOTES

1. Brian Friel, *Translations* (London, 1981), p.29.
2. Laura Hostetler, *Qing Colonial Enterprise: Ethnography and Cartography in Early Modern China* (Chicago, 2001), p.41.
3. Art Ó Maolfabhail, 'An tSuirbhéireacht Ordanáis agus Logainmneacha na hÉireann 1824–34', *Proceedings of the Royal Irish Academy*, 89, C, 3 (1989), p.40.
4. Ibid., p.43.
5. Rev Michael O'Flanagan (ed.) *Ordnance Survey Letters 1834–1841*, 43 volumes typeset and bound in Boole Library, University College Cork, (Bray, 1927–35), Armagh/Monaghan, p.21. These were written in the course of their fieldwork for the survey by John O'Donovan, Eugene O'Curry, George Petrie, Patrick O'Keefe and Thomas O'Connor. *See* Archival Sources for further information. Hereafter abbreviated OSL.
6. OSL, Mayo, Vol.1, p.159.
7. OSL, Meath, pp.3–4.
8. OSL, Sligo, p.17.
9. Ó Maolfabhail, 'An tSuirbhéireacht Ordanáis', p.45.
10. Art Ó Maolfabhail, 'Eoghan Ó Comhraí agus an tSuirbhéireacht Ordanáis', in Pádraig Ó Fiannachta (ed.) *Omós do Eoghan Ó Comhraí* (An Daingean, 1995), p.150.

11. Thomas Davis, *Literary and Historical Essays 1846* (Washington DC, 1998), p.139.
12. Art Ó Maolfabhail, *The Placenames of Ireland in the Third Millennium: Logainmneacha na hÉireann sa Triú Mílaois* (Dublin, 1992), p.17.
13. Ó Maolfabhail, 'An tSuirbhéireacht Ordanáis', p.51.
14. John H. Andrews, '"More Suitable to the English Tongue": The Cartography of Celtic Placenames', *Ulster Local Studies*, 14, 2 (1992), p.11.
15. Ibid., pp.13–14.
16. Angélique Day and Patrick McWilliams (eds), *Ordnance Survey Memoirs of Ireland* (Belfast, 1990–98), 40 volumes, Vol.36, p.61. Hereafter abbreviated OSM.
17. OSL, Cavan/Leitrim, p.109.
18. Ó Maolfabhail, The Placenames of Ireland in the Third Millennium, p.18.
19. Aidan Arrowsmith, Review of 'F. C. McGrath (Post)colonial Drama: Language, Illusion, and Politics (Syracuse, 1999)', *Irish Studies Review*, 9, 1 (2001), p.129.
20. Declan Kiberd, *Inventing Ireland: The Literature of the Modern Nation* (London, 1995), p.620.
21. Maurice Olender, *The Languages of Paradise: Race, Religion and Philology in the Nineteenth Century* (London, 1992), p.61.
22. André Lefevere (ed.), *Translation/History/Culture: A Sourcebook* (London, 1992), p.xi.
23. Lawrence Venuti, *The Translator's Invisibility: A History of Translation* (London, 1995), p.17.
24. Omar S. Al-Shabab, *Interpretation and the Language of Translation* (London, 1996), p.8.
25. Lefevere, Translation, History, Culture, p.1.
26. Friel, *Translations*, p.32.
27. Lefevere, Translation, History, Culture, p.2.
28. David Spurr, *The Rhetoric of Empire: Colonial Discourse in Journalism, Travel Writing and Imperial Administration* (London, 1993), p.4.
29. Anuradha Dingwaney and Carol Maier (eds), *Between Languages and Cultures: Translation and Cross-cultural Texts* (Pittsburgh, 1995), p.4.
30. Lawrence Venuti (ed.), *Rethinking Translation: Discourse, Subjectivity, Ideology* (London, 1992), p.5.
31. Román Álvarez and Carmen-África Vidal (eds), 'Translating: A Political Act', in Román Álvarez and Carmen-África Vidal (eds), *Translation, Power, Subversion* (Clevedon, 1996), p.1.
32. Ovidio Carbonell, 'The Exotic Space of Cultural Translation', in Román Álvarez and Carmen-África Vidal (eds), *Translation, Power, Subversion* (Clevedon, 1996), p.84.
33. Richard Jacquemond, 'Translation and Cultural Hegemony: The Case of French-Arabic Translation', in Lawrence Venuti (ed.), *Rethinking Translation: Discourse, Subjectivity, Ideology* (London, 1992), p.148.
34. T. Tymoczko, 'Translation and Meaning', in F. Guenthner and M. Guenthner-Reutter (eds), *Meaning and Translation: Philosophical and Linguistic Approaches* (London, 1978), p.29.
35. Gísli Pálsson, 'Introduction', in G. Pálsson (ed.), *Beyond Boundaries: Understanding, Translation and Anthropological Discourse* (Oxford, 1993), p.1.
36. Michael Cronin, *Translating Ireland* (Cork, 1996), p.92.
37. Stiofán Ó Cadhla, *Cá bhFuil Éire? Guth an Ghaisce i bPrós Sheáin Uí Ríordáin* (Baile Átha Cliath, 1998), p.41.
38. Pálsson, 'Introduction', p.16.
39. Cronin, *Translating Ireland*, p.92.
40. Ibid., p.105.
41. Ibid., p.107.

42. J. Jorge Klor de Alva, 'Language, Politics, and Translation: Colonialist Discourse and Classic Nahuatl in New Spain', in Rosanna Warren (ed.), *The Art of Translation: Voices from the Field* (Boston, 1989), pp.143–4.
43. Lawrence Venuti, 'Translation as Cultural Politics: Regimes of Domestication in English', *Textual Practice*, 7, 2 (1993), p.209.
44. Spurr, *Rhetoric of Empire*, p.28.
45. Thomas Richards, *The Imperial Archive: Knowledge and the Fantasy of Empire* (London, 1993), p.1.
46. W.F.H. Nicolaisen, 'The Past as Place: Name, Stories, and the Remembered Self', *Folklore 102*, i (1991), p.13.
47. Eric Cheyfitz, *The Poetics of Imperialism: Translation and Colonization from the Tempest to Tarzan* (Philadelphia, 1997), p.48.
48. Ibid., p.105.
49. John H. Andrews, *A Paper Landscape: The Ordnance Survey in Nineteenth-century Ireland* (Oxford, 1975), p.119.
50. Emyr Estyn Evans, *Irish Folk Ways* (London, 1957), p.28.
51. Máirtín Ó Cadhain, 'Béaloideas', in Seán Ó Laighin (ed.) *Ó Cadhain i bhFeasta* (Dublin, 1990), p.160.
52. John O'Donovan, *The Genealogies, Tribes and Customs of Hy-Fiachrach, Commonly called O'Dowda's Country. Now first published from the Book of Leacan, in the library of the Royal Irish Academy, and from the genealogical manuscript of Duald MacFibis, in the library of Lord Roden; with a translation and notes, and a map of Hy-Fiachrach*, (Dublin, 1844), p.204.
53. Ibid., p.453.
54. Tim Robinson, *Stones of Aran: Pilgrimage* (London, 1989), pp.90–1.
55. Estyn Evans, *Irish Folk Ways*, p.29.
56. Cheyfitz, *Poetics of Imperialism*, p.56.
57. Ibid., p.58.
58. Ibid., p.112.
59. Tejaswini Niranjana, *Siting Translation: History, Post-structuralism and the Colonial Context* (Berkeley, 1992), p.2.
60. Ibid., p.13.
61. Olender, *Languages of Paradise*, p.44.
62. Niranjana, *Siting Translation*, p.70.
63. Johannes Fabian, 'Presence and Representation: The Other in Anthropological Writing', *Critical Inquiry*, 16 (1990), p.755.
64. Mahasweta Sengupta, 'Translation as Manipulation: The Power of Images and Images of Power', in Anuradha Dingwaney and Carol Maier (eds), *Between Languages and Cultures; Translation and Cross-cultural Texts* (Pittsburgh, 1995), p.159.
65. Niranjana, *Siting Translation*, p.81.
66. Ibid., p.186.
67. Sengupta, 'Translation as Manipulation', p.165.
68. Lance Hewson and Jacky Martin, *Redefining Translation: The Variational Approach* (London, 1991), p.136.
69. Ibid., p.137.
70. Nollaig Ó Muraíle, 'Seán Ó Donnabháin, "an Cúigiú Máistir"', in Ruairí Ó hUiginn (ed.) *Scoláirí Gaeilge, Léachtaí Cholm Cille XXVII* (Maynooth, 1997), p.41.
71. Rev. Paul Walsh (ed.), *The Placenames of Westmeath* (Dublin, 1957), p.vii.
72. Andrews, *Paper Landscape*, p.125.
73. Ó Maolfabhail, 'Eoghan Ó Comhraí', p.176.
74. OSL, Wexford, Vol.2, p.21.
75. Patricia Boyne, *John O'Donovan* (Kilkenny, 1987), p.4.

76. OSL, Wexford, Vol.2, p.50.
77. OSL, Waterford, p.3.
78. Johannes Fabian, 'Remembering the Other: Knowledge and Recognition in the Exploration of central Africa', *Critical Inquiry*, 26 (Autumn 1999), p.55.
79. OSL, Fermanagh, p.15.
80. OSL, Londonderry, p.38.
81. OSL, Londonderry, pp.38–9.
82. OSL, Fermanagh, p.14.
83. OSL, Donegal, p.20.
84. OSL, Londonderry, p.63.
85. OSL, Londonderry, pp.39–41.
86. Spurr, *Rhetoric of Empire*, p.24.
87. OSL, Londonderry, p.35.
88. OSL, Carlow, p.94.
89. OSL, Donegal, p.32.
90. OSL, Westmeath, Vol.2, p.61.
91. OSL, Kildare, p.11.
92. OSL, Sligo, p.5.
93. OSL, Mayo, Vol.1, p.178.
94. Tim Robinson and Liam Mac Con Iomaire, *A Twisty Journey Mapping South Connemara: Camchuairt Chonamara Theas* (Dublin, 2002), p.78.
95. Ibid., p.50.
96. OSL, Mayo, Vol.1, p.91.
97. OSL, Donegal, p.36.
98. OSL, Armagh/Monaghan, p.37.
99. OSM, 36, p.60.
100. Ó Maolfabhail, 'An tSuirbhéireacht Ordanáis', p.45.
101. OSL, Londonderry, p.47.
102. OSL, Roscommon, Vol.1, p.68.
103. OSM, 28, p.64.
104. OSL, Roscommon, Vol.2, p.15.
105. OSL, Sligo, p.37.
106. OSL, Fermanagh, p.71.
107. OSL, Galway, Vol.1, p.161.
108. OSM, 36, p.61.
109. OSL, Londonderry, p.147.
110. OSL, Westmeath, Vol.2, p.55.
111. OSL, Fermanagh, p.54.
112. OSL, Cavan/Leitrim, p.107.
113. OSL, Donegal, p.116.
114. OSL, Cavan/Leitrim, p.48.
115. OSL, Armagh/Monaghan, p.51.
116. OSM, 28, p.66.
117. Al-Shabab, Interpretation, p.35.
118. Ibid., p.44.
119. Ó Maolfabhail, 'An tSuirbhéireacht Ordanáis', p.60.
120. Fabian, 'Presence and Representation', p.759.
121. Al-Shabab, *Interpretation*, p.39.
122. Venuti, *The Translator's Invisibility*, p.20.
123. OSM, 4, p.1.
124. OSL, Fermanagh, p.73.
125. OSL, Longford, pp.4–5.
126. W.H.F. Nicolaisen, 'Place-name Legends: An Onomastic Mythology', *Folklore*, 87, ii (1976), p.151.

Select Bibliography

Abrahams, Roger D., 'Phantoms of Romantic Nationalism in Folkloristics', *Journal of American Folklore*, 106, 419 (1993), 3–37.

Al-Shabab, Omar S., *Interpretation and the Language of Translation: Creativity and Convention in Translation* (London: Janus, 1996).

Álvarez, Román and Carmen-África Vidal (eds), *Translation, Power, Subversion* (Clevedon/Philadelphia/Adelaide: Multilingual Matters Ltd., 1996).

Amin, Shahid, 'Cataloguing the Countryside: Agricultural Glossaries from Colonial India', in Peter Pels and Oscar Salemink (eds), *Colonial Ethnographies*, a special issue of *History and Anthro-pology*, 8, 1–4 (Harwood Academic Publishers, 1994), pp.35–53.

Anderson, Benedict, *Imagined Communities: Reflections on the Origin and Spread of Nationalism* (New York: Verso, 1983).

Andrews, John H., *History in the Ordnance Survey Map* (Dublin: Ordnance Survey Office, 1974).

Andrews, John H., *A Paper Landscape: the Ordnance Survey in Nineteenth Century Ireland* (Oxford: Clarendon Press, 1975).

Andrews, John H., Kevin Barry and Brian Friel, 'Translations and A Paper Landscape: Between Fiction and History', *Crane Bag*, 7, 2 (1983), pp.118–24.

Andrews, John H., '"More Suitable to the English Tongue": The Cartography of Celtic Placenames', The Ordnance Survey and the Local Historian, *Ulster Local Studies*, 14, 2 (1992), pp.7–21.

Andrews, John H., 'Notes for a Future Edition of Brian Friel's Translations', *The Irish Review*, 13 (Winter 1992/3), pp.93–106.

Andrews, John H., *Shapes of Ireland: Maps and their Makers 1564–1839* (Dublin: Geography Publications, 1997).

Anttonen, Pertti, *Tradition through Modernity: Postmodernism and the Nation State in Folklore Scholarship* (Helsinki, Finnish Literature Society, 2005).

Ardener, Edwin, 'Comprehending Others', in Malcolm Chapman (ed.), *Edwin Ardener: The Voice of Prophecy and Other Essays* (Oxford: Basil Blackwell, 1989).

Ardener, Edwin, 'The Voice of Prophecy: Further Problems in the Analysis of Events', in Malcolm Chapman (ed.), *Edwin Ardener: The Voice of Prophecy and Other Essays* (Oxford: Basil Blackwell, 1989).

Arrowsmith, Aidan, Review of 'F.C. McGrath, (Post) Colonial Drama: Language, Illusion, and Politics (Syracuse: Syracuse University Press, 1999)', *Irish Studies Review*, 9, 1 (2001), p.129.

Asad, Talal (ed.), *Anthropology and the Colonial Encounter* (London: Ithaca Press, 1973).

Ashcroft, Bill, Gareth Griffits and Helen Tiffin(eds), *The Post-Colonial Studies Reader* (London: Routledge, 1995).

Ashcroft, Bill, *et al.* ditto (eds), *Key Concepts in Post-Colonial Studies* (London and New York: Routledge, 1998).

Atchison, J., 'Eton Vale to Bamaga – Place, Geographical Names and Queensland', *Queensland Geographical Journal*, 5 (1990), 1–27.

Aunger, Robert, 'On Ethnography: Storytelling or Science?' *Current Anthropology*, 36, 1 (1995), pp.97–130.

Select Bibliography

Baker, Donald W.A., *The Civilised Surveyor: Thomas Mitchell and the Australian Aborigines* (Melbourne: Melbourne University Press, 1997).
Barnard, Alan and Jonathan Spencer (eds), *An Encyclopedia of Social and Cultural Anthropology* (London: Routledge, 1996).
Barthes, Roland, *Mythologies* (London: Paladin 1973).
Bassnett-McGuire, Susan, *Translation Studies* (London and New York: Methuen, 1980).
Beattie, John, *Other Cultures: Aims, Methods and Achievements in Social Anthro-pology* (London: Routledge & Kegan Paul, 1966).
Bennett, James, 'Science and Social Policy in Ireland in the mid-Nineteenth Century', in Peter J. Bowler and Nicholas Whyte (eds), *Science and Society in Ireland: the Social Context of Science and Technology in Ireland 1800–1950* (Belfast: Institute of Irish Studies, 1997), pp.37–47.
Beverly, John, 'Theses on Subalternity, Representation, and Politics', *Postcolonial Studies*, 1, 3 (1998), pp.305–19.
Bhabha, Homi K., 'Signs Taken for Wonders: Questions of Ambivalence and Authority Under a Tree Outside Delhi, May 1817', in Henry Louis Gates Jr. (ed.) *'Race', Writing, and Differ-ence* (Chicago and London: Chicago University Press, 1985).
Bhabha, Homi K., *The Location of Culture* (London/NewYork: Routledge, 1994).
Bowler, Peter J. and Nicholas Whyte (eds), *Science and Society in Ireland: the Social Context of Science and Technology in Ireland 1800–1950* (Belfast: Institute of Irish Studies, 1997).
Boyne, Patricia, 'Letters from the County Down: John O'Donovan's First Field Work for the Ordnance Survey', *Studies*, (Summer 1984), pp.106–16.
Boyne, Patricia, *John O'Donovan (1806–1861): A Biography* (Kilkenny: Boethius, 1987).
Brantlinger, Patrick, 'Victorians and Africans: the Genealogy of the Myth of the Dark Continent', in Henry Louis Gates Jr. (ed.), *'Race', Writing, and Difference* (Chicago and London: Chicago University Press, 1986).
Brett, David, *The Construction of Heritage* (Cork: Cork University Press, 1996).
Browne, John P., 'Wonderful Knowledge: the Ordnance Survey of Ireland', *Éire-Ireland*, (Spring 1985), pp.15–27.
Budick, Sanford and Wolfgang Iser (eds), The Translatability of Cultures: *Configurations of the Space Between* (California: Stanford University Press, 1996).
Burke, Peter, *Popular Culture in Early Modern Europe* (Aldershot: Scolar Press, 1994).
Burnett, Graham, 'Trig Points', in Rev. Matthew Edney, *Mapping an Empire* (Chicago: Chicago University Press, 1997); *The Times Literary Supplement*, 20 Feb. 1998.
Byrnes, Giselle M., '"The Imperfect Authority of the Eye": Short-land's Southern Journey and the Calligraphy of Colonisation', in Peter Pels and Oscar Salemink (eds), *Colonial Ethnographies, a special edition of History and Anthropology*, 8, 1–4 (Harwood Academic Publishers, 1994), pp.207–35.
Caerwyn Williams, J.E. and Máirín Ní Mhuiríosa, *Traidisiún Liteartha na nGael* (Baile Átha Cliath: An Clóchomhar Tta, 1979).
Cairns, David and Shaun Richards, *Writing Ireland: Colonialism, Nationalism and Culture* (Manchester: Manchester University Press, 1988).
Carbonell, Ovidio, 'The Exotic Space of Cultural Translation', in Román Alvarez and Carmen África Vidal (eds), *Translation, Power, Subversion* (Cleavedon/Philadelphia/Adelaide: Multi-lingual Matters, 1996).
Carr, Edward H., *What is History?* (London: Penguin, 1961).
Carroll, Clare and Patricia King (eds), *Ireland and Postcolonial Theory* (Cork: Cork University Press, 2003).
Carter, Samuel and Anna Maria Hall, *Hall's Ireland: Mr and Mrs Hall's Tour of 1840*, edited by Michael Scott (London: Sphere Books Limited, 1984 [1841]).

Castle, Gregory (ed.), *Postcolonial Discourses: An Anthology* (Massachusetts: Blackwell, 2001).
Chapman, Malcolm, *The Gaelic Vision in Scottish Culture* (Montreal: McGill-Queen's University Press, 1978).
Chapman, Malcolm (ed.), *Edwin Ardener: The Voice of Prophesy* and *Other Essays* (Oxford/NewYork: Basil Blackwell, 1989).
Chapman, Malcolm, *The Celts: the Construction of a Myth* (New York: St. Martin's Press, 1992).
Chesney, Helena C.G., 'Enlightenment and Education', in John Wilson Foster (ed.), *Nature in Ireland: A Scientific and Cultural History* (Dublin: Lilliput Press, 1997), pp.367–87.
Cheyfitz, Eric, *The Poetics of Imperialism: Translation and Colonization from the Tempest to Tarzan* (Philadelphia: University of Pennsylvania Press, 1997).
Cleary, Joe, 'Misplaced Ideas? Colonialism, Location and Dislocation in Irish Studies', in Claire Connolly (ed.), *Theorizing Ireland* (New York: Palgrave Macmillan, 2003).
Clifford, James and George E. Marcus (eds), *Writing Culture: The Poetics and Politics of Ethnography* (Berkeley, CA: University of California Press, 1986).
Cohn, Bernard S., *An Anthropologist Among the Historians* (New York: Oxford University Press, 1990).
Cohn, Bernard S., *Colonialism and its Forms of Knowledge* (Princeton: Princeton University Press, 1996).
Comaroff, John and Jean Comaroff, *Ethnography and the Historical Imagination* (Boulder/San Francisco/Oxford: Westview Press, 1992).
Connolly, Claire (ed.), *Theorizing Ireland* (New York: Palgrave Macmillan, 2003).
Cooper, Frederick and Ann L. Stoler (eds), *Tensions of Empire: Colonial Cultures in a Bourgeois World* (Berkeley, CA: University of California Press, 1997).
Crary, Jonathan, *Techniques of the Observer: On Vision and Modernity in the Nineteenth Century* (London: MIT Press, 1996).
Cronin, Michael, *Translating Ireland: Translation, Languages, Cultures* (Cork: Cork University Press, 1996).
Cronin, Michael, *Irish in the New Century: An Ghaeilge san Aois Nua* (Dublin: Cois Life, 2005).
Crowley, Tony, '"The Struggle Between the Languages": The Politics of English in Ireland', *Bullán: An Irish Studies Journal*, 5, 2 (Winter/Spring 2001), pp.5–21.
Cuddon, J.A., *A Dictionary of Literary Terms* (London: Penguin, 1979).
Cullen, Fintan, *Visual Politics: the Representation of Ireland 1750–1930* (Cork: Cork University Press, 1997).
Cumann Logainmneacha, An, *Logainmneacha as Paróiste na Rinne Co. Phort Láirge* (Dublin: An Cumann Logainmneacha, 1975).
Cunningham, Bernadette, *The World of Geoffrey Keating: History, Myth and Religion in Seventeenth-Century Ireland* (Dublin: Four Courts Press, 2004).
Cunningham, J.B., 'The Letters of John O'Donovan in County Fermanagh: Dogs, Turkeycocks and Ganders', *Ulster Local Studies*, 14, 2 (1992), pp.22–39.
Daly, Mary E., *The Spirit of Earnest Inquiry: the Statistical and Social Inquiry Society of Ireland 1847–1997* (Dublin: Statistical and Social Inquiry Society of Ireland, 1997).
Davis, Thomas, *Literary and Historical Essays 1846* (Washington DC Woodstock Books, 1998).
Day, Angélique, '"Habits of the People": Traditional Life in Ireland, 1830–1840, as Recorded in the Ordnance Survey Memoirs', *Ulster Folklife*, 30 (1984), pp.22–36.
De Alva, J. Jorge Klor, 'Language, Politics and Translation: Colonialist Discourse and Classic Nahuatl in New Spain', in Rosanna Warren (ed.), *The Art of Translation: Voices from the Field* (Boston: Northeastern University Press, 1989).

De Paor, Máire, 'Irish Antiquarian Artists', in Adele M. Dalsimer (ed.) *Visualizing Ireland: National Identity and the Pastoral Tradition* (London: Faber and Faber, 1993).
De Rohan-Csermak, G., 'La Première Apparition du Terme "ethnologie"', *Ethnologia Europaea*, 1, 3 (1967), p.170.
De Valéra, Ruaidhrí, 'Seán Ó Donnabháin agus a Lucht Cúnta', *The Journal of the Royal Society of Antiquaries of Ireland*, 79 (1949), pp.146–59.
Delano-Smith, Catherine and Roger Kain, *English Maps: A History* (London The British Library, 1999).
Derrida, Jacque, 'Des Tours de Babel', in Rainer Schulte and John Biguenet (eds), *Theories of Translation* (Chicago: Chicago University Press, 1992).
Dingwaney, Anuradha and Carol Maier (eds), *Between Languages and Cultures: Translation and Cross-Cultural Texts* (Pittsburgh and London: University of Pittsburgh Press, 1995).
Doherty, Gillian, *The Irish Ordnance Survey: History, Culture and Memory* (Dublin: Four Courts Press, 2004).
Dorson, Richard M., *The British Folklorists: A History* (London: Routledge & Kegan Paul, 1968).
Dundes, Alan (ed.), *International Folkloristics: Classic Contributions by the Founders of Folklore* (Oxford: Rowman & Littlefield Publishers Inc., 1999).
Edney, Matthew, 'Mathematical Cosmography and the Social Ideology of British Cartography 1780–1820', *Imago Mundi*, 46 (1994), pp.101–16.
Edney, Matthew, *Mapping an Empire: the Geographical Construction of British India, 1765–1843* (Chicago and London: University of Chicago Press, 1997).
Edney, Matthew, 'Reconsidering Enlightenment Geography and Map-Making: Reconnaissance, Mapping, Archive', in W.J. Withers and D.N. Livingstone (eds), *Geography and Enlightenment* (Chicago: University of Chicago Press, 1999).
Edwards, Ruth Dudley, 'Preliminary Report on the Ordnance Survey Manuscripts', *Analecta Hibernica*, 23 (1966), pp.272–96.
Ellis, Steven G., 'Writing Irish History: Revisionism, Colonialism, and the British Isles', *The Irish Review*, 19 (Spring/Summer 1996), 1–22.
Englander, David and Rosemary O'Day, *Retrieved Riches: Social Investigation in Britain 1840–1914* (Aldershot: Ashgate, 1998).
Eriksen, Thomas and Finn S. Nielsen, *A History of Anthropology* (London: Pluto Press, 2001).
Estyn Evans, E., *Irish Folkways* (London and New York: Routledge, 1957).
Estyn Evans, E., *Ireland and the Atlantic Heritage: Selected Writings* (Dublin: Lilliput Press, 1996).
Fabian, Johannes, *Time and the Other: How Anthropology Makes Its Object* (New York: Columbia University Press, 1983).
Fabian, Johannes, 'Presence and Representation: the Other in Anthropological Writing', *Critical Inquiry*, 16 (1990), pp.753–72.
Fabian, Johannes, 'Remembering the Other: Knowledge and Recognition in the Exploration of Central Africa', *Critical Inquiry*, 26 (Autumn 1999), pp.49–69.
Fanon, Frantz, *The Wretched of the Earth* (London: Penguin, 1990).
Ferro, Marc, *Colonization: A Global History* (London and New York: Routledge, 1997).
Fitzpatrick, David, 'Ireland and the Empire', in Andrew Porter (ed.), *The Oxford History of the British Empire, Volume 111: The Nineteenth Century* (Oxford: Oxford University Press, 1999), pp.494–521.
Foster, John Wilson, *Fictions of the Irish Literary Revival: A Changling Art* (Gill & Macmillan: Syracuse University Press, 1987).
Foster, John Wilson, *Nature in Ireland: A Scientific and Cultural History* (Dublin: Lilliput Press, 1997).

Foster, Roy, *Modern Ireland 1600–1972* (London: Penguin, 1988).
Friel, Brian, *Translations* (London: Faber and Faber, 1981).
Frykman, Jonas and Orvar Löfgren, *Culture Builders: A Historical Anthropology of Middle-class Life* (New Jersey: Rutgers University Press, 1996).
Frykman, Jonas and Orvar Löfgren (eds), *Force of Habit: Exploring Everyday Culture* (Lund: Lund University Press, 1996).
Fulford, Tim and Peter J. Kitson, *Romanticism and Colonialism: Writing and Empire, 1780–1830* (Cambridge: Cambridge University Press, 1998).
Gailey, Alan, 'Folk-life Study and the Ordnance Survey Memoirs', in Alan Gailey and Daithí Ó hÓgáin (eds), *Gold Under the Furze: Studies in Folk Tradition* (Dublin: The Glendale Press, 1982).
Gailey, Alan and Daithí Ó hÓgáin (eds), *Gold Under the Furze: Studies in Folk Tradition* (Dublin: The Glendale Press, 1982)
Gantz, Jeffrey, *Early Irish Myths and Sagas* (London: Penguin, 1981).
Gates, Henry Louis Jr. (ed.), *'Race', Writing, and Difference* (Chicago and London: Chicago University Press, 1985).
Geddes, Arthur, 'Scotland's "Statistical Accounts" of Parish, County and Nation: c. 1790–1825 and 1835–1845', *Scottish Studies*, 3, 1 (1959), pp.17–29.
Geertz, Clifford, *Works and Lives; the Anthropologist as Author* (Stanford: Stanford University Press, 1988).
Gibbons, Luke, *Transformations in Irish Culture* (Cork: Field Day and Cork University Press, 1996).
Glassie, Henry, *Irish Folktales* (London: Penguin, 1985).
Godlewska, Anne and Neil Smith (eds), *Geography and Empire* (Oxford, UK and Cambridge, MA: Blackwell, 1994).
Graham, Colin, 'Post-Colonial Theory and Kiberd's "Ireland"', *The Irish Review*, 19 (Spring/Summer 1996), pp.62–7.
Graham, Colin, *Deconstructing Ireland: Identity, Theory, Culture* (Edinburgh: Edinburgh University Press, 2001).
Guenthner, F. and M. Guenthner-Reutter (eds), *Meaning and Translation: Philosophical and Linguistic Approaches* (London: Duckworth, 1978).
Hadfield, Andrew and John McVeagh (eds), *Strangers to That Land: British Perceptions of Ireland from the Reformation to the Famine* (London: Colin Smythe, 1994).
Hall, Catherine, *Civilising Subjects: Metropole and Colony in the English Imagination, 1830–1867* (Cambridge: Polity Press, 2002).
Hamer, Mary, 'Putting Ireland on the Map', *Textual Practice*, 3 (1989), pp.184–201.
Harley, John B., 'Irish Geography and the Early Ordnance Survey: A Review', *Irish Geography*, IX (1976), pp.137–42.
Harley, John B., 'Silences and Secrecy: the Hidden Agenda of Cartography in Early Modern Europe', *Imago Mundi*, 40 (1988), pp.57–76.
Herrity, Michael (ed.), *Ordnance Survey Letters Meath* (Dublin: Four Masters Press, 2001).
Herrity, Michael (ed.), *Ordnance Survey Letters Dublin* (Dublin: Four Masters Press, 2001)
Hewson, Lance and Jacky Martin, *Redefining Translation: The Variational Approach* (London and New York: Routledge, 1991).
Hindley, Reg, *The Death of the Irish Language* (London and New York: Routledge, 1990).
Hirsch, Eric and Michael O'Hanlon, *The Anthropology of Landscape: Perspectives on Place and Space* (Oxford: Oxford University Press, 1995).

Hodson, Yolande, *Popular Maps: The Ordnance Survey Popular Edition One-inch Map of England and Wales, 1919–1926* (London, Charles Close Society, 1999).
Honko, Lauri, 'The Folklore Process', *Folklore Fellows' Summer School* (Turku: FFSS, 1991), pp.25–47.
Hostetler, Laura, *Qing Colonial Enterprise: Ethnography and Cartography in Early Modern China* (Chicago and London: University of Chicago Press, 2001).
Hughes, A.J., '"Descriptive Remarks" for the County Armagh Portion of Creggan Parish Contained in the Nineteenth-Century Ordnance Survey Name-Books', *Seanchas Ard Mhacha*, 15, 1 (1992), pp.97–112.
Hultkrantz, Åke, *General Ethnological Concepts* (Copenhagen: Rosenkilde and Bagger, 1960).
Jacob, Christian, 'Toward a Cultural History of Cartography', *Imago Mundi*, 48 (1996), pp.191–7.
Jacquemond, Richard, 'Translation and Cultural Hegemony: The Case of French-Arabic Translation', in Lawrence Venuti (ed.), *Rethinking Translation: Discourse, Subjectivity, Ideology* (London and New York: Routledge, 1992).
Jordan, Glenn, 'Flight from Modernity: Time, the Other and the Discourse of Primitivism', *Time & Society*, 4, 3 (1995), pp.281–303.
Joshi, Svati, *Rethinking English: Essays in Literature, Language, History* (Delhi: Oxford University Press, 1994).
Judd, Denis, *Empire: The British Imperial Experience, from 1765 to the Present* (London: HarperCollins, 1996).
Kanneh, Kadiatu, '"Africa" and Cultural Translation: Reading Difference', in Keith Ansell Pearson, Benita Parry and Judith Squires (eds), *Cultural Readings of Imperialism: Edward Said and the Gravity of History* (London: Lawrence and Wishart, 1997).
Kiberd, Declan, *Inventing Ireland: the Literature of the Modern Nation* (London: Jonathan Cape, 1995).
Kiely, Benedict, *Poor Scholar: A Study of William Carleton* (Dublin: Talbot Press, 1972).
Knight, Roger, 'Colonialism and its Forms of Knowledge', a review of Bernard S. Cohn's work of the same title, *Postcolonial Studies*, 1, 3 (1998), pp.435–40.
Kuklick, Henrika, *The Savage Within: the Social History of British Anthropology, 1885–1945* (Cambridge: Cambridge University Press, 1991).
Kuper, Adam, *Culture: The Anthropologists' Account* (Cambridge, MA: Harvard University Press, 1999).
Kuper, Adam, *The Invention of Primitive Society: Transformation of an Illusion* (London: Routledge, 1988)
Lebow, Richard Ned, *White Britain and Black Ireland: The Influence of Stereotypes on Colonial Policy* (Philadelphia: Institute for the Study of Human Issues, 1976).
Leerssen, Joep, *Remembrance and Imagination: Patterns in the Historical and Literary Representation of Ireland in the Nineteenth Century* (Cork: Cork University Press with Field Day, 1996).
Leerssen, Joep, *Mere Irish and Fíor-Ghael: Studies in the Idea of Irish Nationality, its Development and Literary Expression Prior to the Nineteenth Century* (Cork: Cork University Press with Field Day, 1996).
Lefevere, André, *Translation/History/Culture* (London and New York: Routledge, 1992).
Liebenberg, Elri, 'Mapping British South Africa: The Case of G.S.G.S. 2230', *Imago Mundi*, 49 (1997), pp.129–42.
Livingstone, David N., 'Climate's Moral Economy: Science, Race and Place in Post-Darwinian British and American Geography', in Anne Godlewska and Neil Smith (eds), *Geography and Empire* (Oxford UK and Cambridge, MA: Blackwell, 1994).

Lloyd, David, *Anomalous States: Irish Writing and the Post-Colonial Moment* (Dublin: Lilliput Press, 1993).
Lotman, Juri. M., 'On the Metalanguage of a Typological Description of Culture', *Semiotica*, 14, 2 (1975), pp.97–123.
McDonough, Terence, *Was Ireland a Colony? Economics, Politics and Culture in Nineteenth-Century Ireland* (Dublin: Irish Academic Press, 2005).
McEvoy, John, *Statistical Survey of the County of Tyrone; With Observations on the Means of Improvement. Drawn up in the years 1801, and 1802, for the consideration, and under the direction of the Dublin Society* (Dublin: Graisberry and Campbell, 1802).
McGrane, Bernard, *Beyond Anthropology: Society and the Other* (New York: Columbia University Press, 1989).
MacLaughlin, Jim, *Re-imagining the Nation-State: The Contested Terrains of Nation Building* (London/Sterling/Virginia: Pluto Press, 2001).
MacLeod, Roy, 'On Science and Colonialism', in Peter J. Bowler and Nicholas Whyte (eds), *Science and Society in Ireland: the Social Context of Science and Technology in Ireland 1800–1950* (Belfast: Institute of Irish Studies, 1997), pp.1–19.
McWilliams, P. and Angelique Day (eds), *Ordnance Survey Memoirs of Ireland, Vol.39, Parishes of County Donegal II*; Mid, West and South Donegal (Belfast: Institute of Irish Studies, 1997).
Memmi, Abert, *The Colonizer and the Colonized* (London: Earthscan Publications, 1990).
Monmonier, Mark, *How to Lie With Maps* (Chicago and London: University of Chicago Press, 1991).
Motzkin, Gabriel, 'Memory and Cultural Translation', in Sanford Budick and Wolfgang Iser (eds), *The Translatability of Cultures* (California: Stanford University Press, 1996).
Mudimbe, V.Y., *The Invention of Africa: Gnosis, Philosophy, and the Order of Knowledge* (London: James Curry Indiana University Press, 1988).
Murphy, John A., *The College: A History of Queens University College Cork* (Cork: Cork University Press, 1995).
Murray, Damien, *Romanticism, Nationalism and Irish Antiquarian Societies, 1840–80* (Maynooth: Maynooth Monographs, 2000).
Murray, Peter, *George Petrie 1790–1866: The Rediscovery of Ireland's Past* (Cork: Gandon Editions, 2004).
Myers, Kevin, 'An Irishman's Diary', *The Irish Times*, 17 Dec. 2003, p.17.
NicCraith, Máiréad, *Malartú Teanga: An Ghaeilge i gCorcaigh sa Naoú hAois Déag* (Bremen: Cumann Eorpach Léann na hÉireann, 1993).
Nicolaisen, W.F.H., 'Place-Name Legends: An Onomastic Mythology', *Folklore*, 87, ii (1976), pp.146–59.
Nicolaisen, W.F.H., 'The Past as Place: Names, Stories, and the Remembered Self', *Folklore*, 102, i (1991), pp.3–15.
Niranjana, Tejaswini, *Siting Translation: History, Post-Structuralism, and the Colonial Context* (Berkeley/Los Angeles/Oxford: University of California Press, 1992).
Niranjana, Tejaswini, 'Translation, Colonialism and the Rise of English', in Svati Joshi (ed.), *Rethinking English: Essays in Literature, Language, History* (Delhi: Oxford University Press, 1994).
Nolan, William, *Sources for Local Studies* (Nolan: Blackrock Printers, 1982).
Nolan, William, 'Introduction', John O'Donovan and Eugene O'Curry, *The Antiquities of County Clare* (Clare: Clasp Press, 2003).
Ó Buachalla, Breandán, 'Foreword', *Foras Feasa ar Éirinn: the History of Ireland by Geoffrey Keating*, Vol.1, trans. David Comyn (London: Irish Texts Society, 1987 2nd edn).

Ó Cadhain, Máirtín, 'Béaloideas', in Seán Ó Laighin (ed.), *Ó Cadhain i bhFeasta* (Dublin: An Clóchomhar, 1990).

Ó Cadhla, Stiofán, *Cá bhFuil Éire? Guth an Ghaisce i bPrós Sheáin Uí Ríordáin* (Baile Átha Cliath: An Clóchomhar, 1998).

Ó Cadhla, Stiofán, 'Mapping a Discourse: Irish Gnosis and the Ordnance Survey 1824–1841', in Jamie Saris and Steve Coleman (eds) *Culture, Space and Representation*, a special issue of *Irish Journal of Anthropology*, 4 (1999), pp.84–109.

Ó Cadhla, Stiofán, *The Holy Well Tradition: the Pattern of St Declan, Ardmore, County Waterford, 1800–2000* (Dublin: Four Courts Press, 2002).

Ó Cíosáin, Niall, 'Introduction', in *Poverty Before the Famine: County Clare 1835: First Report of His Majesty's Commissioners for Inquiring into the Condition of the Poorer Classes in Ireland* (Clare: Clasp Press, 1996).

Ó Coileáin, Seán, 'Place and Placename in Fianaigheacht', *Studia Hibernica*, 27 (1993), pp.45–60.

O'Connor, John, *The Workhouses of Ireland: The Fate of Ireland's Poor* (Dublin: Anvil Books, 1995).

Ó Crualaoich, Gearóid, *The Book of the Cailleach: Stories of the Wise Woman Healer* (Cork: Cork University Press, 2003).

O'Curry, Eugene, *Lectures on the Manuscript Materials of Ancient Irish History* (New York: Burt Franklin, 1861).

Ó Danachair, Caoimhín, 'The Progress of Irish Ethnology, 1783–1982', *Ulster Folklife*, 29 (1983), pp.3–17.

O'Donovan, John, *The Genealogies, Tribes, and Customs of Hy-Fiachrach, commonly called O'Dowda's country. Now first published from the Book of Leacan, in the library of the Royal Irish Academy, and from the genealogical manuscript of Duald MacFirbis, in the library of Lord Roden; with a translation and notes, and a map of Hy-Fiachrach*, (Dublin: Irish Archaeological Society, 1844).

O'Donovan, John, *The Antiquities of the County of Kerry* (Cork: Royal Carbery Books, 1983).

O'Donovan, John and Eugene Curry, *The Antiquities of County Clare* (Clare: Clasp Press, 2003, 2nd Edn).

Ó Duilearga, Séamus, 'Notes on the Oral Tradition of Thomond', *The Journal of the Royal Society of Antiquarians of Ireland*, papers in honour of Liam Price, 95 (1965), pp.133–47.

Ó Giolláin, Diarmuid, *Locating Irish Folklore: Tradition, Modernity, Identity* (Cork: Cork University Press, 2000).

O'Halloran, Clare, *Golden Ages and Barbarous Nations: Antiquarian Debate and Cultural Politics in Ireland c. 1750–1800* (Cork: Cork University Press, 2004).

Ó hÓgáin, Daithí, 'Béaloideas – Notes on the History of a Word', *Béaloideas*, 70 (2002), pp.83–98.

Ó Maolfabhail, Art, 'An tSuirbhéireacht Ordanáis agus Logainmneacha na hÉireann 1824–34', *Proceedings of the Royal Irish Academy*, 89(C)(3) (1989), pp.36–66.

Ó Maolfabhail, Art, 'Éadbhard Ó Raghallaigh, Seán Ó Donnabháin agus an tSuirbhéireacht Ordanáis 1830–4', *Proceedings of the Royal Irish Academy*, 91(C)(4) (1991), pp.73–103.

Ó Maolfabhail, Art, *The Placenames of Ireland in the Third Millennium: Logainmneacha na hÉireann sa Triú Mílaois* (Dublin: Ordnance Survey, 1992).

Ó Maolfabhail, Art, 'Eoghan Ó Comhraí agus an tSuirbhéireacht Ordanáis', in Pádraig Ó Fiannachta (ed.), *Ómós do Eoghan Ó Comhraí* (An Daingean: An Sagart, 1995).

Ó Muraíle, Nollaig, 'Seán Ó Donnabháin, "An Cúigiú Máistir"', in Ruairí Ó hUiginn (ed) *Scoláirí Gaeilge, Léachtaí Cholm Cille XXVII*, (Maynooth: An Sagart, 1997), pp.11–82.

O'Toole, Fintan, 'Going Native', in *Black Hole, Green Card: the Disappearance of Ireland* (Dublin: New Island Books, 1994).
Ó Tuama, Seán, *Cúirt, Tuath agus Bruachbhaile* (Baile Atha Cliath: An Clóchomhar, 1991).
Olender, Maurice, *The Languages of Paradise: Race, Religion, and Philology in the Nineteenth Century* (London and Cambridge, MA: Harvard University Press, 1992).
Ordnance Survey of Ireland. An Illustrated Record of Ordnance Survey in Ireland (Dublin: Ordnance Survey of Ireland, 1991).
Pálsson, Gísli (ed.), *Beyond Boundaries: Understanding, Translation and Anthropological Discourse* (Oxford: Berg, 1993).
Pearson, Keith A. Benita Parry, Judith Squires, *Cultural Readings of Imperialism: Edward Said and the Gravity of History* (London: Lawrence and Wishart, 1997).
Pels, Peter, 'The Construction of Ethnographic Occasions in Late Colonial Uluguru', in Peter Pels and Oscar Salemink (eds), *Colonial Ethnographies* (Harwood Academic Publishers) a special issue of *History and Anthropology*, 8, 1–4 (1994), pp.321–51.
Pels, Peter, 'The Anthropology of Colonialism: Culture, History and the Emergence of Western Governmentality', *The Annual Review of Anthropology*, 26 (1997), pp.163–83.
Pels, Peter, 'Professions of Duplexity: A Prehistory of Ethical Codes in Anthropology', *Current Anthropology*, 40, 2 (1999), pp.101–36.
Pels, Peter and Oscar Salemink (eds), *Colonial Ethnographies*, a special issue of *History and Anthropology*, 8, 1–4 (USA Harwood Academic Publishers, 1994).
Pels, Peter and Oscar Salemink, 'Five Theses on Ethnography as Colonial Practice', in Peter Pels and Oscar Salemink (eds), *Colonial Ethnographies* (Harwood Academic Publishers) a special issue of *History and Anthropology*, 8, 1–4 (1994), pp.1–35.
Perry Curtis Jr., Lewis, *Apes and Angels: the Irishman in Victorian Caricature* (Washington and London: Smithsonian Institution Press, 1997).
Phillips, C.W., *Archaeology in the Ordnance Survey 1791–1965* (London, The Council for British Archaeology, 1980).
Porter, Andrew (ed.), *The Oxford History of the British Empire, Vol.111: the Nineteenth Century* (Oxford: Oxford University Press, 1999).
Pratt, Mary L., 'Scratches on the Face of the Country: or, What Mr Barrow Saw in the Land of the Bushmen', *Critical Inquiry*, 12 (1985), pp.119–43.
Prunty, Jacinta, *Maps and Map-Making in Local History* (Dublin: Four Courts Press, 2004).
Raheja, Gloria Goodwin, 'Caste, Colonialism, and the Speech of the Colonized: Entextualization and Disciplinary Control in India', *American Ethnologist*, 23, 3 (1996), pp.494–513.
Raheja, Gloria Goodwin, 'The Ajaib-Gher and the Gun Zam-Zammah: Colonial Ethnography and the Elusive Politics of "Tradition" in the Literature of the Survey of India', *South Asia Research*, 19, 1 (1999), pp.29–51.
Richards, Thomas, *The Imperial Archive: Knowledge and the Fantasy of Empire* (London: Verso, 1993).
Robinson, Douglas, *Translation and Taboo* (Illinois: Northern Illinois University Press, 1996).
Robinson, Tim, *Stones of Aran: Pilgrimage* (London: Penguin, 1989).
Robinson, Tim and Liam Mac Con Iomaire, *A Twisty Journey Mapping South Connemara: Camchuairt Chonamara Theas* (Dublin: Coiscéim, 2002).
Royle, Stephen, 'Irish Manuscript Census Records: A Neglected Source of Information', *Irish Geography*, IXI (1978), pp.110–23.

Sahlins, Marshall, *Islands of History* (London and Chicago: Chicago University Press, 1985).
Said, Edward W., *Culture and Imperialism* (London: Chatto & Windus, 1993).
Said, Edward W., *Orientalism: Western Conceptions of the Orient* (London: Penguin, 1978).
Said, Edward W., 'Afterword. Reflections on Ireland and Post-colonialism', in Claire Carroll and Patricia King (eds), *Ireland and Postcolonial Theory* (Cork: Cork University Press, 2003).
Schacker, Jennifer, *National Dreams: The Remaking of Fairy Tales in Nineteenth-Century England* (Philadelphia: University of Pennsylvania Press, 2003).
Schaffer, Kay, 'Colonizing Gender in Colonial Australia: The Eliza Fraser Story', in Gregory Castle (ed.), *Postcolonial Discourses: An Anthology* (Massachusetts: Blackwell, 2001).
Schama, Simon, *Landscape and Memory* (London: HarperCollins Publishers, 1995).
Schulte, Rainer and John Biguenet (eds), *Theories of Translation: An Anthology of Essays from Dryden to Derrida* (Chicago: Chicago University Press, 1992).
Scríbhneoirí Ban Ros Muc, *Idir Mná* (Conamara: Pléarácha Chona-mara Teo, 1995).
Sengupta, Mahasweta, 'Translation as Manipulation: the Power of Images and Images of Power', in Anuradha Dingwaney and Carol Maier (eds), *Between Languages and Cultures* (Pittsburgh: University of Pittsburgh Press, 1995).
Shanin, Teodor (ed.), *Peasants and Peasant Societies* (London: Penguin, 1998).
Sherzer, Joel, *Verbal Art in San Blas: Kuna Culture Through its Discourse* (Cambridge: Cambridge University Press, 1990).
Smith, Gillian, '"An Eye on the Survey": Perceptions of the Ordnance Survey in Ireland 1824–1842', *History Ireland*, (Summer 2001), pp.37–41.
Smith, Linda Tuhiwai, *Decolonizing Methodologies: Research and Indigenous Peoples* (London and New York: Zed Books Ltd., 1999).
Spivak, Gayatri Chakravorty, 'Can the Subaltern Speak?', in Bill Ashcroft et al. (eds), *The Post-Colonial Studies Reader* (London: Routledge, 1994).
Spurr, David, *The Rhetoric of Empire: Colonial Discourse in Journalism, Travel Writing, and Imperial Administration* (London: Duke University Press, 1993).
Stagl, Justin, *A History of Curiosity: the Theory of Travel 1550–1800* (London and New York: Routledge, 1995).
Stevenson, David, 'Cartography and the Kirk: Aspects of the Making of the First Atlas of Scotland', *Scottish Studies*, 26 (1982), pp.1–12.
Stocking, George W. Jr., 'What's in a Name? The Origins of the Royal Anthropological Institute (1837–71)', *Man: the Journal of the Royal Anthropological Institute*, 6, 3 (1971), pp.369–90.
Stocking, George W. Jr., *Victorian Anthropology* (NewYork: The Free Press, 1987).
Stoeltje, Christie Beverly J., 'The Self in "Fieldwork": A Methodological Concern', *Journal of American Folklore*, 112, 444 (1999), pp.158–82.
Swift, Cathy, 'John O'Donovan and the Framing of Early Medieval Ireland in the Nineteenth Century', *Bullán: An Irish Studies Journal*, 1, 1 (1994), pp.91–103.
Thomas, Nicholas, *Colonialism's Culture* (Cambridge: Polity Press, 1994).
Thompson, Stith, *Four Symposia on Folklore* (Westport, CT: Greenwood Press, 1953).
Tierney, Michael, 'Eugene O'Curry and the Irish Tradition', *Studies,* (Winter 1962.), pp.449–62.
Tonkin, Elizabeth, Maryon McDonald and Malcolm Chapman (eds), *History and Ethnicity* (London and New York: Routledge, 1989).
Turnbull, David, 'Cartography and Science in Early Modern Europe: Mapping the Construction of Knowledge Spaces', *Imago Mundi*, 48 (1996), pp.5–24.

Van Maanen, John, *Tales of the Field: On Writing Ethnography* (Chicago: The University of Chicago Press, 1988).
Venuti, Lawrence, *Rethinking Translation: Discourse, Subjectivity, Ideology* (London: Routledge, 1992).
Venuti, Lawrence, 'Translation as Cultural Politics: Regimes of Domestication in English', *Textual Practice*, 7, 2 (1993), pp.208-23.
Venuti, Lawrence, *The Translators Invisibility: A History of Translation* (London and New York: Routledge, 1995).
Vicziany, Marika, 'Imperialism, Botany and Statistics in Early Nineteenth-Century India: the Surveys of George Buchanan 1762-1829', *Modern Asian Studies*, 20, 4 (1986), pp.625-60.
Walsh, Rev. Paul, *The Placenames of Westmeath* Dublin: Institute for Advanced Studies, 1957).
Warren, Rosanna, *The Art of Translation: Voices from the Field* (Boston: North-eastern University Press, 1989).
Welch, Robert, *A History of Verse Translation from the Irish 1789-1897* (London: Colin Smythe, 1988).
Welch, Robert, *The Oxford Companion to Irish Literature* (Oxford: Clarendon Press, 1996).
West, Andrew, 'Writing the Nagas: A British Officers' Ethnographic Tradition', in Peter Pels and Oscar Salemink (eds) *Colonial Ethnographies* (Harwood Academic Publishers), a special issue of *History and Anthropology*, 8, 1-4 (1994), pp.55-89.
Whelan, Kevin, 'Beyond a Paper Landscape – John Andrews and Irish Historical Geography', in Kevin Whelan and F.H.A Aalen (eds), *Dublin City and County: From Prehistory to Present* (Dublin: Geography Publications, 1992), pp.379-424.
Whyte, Nicholas, *Science, Colonialism and Ireland* (Cork: Cork University Press, 1999).
Williams, Raymond, *Keywords: A Vocabulary of Culture and Society* (London: Fontana Press, 1976).
Withers Charles W.J. and David N. Livingstone (eds), *Geography and Enlightenment* (Chicago: University of Chicago Press, 1998).
Wolfe, Patrick, '"White Man's Flour": Doctrines of Virgin Birth in Evolutionist Ethnogenetics and Australian State-formation', in Peter Pels and Oscar Salemink (eds), *Colonial Ethnographies* (Harwood Academic Publishers) a special issue of *History and Anthropology*, 8, 1-4 (1994), pp.165-205.
Wood, Denis, *The Power of Maps* (London: Routledge, 1992).
Wooding, Jonathan M., *The Otherworld Voyage in Early Irish Literature* (Dublin: Four Courts Press, 2000).
Young, Robert J.C., *Postcolonialism: An Historical Introduction* (Massachusetts: Blackwell, 2001).

Index

Abbey Gormigan 186
Abbey of Ballintober 198, 199
ab origine 203
aborigines, aboriginal 103–6, 119, 138, 160, 166, 173, 203, 205–14, 236
Aborigines Protection Society of London 52
Abraham 54, 55
academia, Irish 92
'Académie Celtique', Mémoirs of 55
accent 104, 106, 116, 128, 194, 196
acculturation 115
accuracy 75, 77, 148
Achebe 41
Act of Union, the 53, 54, 62, 73, 85, 88
administration, imperial 88
aesthetics 222
   elitist 108; modern 184
Africa 88, 121
   travellers to 60; explorers of 111
Africanism 83
Africanoid 43
Africans 40
Aghacon 195
Aghadowney, Parish of, Co Derry 113, 210
Aghagallon 1
Aghalurcher, Parish of, Co Fermanagh 116, 120, 246–7
Aghanloo, Parish of, Co Derry 114, 125, 206
Aghalurcher, Co Fermanagh 110
Ahanloo, Parish of, Co Derry 127
Aher, David 135
Ahern 133
aide-mémoire 102
Aileach 151
Alexander 4
Alexandria 13, 178
allegory, allegorization 41
Al-Shabab 224, 244
Álvarez 74
'*am Faoilteach problem*', the 76–83
America 14, 41, 85, 197, 231
American Ethnological Society, the 52
American Indians 40, 179
American natives 53
Andrews, John H. 9, 19, 25, 28, 34, 73, 107, 137, 221, 235
Anglicization 14–5, 26, 28, 47, 73, 81, 83, 88, 91, 92, 148, 172, 181, 189, 196, 202, 204, 214, 218–50
anglocentric, anglocentrism 3, 171
Anglo-Irish 54, 122
Anglo-Norman(s, the) 193, 198
anglophile, anglophilic 47, 227
anglophone 8, 174, 232

(Anglo-)Saxons 43, 176
anthropocentric, humanism 102
anthropology, anthropological 10, 13, 32, 42, 62
   name coined 44
Anthropological Institute of Great Britain and England 52
Anthropological Society of London 52
anthropologist(s) 26, 188, 190, 202, 229–33
Antipodes, the 13
Antiquarian Department, the 21, 23, 25, 27, 103, 107, 136, 146, 170, 220, 232, 238, 240
antiquarian, antiquarianism 4, 13– 35, 45, 46, 47, 50, 57, 63, 67, 83, 89, 142, 144, 148, 149, 153, 154, 173, 174, 175, 202, 228
   Vallancey school of, the 174–7
antiquaries 48, 54, 55, 78, 90, 145, 154, 170, 171, 172, 177, 178, 198, 212
antiquities 4, 17, 20, 33, 42, 48, 51, 53, 54, 96, 147, 176, 203
Apocalypse 137
Antrim Memoirs 29
Arabic, translations from 78
Aran Islands 26, 188, 191, 231
Aranmore 188
archéocivilisation 53, 170
archaeology, archaeological, archaeolologist(s) 28, 32, 53, 78, 227
archaize, archaization 227, 243, 250
archive, archives 32, 65, 83, 102
   colonial 93
Ardee 136
aristocracy 114
arrivistes 91
ars apodemica 55
ars memorativa 102
Artrea, Parish of, Co Derry 118, 120, 140
artefacts 6, 54, 233
Ashcroft 174
Asia 88
Assfeenan 160
assimilation 96, 172, 200, 250
Atateemore 26
Athboy 185
Athlone 164
Aughlish 104
Aunger 75
Australasia 88
Australia 56, 81
autobiography, genre of 102

Babbage, Charles 50
Bailey, C., Lieutenant 108
baile biataigh 230

266 Index

baile fearainn 230
ballad, political 23
Ballina, Co Mayo 219
Ballinvicar 161
ballybetagh 230, 231
Ballybofea 160
Ballybriest Glen 191
Ballyconnell 111
Ballyheelan 243
Ballyhillan 243
Ballymartin, Parish of, Co Antrim 109, 138
Ballymullins district 104
Ballynadrimney, Parish of, Co Kildare 150
Ballynascreen, Parish of, Co Derry 114, 117, 122, 155
Ballyrobert, Grange of, Co Antrim 109, 115
Ballysadare 189
Ballyscullion, Grange of, Co Antrim 126
Ballywater, Grange of 138
Balteagh spade, the 125
Banagher, Parish of, Co Derry 106, 142, 157
Bantry Commons, Co Wexford 137
barbarism, barbarity 67, 76, 84, 221, 237
  vs civilization 42, 250; vs correctness 221
barbarous, barbarian 14, 15, 243, 244
Barons of Tooráá, the 190
Barony of Ida 236
Barony of Ofa 236
Barrett, John 127
barriers, cultural 31
Bartlett, Richard 15
Beal tinne 59
Bealach an Chaoláin 243
béaloideas 2, 47, 48; *see* folklore
Bealtaine/ Bealtinne 66
béarla 41; *see* language, English
Beds of Dermot, the 159
behaviour 85, 113, 141
  habitual 59; sexual 127
bélaiteas 47; *see* folklore
Beauford, William 176, 177
Beaufort, Louisa C. 146; *see* Miss Beaufort
Beaufort, Daniel Augustus 102
Beneda glen, the 119
Bengal 51, 57
Bennada Glen151, 165
Bennett, George A., Lieutenant 19, 104, 117, 203
Beranger, Gabriel 42
Berlin 32
Berry, Corporal 137, 241
Betham, Sir William 54–5, 146
Bhabha 94, 204
bilingualism 2, 3, 92, 105
Bin Bulbin (mountain/ fort) 162,196
Bínn Bulbain 196
Black Hole of Knockfierna, the 133
Blackwater 199
Blackwoods 86
Bleakley, J. 113, 123, 127, 138
Bloody Foreland 160
Bloomfield, Major 190

Blount, Charles 15
blue books, the 50
Board of Ordnance18, 220
  Master General of 1
Boazio, Baptista, mapmaker 15
Bogoidigh 194
Boldero, W. G., Captain 25
Book of Common Prayer, the 105
Book of Revelations, the 137
Bopp, Franz
  The Conjugation System of the Sanskrit Language 180
Boteler, R., Lieutenant 114, 241
boundaries 16, 18, 24, 34, 103, 133, 137, 172, 221, 232, 241
  cultural and racial 119; linguistic 245; national 59
Boundary Commission 18
Boundary Surveyors 135, 241, 242, 243
bourgeois, ideas 67
bourgeoisie 66, 114
Bourne, Henry, *Antiquitates Vulgares* 54
Bovevagh, Parish of, Co Derry 106, 111, 142
Boyle, James 109, 114, 115, 117, 118, 119, 120, 122, 124, 126, 138, 139, 141, 143, 188, 189
Boyne, John 73
Brackagh Glen 191
Bran, *Voyage* 13
Brand, Reverend Joseph, *Popular Antiquities* 54
Brantlinger 121
Brash, Richard R. 147
Brendan, *Navigatio* 13
Brewer, James, *Beauties of Ireland* 17
Brian, time of 158
Brian Man Mullen 164
Bristol 43
Brissue-na-scaribheshoilse 207
British Act of Union with Ireland 84
British Association 23
  for the Advancement of Science 49, 50
British East India Company 56
British Museum 22
British Union 54
Brett 107
Brooke, Charlotte, 174
  Reliques of Ancient Irish Poetry 54
Buchanan, Francis 51
Buchanan, George 54
Buck, Henry 221, 241
buí 241
buildings 20, 26
  Anglo-Norman 198
Bunting, Edward 174
'Bureau de Statistique' 49
bureaucracy, imperial 220
Bush, John, *Hibernia Curiosa* 57
Busteed, Thomas 22

cabin, the, as sign of Irishness 164
Caddamstown, Parish of, Co Kildare 150
Caesar 4

Index

caisiol 159
calendar(s) 123, 219
  Gregorian 76; popular 139; Standard English 76; vernacular 190
Camden, William, Britannia 54
camera, the 33
Camlin, Parish of, Co Down 111
campaigns, military 15
Campbell, Iain, of Islay' 45
Campion, *History of Ireland* 219
Canada Act 84
Cape, peoples of 88
capitalism 16
Carbry's Island 198
Carleton, William 96, 199
Carlow 187
carn 136
Carnanbare 154
Carmnoney, Parish of, Co Antrim 139
Caroline, William 235
Carr 9
cartography, cartographic 5, 6, 10, 13, 14, 15, 21, 28, 33–5, 73–97, 103, 174, 218, 223, 233; *see* map
Cashel, wall of 209
Castle Caldwell 32
Castle Dillon, Loughgall 108
Castlebellingham 192
Castlecomer, Co Kilkenny 164
Castlekeeran, Church of 199
catalogue, cataloguing 20, 21, 54
cathair 159
Catholic Emancipation 23
  movement for 17
Catholic University 22
Catholics 45–6, 141
Cavan 138, 147
Ceathramhadh na Madadh 196
Celt, Celtic 43, 160, 164, 222
Celtic Fringe, the 41, 55, 124
Celtic revival 46
Celtic Society, the 46
Celticism 56, 77, 83, 89
Celticist(s) 47, 124, 142, 178, 233
Celtomania 56
census 51
  Belgian 50
Central Statistics Office 1
cess (Collector) 92, 181, 206
character 59, 124, 127, 163
Charter of Londonderry 240
Charter of Newry 27
Chaytor, John, Lieutenant 112, 125, 139, 141, 191, 241
Chesney 62
Chetwood, William Rufus, *A Tour Through Ireland* 57
Chieftain Fights 163
Church of England 242
churches 149, 152, 176
  Scottish 105
Civil Survey, the 49, 86, 219

civilization 10, 17, 25, 30–1, 41, 42, 47, 59, 66, 92, 93–4, 97, 110, 121, 122, 133–66, 158, 160, 163, 166, 166, 173, 179, 223, 231, 235, 250
  vs savagery 173; *see* barbarism; and environment/ climate 111; debate about 56; earlier phases, reconstruction of 48; signs of 105; standards of 67
Claddagh 188
Claggan Glen 191
clans, tribal 119
Clare 204; see County Clare
class 45, 84, 107, 113, 120, 126, 126, 127, 156, 230
classifying, classification 32, 41, 62, 65, 104, 175
cleanliness 66, 121, 125, 126, 127, 129
  moral problem of 119
Cleary 85
Clifford 82
Cloch an Ghadaí 248
Cloch an Phoill 210
clock, the 114; *see* time
Cloghagaddy 248; *see* Thief's Stone, the 248
Cloghaneely, Parish of 160
Clogher, Parish of, Co Tyrone 121, 123, 140, 185
Clondermot, Parish of 206
Clonmany, Parish of, Co Donegal 140, 160
Co Antrim 109, 114, 115, 117, 119, 120, 122, 124, 125, 126, 127, 138, 139, 141, 143, 221
Co Armagh 108
Co Carlow 144, 146, 156, 175, 210, 238
Co Clare 137, 184
Co Derry 1, 20, 105, 111, 113, 114, 117, 118, 119, 120, 122, 123, 125, 127, 128, 133, 135, 136, 140, 142, 150, 152, 155, 165, 191, 192, 196, 206, 207, 208, 210, 211, 222, 238, 241, 242
Co Donegal 111, 112, 127, 140, 160, 161, 163, 164, 184, 185, 190, 191, 199, 205, 206, 207, 209, 243
Co Down 110, 111, 117, 135, 139, 203
Co Dublin 191
Co Fermanagh 110, 111, 116, 120, 139, 163, 185, 190, 206, 209, 243, 246, 247
Co Galway 149, 187
Co Kerry 149, 161
Co Kildare 150, 181, 194, 211, 238
Co Kilkenny 137, 164, 173, 177, 182, 236
Co Laois 186, 199
Co Longford 156, 183
Co Louth 185, 191, 194, 211
Co Mayo 175, 188, 191, 198, 211, 219, 238
Co Meath 156, 162, 183, 184, 187, 199, 210, 220
Co Monaghan 206, 243
Co Offaly 187, 193, 195, 199
Co Roscommon 156, 159, 180, 181, 183, 191, 193, 200, 240, 241
Co Sligo 161, 182, 188, 191, 194, 196
Co Tyrone 105, 110, 121, 140

Co Waterford 119
Co Westmeath 157, 164, 183, 185, 193,195, 197, 238
Co Wexford 137, 235
Co Wicklow 189
cock-fighting 123, 140
code, cultural 76
Cohn 87
coimgne (comheagna) 47
Colby, Thomas, Colonel 18, 19, 25, 28, 49, 50, 59, 126, 219, 245
  Instructions for the Interior Survey of Ireland 218
Coleridge, Samuel Taylor 84
Colgan 153, 220
collecting, collection, collector(s) 4, 16, 25, 29, 32, 33, 44, 45, 50, 57, 61, 94, 102, 156
Colleges (Ireland) Act, the 46, 51
colo 128
colonia 128
colonial, colonialist, colonialism 9, 10, 14, 22, 31, 34, 35, 41, 43, 46, 53, 56, 59, 74, 81, 83, 84, 86, 87, 96, 114, 122, 127, 129, 145, 157, 203, 205, 232
  vs imperialism 86; cultural 33; regime 79, 89, 96, 184
  concept of 90; critique of 93; culture of 73–97; interpretation of 89
colonization 15, 42, 85, 88, 91, 129, 229
  invoked cause of improvement 115, 116; justification for 120
colonized, the 3, 29, 84
colonizer 17, 29, 84, 89, 156, 203, 204, 223, 226, 234
  civilizing 121; values and aesthetics of 129
colony 8, 88, 89, 93–4, 126, 184, 203
Colum Cille 13
Columbus 41
commentaries 27, 48
commission, commissioners 25, 45, 94
committee on public income and expenditure 19
comparative philology 180
compilation, encyclopaedic 102–3
Cona 142
Conloch 142
Conmaicne Mara 184
Connacians, the 196
Connaught 90, 163, 196, 207
Connemara 4, 6, 134, 184, 239
  An Spidéal 26
conquest 14, 15
  justification of 94; Elizabethan 14–5; Norman 14
construction, social 33, 65
consumerism 16
Continent, the 13, 17
control 34, 77, 234
  governmental 19, 47
controversy, religious and political 24
Conwall 206, 207
Conyngham, William Burrton 42

Copernicus 13
Cook, James, Captain 42, 134
Coote, Charles
  Statistical Survey of the Queen's Country 176
Corick 191
Cork 1, 51, 57
Cork Cuvierian Society for the Cultivation of Sciences 46
Cormac 153
Cormac MacAirt 151
Corofin 62
cosmology, popular 247
cottage, the 119
countryside 4, 27, 119, 165, 231
County Clare 22, 61, 62
County Roscommon 81
Cranfield Holy Well 144
craniology 42
Crary 30
Creagh, the, Co Derry 118
creative, creativity 148, 150, 152, 224, 234
Croker, Thomas Crofton, 133, 134, 142–3, 174, 175, 199
  Researches in the South of Ireland 17, 40
  Fairy Legends and Traditions of the South of Ireland 143
cromlechs, archaeological account of 159
Cromwell, Oliver 15, 240
Cromwell, T. K., *Excursion through Ireland* 17
Cronin 227
Crooke, William 59
Crossmolina 191
Crown Bridge 139
Cuarsan 198
Cuchulin 142
Cuddon 102
cúige 230, 131
Cullen 40
culture, cultures 2, 3, 4, 8, 18, 35, 47, 58, 77, 82, 90, 96, 106, 113, 162, 195, 218–50
  Celtic 56; dying 94; elite 153; European 14, 107; indigenous 47, 203; Irish 21, 28, 56, 77, 94, 138, 146, 155, 166, 170, 172, 175–6; defining of 96, 144; material 117, 125; native 4, 213; oral 123; Oriental 56; peasant 247; popular 5, 20, 144, 152; primitive 16, 142, 235 – see primitivism; primordial 202; receptor/ target vs source 226, 233, 234; traditional 123; vernacular 5, 7, 29, 117, 123, 135, 138, 144, 155, 166, 185, 212, 234, 239; visual 107; Western 108
  of colonialism 73–97; appropriation of 228 ; interpretation of 10; contact between 134; vs nature 173
curiosity 13, 41, 73, 78, 155, 173
Curristown, Co Meath 210
custom, customs 48, 51, 57, 60, 63, 147, 157, 166
  and beliefs 49; see manners, habits
Cuvier, Georges 46

Dagda 178, 180

Daguerre, Louis 184
dancing 140, 208
  masters vs farmers 178–80
Danes, the 159, 238
data 9, 29, 166
  collection of 75, scientific 50; data pilgrimage 57, 136
Davis, Thomas 221
Dawson, Robert Kearsley, Captain/ Lieutenant 29, 119
Day 73
de Chevannes, Alexandre César 52
de Gérando, Joseph Marie 49
  Considérations sur les méthodes à suivre dans l'observation des peuples sauvages 55
De hÓir 235
de la Ramée, Pierre 52; *see* Petrus Ramus
de Santarém, Viscount 33
Declinatio Rustica 103
Deire 152
Delves-Broughton, Lietenant 238
Delves-Broughton, W. E., Lieutenant 140, 207
demesnes 34, 242
demos, modern terms derived from 53
Department of Science and Arts, the 19
Dermot, warrior 151
Derrida 91, 218–50
Derry 104, 145, 152, 206, 208, 218–50
Derry Memoirs 29
Derrygonneely 140
Derrysherridan 159
description 2, 27, 33, 40, 41, 44, 45, 48–55, 64, 96, 107, 108, 112, 114, 115, 120, 122, 123, 152, 162, 164, 196, 218, 245
  archaeological 28; cultural 3, 4, 15–6; ethnographic 15–6, 108, 123; evolutionary 32; romantic 108; territorial 17
desert, cultural 157
Desertegney, Parish of, Co Donegal 111, 112
Desertlyn, Parish of, Co Derry 128
Desertmartin, Parish of, Co Derry 117, 122, 128, 136
Devereaux, Robert 15
Devil's Bit, the 187
diaries 61, 86, 102, 189; *see* field-diary
dictation 241, 243
Dicuil 13
Dinnsheanchus 159
Dinnshenchas, the 152
diorama, the 184
discourse 5, 6, 9, 34, 35, 62, 64, 76, 112–13, 122, 143, 145, 224, 230
  academic 7; Celticist 96; colonialist 20, 67, 83, 84, 89, 93, 94, 97, 106, 110, 122, 153, 155, 160, 173, 178, 204, 212, 213, 227, 228, 234, 235; cultural 7; ethnographic 8, 124, 148; ethnological 95; evolutionary 20, 78, 90, 113, 124, 203; folkloristic 148; historical 8; scientific 76; vernacular 141, 146, 148, 151, 152
Disraeli, Benjamin 86
disturbances, agrarian 137

divisions, territorial 229, 230
Doherty 91
dominance, linguistic and cultural 91
domination 81, 85, 228
Donacavey, Co Tyrone 123
Donagh, Parish of, Co Donegal 140
Donaghpatrick 220
Domaghmore 206
Donegal 15, 23, 66, 121, 239
Donegal Memoirs 29
Doonaha 22
Dorson 45
Doyle, Roddy, *The Commitments* 43
Dowds 119
Down, County of 28
Down Survey, the 15, 49, 86, 102, 219, 240
Downes, G. 105, 207
Downing Street 25
Draper's Company 114, 128
drawing, tuition in 24, 182
drinking 128
  at funerals 123
Drinkwater, John Elliot 50
Drogheda 136
Dromin, Parish of, Co Louth 211
Drumachose, Parish of, Co Derry 105, 125, 135
Drumgath, Parish of, Co Down 117
Drummaul, Parish of, Co Antrim 119, 141
Drummond 135
Drummond, Thomas, Lieutenant 20
Drumshambo 207
Drumsna 207
Dublin 1, 20, 23, 27, 46, 57, 59, 136, 149, 171, 183, 186
  Coal Quay 163; Grafton Street 20; Great Charles Street 27, 196; Phoenix Park 19
Dublin Corporation, the 223
Dublin Penny Journal 27
Dublin Society, the 19, 176; *see* Royal Dublin Society
Dublin Statistical Society, the 49
Dublin University Magazine 56, 175
Duke of Cumberland, the 15
Duke of Wellington, the 18, 19
  government of 23
du Noyer, Georges Victor 26, 107, 182
Dún Laoghaire 223
Dunaghy, Parish of, Co Antrim 120, 125
Dunboe, Parish of, Co Derry 123
Duneane, Parish of, Co Antrim 122, 126, 127, 143
Dunfierth, Parish of, Co Kildare 150
Dungiven, Parish of, Co Derry 106, 119, 125, 128, 133, 136, 142, 151
Dunglow 161, 185
Dunleer 185
Dunquin 161
Dunree Fort 111
Durnford, E. W., Lietetnant 111, 247
Dutton 177
  Statistical Survey of the County of Galway 176

Eagleton, Terry 5
Early English Inquisitions 240
early modern 16
Easter Rising, the 27
economy, agricultural 117, political 87
Edgeworth, Maria 105
Edgeworth, William 3, 135
Edinburgh 53, 179
Edinburgh Review 40, 44, 86
Edney 107, 108
education 45–6, 57, 63, 73, 94, 107, 113, 115, 126, 163, 166, 232
Edwards, W. F. M. 52
eile 13
Éire 57
Elizabethans, the 41
elves 143
Emmet, Robert 40
Empire, the 4, 7, 8, 14, 15, 16, 17, 21, 22, 23, 31, 34, 57, 59, 60, 61, 64–5, 74, 77, 82, 83–97, 134, 135, 136, 137, 212, 228, 229
encyclopaedism, encyclopaedic 34, 55, 66, 102–3, 202
England 13, 18, 49, 60
  colonial 32; and Ireland, difference between 3, 43, political and cultural relations 17
English, the 14, 125
English Law Deeds 219
Englishness 43, 59, 111, 119
English rule, the 15
Enlightenment, the 31, 34
Enniskillen, Co Fermanagh 112, 114, 139
Ennismurray 42
Ennistymon 62
entertainment, popular 110
Eriksen 41
Erin, tales of 142
Erne-head Lake, the 249
Errigal 152
Erse 103; *see* language(s), Irish
erudition 77
Essex's massacre 2
ethnocentric 33, 34, 120
ethnocentrism 81
ethno-folklore 60
Ethnological Society of London, the 44, 52
ethnologia 52
ethnologie, appearance of 52
ethnology, ethnological, ethnologist 3, 4, 8, 10, 11, 16, 26, 31–3, 40–67, 141, 144, 146, 163, 171, 190
  term use 52; theory of 81; *see* ethnography, folklore
ethnográphia 52
ethnographie, appearance of 52
ethnography, ethnographical, ethnographer/ist 2, 3, 4, 5, 6, 8, 17, 20, 21, 24, 25, 31, 41, 48, 52–3, 73–97, 103, 107, 120, 212, 223, 233, 245
  colonial 8, 124; origins of 56; Irish, path breakers in 178; ethnographer as mediator 226, personality of 75; *see* ethnology, folklore

ethnos 44–5
  ethnos-terms 52
etymologist(s) 80, 173, 220, 235
etymology 26, 152
Europe 20, 41, 52, 53, 56
European grand tour 107
Evans, Estyn 9, 231
Everest, George 51, 58
evidence 5, 178
  documentary 9; primary 6
evolution of man 120
evolutionism 63, 96
exactness 78; *see* accuracy
exile 13
exoticism 78
expansion
  capitalist 87, 96; colonial 84; economic 87
expansionism 85
expedition 13, 41–2, 47
expertise 54
exploitation 120
exploration, explorer(s) 32, 55, 57, 73, 102, 121, 160, 171, 172, 203

Fabian 34, 45, 52, 60, 111, 127, 166, 233, 236
Fabri, Johann Ernst 52
'fack' 125
'Fadh' surname 119
fair 140, 190
fairies 6, 133, 141, 142, 143, 144, 147, 148, 157, 175, 176
Fairy Legends and Traditions of the South of Ireland 17
Faithleg, Co Waterford 236
famines of the 1890s 6
Fanon 92
Faraday Wheel, the 110
Faroe Islands, the 13
Farrer, James Anson 60
fashions 30
Father Ted 5
Faughallstown 157
Faughanvale, Parish of 122, 206
Feary, James 151
Fellenberg, Philipp Emanuel von 126
fence, the 231
Fenian tales/ cycle 237
  characters of, in place-names formation 152; *see* place names
Fenwick, Lieutenant 81
Ferguson, Sir Samuel 27, 175
  Congal: a Poem in Five Books 175
  Hibernian Knights Entertainments 175
feudalism, colonial 86
Fiannaíocht 142, 237
field, fieldwork 4, 10, 20, 22, 27, 28–9, 62, 170–214, 184––90, 192, 205, 212
  experience 194
Field Day Theatre Company of Derry, Guild Hall 218
field-diary 62, 189
field-notes 106, 189, 195

fieldworkers 10, 24, 26, 63, 82, 86, 135, 144, 165, 171, 173, 182, 189, 190, 195, 207, 213, 241
Fin Mac Cool 152, 154, 238
Fin McCoul 141
Finn MacCool 246
Fiodh Ard 236
Fion MacComhal 142
Fionn Mac Cumhaill 152
Fionn MacCool 151
fios 134; *see* power, supernatural
Fir Bolg 202
First Report of his Majesty's Commissioner for Inquiring into the Condition of the Poorer Classes in Ireland of 1835 51
Fitzpatrick 88
Flannery 235
Flood, Henry 47
folklore 3, 4, 6, 8, 16, 17, 45, 50, 51, 52, 61, 65, 74, 94, 96, 133, 138, 139, 141, 146, 147, 156, 170, 171
  emergence of 40–67; term coined 53; Irish 2, 4, 7, 11, 26, 29, 83, 123, 173, 174; *see* ethnology, ethnography
Folklore Comission, the 170
Folklore of Ireland Society, the 61
folklorist(s) 4, 26, 188
folkloristics 48, 56
foreigners, British officers as 112
fort 159
Foucault, Michel 34
France 88
Frederick III 32
Friel, Brian 221, 223, 224
  Translations 9, 73, 218, 222
Frizell, Peggy 1
Frykman 119, 156

Gaédhil nobility 163
Gaeltacht 162
Gael, Gaelic 4, 124, 238
  law 15
Gaeldom 3
Gaelic Society of Ireland, the 46
Gailey 73
Galbally 5
Gall, Franz 42
Gal(l)oon, Co Fermanagh 247, 248
Galway 14, 181, 187
game of common 140
Gambian River, the 40
Garvagh 150, 210
Gatterer, Johann Christoph 45
Gaul, Son of Morni 142
Gatterer, Johann Christoph 45
gaze, the 107, 108, 204
Geeadore 160
Geertz 82
genealogy 157
General Post Office, the 27
genre 16, 41, 62, 102, 120, 142
Gentlemen's Stewards 137
gentry, the 34, 96, 242

Geographical Societies, Britain 44
geography, geographic 13, 44, 45
  mythological 152
Gerald of Wales 5, 17, 41, 56, 127; *see* Giraldus Cambrensis
Germany 53
ghosts 141, 144, 149, 157
giants 143, 154, 157, 159
Giraldus Cambrensis 5, 48, 219
  Topographia Hiberniae (Topography of Ireland) 14
Glean Finne 160
Glen Fin 161, 208
Glen Gavlen/ Glengavlen 162, 189
Glen Togher 162
Glenavy, Parish of, Co Antrim 117
Glenbush, valley of 118
Glencolumbkille 161
Glenconkeine 155
Glendalough 189
Glenomara, Co Clare 137
Glensoolie 206
Glenuller 237
Glinsce 239
glossaries 63
Goethe, Johann Wilhelm 56
Goldsmith, Oliver 28
Goll Mac Morna 154
Gothic, Gothicism 55
Goulburn, Henry 23, 25
Gough, *Camden* 219
governance 15, 18, 28
government 15, 26, 91, 136
  British 20, 84; colonial 14, 58; imperial 21, 162
grammar 105
Granges 149
Grania 159
Graves, Charles, Rev. Dr 147
Great Trigonometrical Survey of India 58
Greatorex, J., Lieutenant 110, 116, 118, 120, 139–40, 141, 246
Griffith, Arthur 27
Griffith, Richard 18, 137, 221, 241
Griffith's Boundary Survey 166, 229
Grimm, Jacob 49, 63, 143
Grimm, Wilhelm 63, 143
group interviews 191

habits 143, 166
  of the people 10, 40–67
habitus 59
Hackett, William 147
hagiology 141–4, 175, 213
Hall 60
Hall, Samuel Carter, Mr and Mrs 107
Halloween 5
Hamer 34, 35, 73
'Hamish' surname 119
Hammond, Andrew 121
Hancock, William Neilson 22
Hannyngton, T. C. 109
Hardiman, James 27, 124, 181
  Irish Minstrelsy 56

Harley 34
Harvey, James 187
Hawaii 134
hegemony 33, 91, 225
Henry VIII 14, 40
Herder, Johann Gottfried von 56
hereditary, crime 63
heredity 42, 59
Hemans, G. W.,, 122, 126, 127, 143
Hibernian Antiquarian Society, the 42, 46
Hibernian tongue, the 103
Hiberno-Romanesque revival 26
Highlands, Scottish, the 17
Hill of Beinnin, the 160
Hill of Howth, the 20
Hill of Mullaghcormick, the 200
Hill of Wonders, the 210
Hilltown 109
Historical Department, the 21
historical–antiquarian vs quantitative–statistcal 25
historiography 46, 47, 174
history 4, 28, 51, 93, 145, 150, 157, 158, 161, 173, 225
  conjectural 179; Irish 176–7; national 16; of ideas 10
Holy Scriptures 91
holy wells 149, 172
Homer 44
Honan Chapel 26
Hong Kong 88
Honko 65
Horn Head 209
Hostetler 212
Hottentot, the 166, 204
house, thatched, the 119–20
  as sign of Irishness 119
House of Commons 24, 85, British 18
House of Lords, British 53, 85
humanism 102
humanitarianism 85
Hungry Hill 20
Huxley, Thomas H. 59
hybridization, cultural 92
Hyde, Douglas 45

IarConnaught 182
Iberno-Celtic Society, the 46
identity 3, 28, 56, 174, 235
  cultural 19; Irish 10
ideology 10, 62, 63, 96, 121
illness 186
imagination 65, 155
imitation 30, 33
  objective 34
imperialism 6, 83, 85, 90, 225
improvement 108, 110, 112, 113, 114, 118, 122, 126, 161, 162, 163, 232, 249
  and colonization 115, 116
India 33, 41, 51, 58, 118, 136
India Act 84
indigenous 16, 29, 95, 96, 174
Indomania 56

Industrial Revolution, the 60, 113
industrialization 30
inferiority 93, 105, 124, 138, 203
  cultural 64; mental 43; racial 113
informants 172, 190, 191, 193, 195, 199, 205, 207, 208, 213
information 1, 20, 24, 29, 30, 65, 83, 102, 107, 108, 124, 136, 145, 170, 194, 202, 211, 223
Inishowen mountains 113
Inishkeel, Parish of, Co Donegal 127
Inishmacsaint, Parish of, Co Fermanagh 140
Innishcollen 246
Innishmurray island, Co Sligo 161, 162
inquiry 31, 32, 62–3, 179, 182
  disinterested scientific 179; statistical 25; heads of 58, 61, 108, 140
intellectualism 124
intellectuals, German 63
intelligentsia, Irish, the 13
intelligibility 78
interlanguage 250
interpretation 3, 10, 29, 35, 42–3, 47, 74, 224, 234, 237, 246, 248, 250
invasion 57
inventory 29
  cultural 28
Iran 57
Ireland 7, 17, 40, 89, 121, 172
  Africanized 84; ancient and new 35; Anglicized 94, 171; (archaic) primordial 166, 241; Indianized 84; pre-academic pre-independence 226; precolonial civilized vs postcolonial uncivilized 154; rural 51
  Chief Secretary of 23; conquerors of 9; economy of 88; images/ representations of 10, 16, 22, colonialist 17, primitive 65, British market for 22; interpretations of 79; moral depravity of 23; population of 112; scenery of 107
  discourse on 5; Poor Law legislation for 50; unrest in 88; as colony 84; and the Empire 83–97; see England
Irish, the 3, 5, 9, 22, 23, 67, 118, 124, 127, 128, 145, 248
  as barbarians 14; as barbaric/ exotic natives 41; as noble savages 122; as outsiders 59 exotic/ romantic 56; Milesians 177; and the English 60, 106, 177; and the Scots 104, 117, 125, 128, 138, 139, 177
Irish Aborigines, the 103
(Irish) academic disciplines, emergent 153, 179
Irish Academy, the 9, 213; see Royal Irish Academy
Irish acre, the 231
Irish Annals, the 236, 240
Irish Antiquarian Research, the 177
Irish Archaeological Society, the 92
Irish Biblical Society, the 105
Irish Channel, the 88
Irish Constabulary, the 94
Irish cry, the 140
Irish Folklore Commission, the 5, 61, 62

Index

Irish Free State, the 61, 62
Irish Record Office, the 27
Irishness 5, 10, 119, 164, 178, 205, 213
Irwin, Rev. Mr 192
ithir 13

Jack Daly 193
Jacobites, the 15
Jacquemond 78
Jamaica 66
Jervis, Thomas 126
Jews, the 248
Johnson, Samuel 42
  Dictionary 219
Jones, Richard 50
Jones, Sir William 56, 137
Journal of Rebellion, the 219

kaffir, the 166, 204
Kanneh 123
Kavanagh, Patrick 3
Keane, John B. 4
Keating, Geoffrey 46, 48, 159
'keenie' 140
Kells 185
Kelly, D. H., Dr 191, 192
Kennedy 3
Kenny, Peter 189
Keogh 231
Kerkland, M. M. 110
Kerry 51
Kiberd 223
Kilbroney, Parish of, Co Down 109, 135
Kilcar, Co Donegal 161
Kilcooley, Parish of, Co Roscommon 159
Kilcronaghan, Parish of, Co Derry 117, 122
Kildysart Workhouse 22
Kilglass 200, 201, 202
Kilkee 22
Kilkeel 109
Kilkenny 26
Kilkenny Archaeological Society 51
Killcarney 164
Killererin, Parish of 149
Killeshin, Co Carlow 175, 176
Killimor 186
Killymard, Parish of 66
Kilmacumshy 190
Kilronan 188
King, Rev. Gilbert 105
King's County 187
King's Evil, the 127
knowledge 15, 17, 29, 30, 31, 93, 94, 96, 134, 151, 171, 202, 223, 226
  classified 44; cultural 47; dispersed 194; encyclopaedic 34, 147, 179; normative 205; unified 65; vernacular 147, 174, 208
  creation/ production of 34, 225; genre of 62; transmission of 225
Kollár, Adam Frantisek 52
Kuper 31

Labby 145, 208

Lactus Vetuli 249
Lake District 17
Lake of the Calf, the 249
lament for the dead 140
Lancey, W., Lieutenant 66, 105, 111, 112, 140
land 13–35, 134–8
land measure(s) 230–1
landscape 17, 34, 106, 109, 111, 112, 123, 133, 134, 135, 150, 154, 160, 162, 163, 172, 223, 232, 249, 250
  cultural vs natural 110; of symbols 229
landschap 106
Lang, Anderw 60
language, languages 2, 7, 20, 30, 46, 52, 74, 89, 96, 102–29, 160, 170, 225, 226, 237, 250
  and race 17; families of 54
  ancient 160; living 4, 78, 92, 166, 227, 233, 235, 239; target / source 76, 78, 228, 235, 237, 246
  Celtic 54, 160, 179; Czech 52; Dutch 106; English 5, 7, 8, 9, 14–5, 22, 42, 45, 47, 56, 63, 76, 77, 80, 91, 97, 156, 214, 229 – proper names translated into 205; Gaulish 54; Irish 3, 4, 7, 9, 13, 14, 21, 22, 27, 29, 45, 47, 57, 63, 80, 89, 90, 91, 97, 104, 106, 134, 166, 178, 193, 207, 212, 222, 235, 236, 239, 240, revival of 7, Chair in 47, marginalization of 226, *see* Milesian Irish, Erse, Hibernian; Italian 107; Hungarian 52; Latin 27; Scottish 156, 193; Scottish Gaelic 240; Welsh 54
Larcom, Sir Thomas, Major General 18–9, 21, 23, 24, 25, 28, 29, 49, 50, 56, 58, 137, 140, 186, 204, 205, 221, 236, 242, 243, 245
Lawton, James Anthony 49
Layd, Parish of, Co Antrim 141
learning, Irish 21–2, 27, 55, 170
Leckpatrick, Parish of, Co Tyrone 113
Ledwich, Edward 55, 146, 176, 177
Leersen 4, 83, 94
legend(s) 157, 158, 159, 172, 199, 208, 240, 248
legitimacy 84, 223, 224
Leiden 32
leírscáil 13
Leitrim 147, 207
letter(s) 4, 5, 20, 22, 24, 28–9, 32, 83, 86, 91, 110, 138, 146, 147, 156, 160, 164, 166, 170, 171, 172, 184, 187, 189, 203, 212, 223, 245, 247–8
Letterbaily 210
Letterkenny 207
Lewis, Samuel, *Topographical Dictionary* 51
lexicography 26
Leyden, Michael 189–90, 221
Lhuyd, Edward, *Archaeologia Britannica* 54
library 29
life 7, dimensions of 6
  cultural 7, 8, in Ireland 3; economic 63; rural, in Britain 113
lifestyle 20, 60, 117
Ligar, C. W. 113, 114, 119, 123, 125, 128, 133

Limavaddy 206
limelight, invention of 20
Linaskea 110
linguistic difference 106
Lis Molaise, Parish of 186
Lisburn 116
Liscanor Bay, place names in 209
Lissan, Parish of 128, 191
literacy 17
literature, of confutation 173; popular 143
Living Island, the 199
*Loch Gamhn*a 249
Löfgren 119, 156
Loghermore 197
Longfield, Co Tyrone 110, 113
London 5, 15, 16, 19, 32, 57, 80, 85, 128
London Hibernian Society, the 91
London Statistical Society, the 49, 50
Longfield, Co Tyrone 110, 113
Longley 223
Looee, Olympic Games of 220
Loop Head, County Clare 22
Lord Anglesey 1
Lord Bacon 177
Lord Clare 85, 136
Lord Roden 192
Lord Stanley 23
Lord Templetown 109, 115
Lord–Lieutenant 88
Lorum 146
Lotman 95
Lough Boderg 200
Lough Clugacommen 239
Lough Derg 66
Lough Erne 192
Lough Foyle 20
Lough Key 156
Lough Lagan 201
Lough Melvin 163, 206
Lough O'Gara 187
Lough Ree 164, 191
Lough Sheelin 187
Lough Sheelion 147, 159
Loughgall 108
Loughguile, Parish of, Co Antrim 118, 141
Loup, the, Co Derry 118, 119
Luddite natives 2
Lurg, Co Fermanagh 206
Lyle's Hill 109
Lynch, Patrick, *Life of St Patrick* 159
Lyons 238

Macaulay 6
MacFlail, Bostoon 22
Mackenzie, Colin 51
MacLeod 10, 29, 92
Macosquin 210
MacPherson 55
MacShane 200–3
Magaghran 158–9
Maghera, Co Derry 152, 192
Magherafelt estate 119
magic, vernacular 133
Magician Maugraby 178

Magilligan 206
Maguire, Patrick 209–10
Malin, Fanny 1
Malinowski, Bronislaw 62
Mallusk, Grange of, Co Antrim 114
Malthus, Thomas Robert, 63, 179, 213
  Essay on Population 50
Malthusian 177
Manchester 6
Manchester Statistical Society, the 49, 50
Mangan, James Clarence 26
manners 42, 47, 63–4, 105, 116, 122, 157
  and customs 50, 54, 63–4, 121, 225,
  collection of 53, description of 25; and
  language 105
Mansfield, George D. 113
manuals 63
  anthropological 55; encyclopaedic 55
Manus 151, 152
manuscripts 27, 47, 54, 102, 170, 171, 178,
  210, 235
Maoris, the 88
map, mapping 5, 6, 10, 15,16, 17, 18, 19, 20,
  25, 33, 55, 62, 64, 83, 102, 103, 107, 108,
  133, 135, 165, 172, 184, 187, 219, 221,
  223, 225, 227, 233
  of Ireland 107, earliest 13; Grand Jury maps
  219; Norden, Beaufort and Mercator maps
  219; accuracy of 33; *see* cartography
mapa 13
mapmakers 15, 17, 218, 247
mappa 33
Marcus 82
marginal(s), marginality, marginalization 79–80,
  91, 96, 174, 204, 226, 238
Martin, R. 139
Maskelyne, Nevil 57–8
Mason, William Shaw 51, 102
  Parochial Surveys 51
McCloskey, clan of 119
McCloskey, John 104, 105, 106, 117, 122, 142
Me, ancient kingdom of 157
Mead, Margaret 62
meaning, meanings 34, 74, 79, 80, 113, 134,
  135, 152, 175, 196, 208, 227, 240
measurement 20, 103; *see* land measure(s)
meerings 18; *see* boundaries
Memmi 92
memoir, memoirs 5, 10, 20, 24, 33, 58, 73, 83,
  102, 103, 104, 106, 107, 110, 112, 118,
  119, 122, 138, 203, 206, 207, 221, 223,
  245, 246, 247
  writers of 105, 109, 115, 118, 119–20, 123,
  124, 125, 126, 128, 146, 166
  see aide memoire
Merly 208
metalanguage 96
metaphor 11, 34, 76, 222, 226, 227
method 6, 25, 170–214
methodology, methodological 7, 10, 30, 32,
  55, 61, 62, 154, 171, 173, 182, 190, 212,
  220, 241
microscope, the 33, 34

Mídhe 157; *see* Me
Midlands, the 91
Milesian Irish 103
Milesians 160, 177, 190, 193, 203
mimic man, the 203–5
mimicry 21, 170–214
Miss Beaufort 26; *see* Beaufort
Mithchell, Sir Thomas 56, 59
Mobwee Glen 191
modern(ity) 31, 134, 233
 concept of 6–7; Irish 6
modernizing 30, 134, 247
Mohill 207
Molière 177
Molyneux, Sir Thomas, Lieutenant General 108
Monaghan town 190
Monaster Evin, Co Kildare 211
Moneyhanegan 242
Moneyneeny, Co Derry 210
monolingualism 3
Monmonier 33
Montesquieu, Baron de 42
monuments 42
Moore 175
moral philosophy 180
morality 113; *see* rectitude
Morni 142
Mount Brandon 20
mountaineers 103–6
Mountjoy 25
Mountjoy, Baron 15
Mountjoy Barracks 19
Mountjoy House 19
Moy-Iha 161
Muckish 160
Mudge, William 57, 59
Mudimbe 83
Mullaboy 241
Munich 32
Munster 43, 57, 91
museum 29, 32, 102
 ethnographic 32
Museum of Irish Industry, the 19
'Museum voor Volkenkunde' 32
Myres, Kevin 3
Mysore, India 51
mythology 157, 232
 Irish 57

name(s)
 archaic 239; personal 235; name books 244, 245
naming 223
Napoleon Bonaparte 49, 56
národpis 52; *see* ethnography
nation 10, 15, 77, 92, 124, 229
nation-state 16, 61, 89, 126
National Museum of Ireland, the 33
nationalism, nationalist(s) 222, 234
natives 42, 51, 103–6, 122, 135, 161, 223
natural history 31, 44, 48–55, 62, 67
natural philosophy 177

natural sciences 62, 75, 180
Navan 137
Neeny Roe 141
Nephin mountain, the 191
New Statistical Account of Scotland 49
Newbawn, Co Wexford 235
Newry, Co Down 139
Niall of the Nine Hostages 151
Nicholson, Francis 17
Nicholson, Marianne 17
Nicolaisen 229
Niranjana 74, 81, 234, 235
Normans, the 14–5
North West Farming Society, the 66, 105, 113, 120–1, 127

objectivity 34–5, 74, 147, 177, 212
O'Brollaghans, tribe of 145
observation 31, 32, 34, 40, 45, 48, 117, 122, 162, 172, 190, 203, 212
observer, the 30, 31, 34, 60, 62, 66, 96, 106, 108, 110, 124, 137, 154, 204, 233
Ó Buachalla 48
Ó Cadhain, Máirtín 89, 230
O'Carolan, Turlogh 32, 43
oceanus 13
Ó Cíosáin 51
Ó Coileáin 152–3, 180
O'Connell, Daniel 22, 86, 137
O'Connell's Tower 137
O'Connor, Cathal Croibhdhearg 199; *see* O'Connor, Charles the Redhanded
O'Connor, Charles 146, 174
O'Connor, Charles the Redhanded 198–9
 civilization of 198
O'Connor, Roger 177
O'Connor, Thomas 26, 136, 137, 146, 150, 164, 171, 183, 185, 186, 191, 192, 194, 196, 198, 204, 205, 211, 219, 238, 241
O'Curry, Eugene 21–2, 22–6, 45, 61, 62, 137, 146, 164, 165, 171, 175, 181, 182, 191, 204, 205, 210, 237
Ó Danachair 57
Ó Donnabháin, Seán 222; *see* O'Donovan, John
O'Donnell, Owen Hugh 218; *see* O'Donnovan, John
O'Donovan, John 11, 21–2, 26, 27, 28, 32, 42, 45, 46, 48, 56, 61, 62, 73, 77, 81, 122, 137, 138, 139, 142, 143, 144, 145, 146, 147, 150, 151, 152, 153, 155, 156, 157, 160, 164, 165, 166, 170, 171, 172, 173, 174, 175, 176, 177, 178, 179, 180, 181, 182, 184, 185, 186, 188, 189, 190, 191, 191, 192, 194, 195, 196, 197, 198, 199, 200, 202, 204, 205, 206, 208, 209, 211, 212, 218, 219, 220, 221, 222, 227, 231, 232, 235, 236, 238, 240, 241, 242, 244, 245, 247, 248
 records quarried by 219; translation strategy of 236–7
 Annals of the Four Masters 77, 219
 The Annals of Kilronan 219
 The Book of Clann Firbis 219

The Book of Fenagh 219
The Book of Kilkenny 219
The Book of Lecan 219
The Book of Lismore 219
The Tripartite Life of St Patrick 219
Leabhar na gCeart (book of rights) 219
Ó Duilearga, Séamas 61, 62, 156, 170, 204
O'Felme, Tulach 146
O'Flanagan, Father Michael 27
O'Flanagan, Theophilus 174
O'Flynn, Edmund 211
Ó Giolláin 32, 173
O'Gorman, Maurice 47, 174
O'Halloran 56
   General History of Ireland 54
   Ierne Defended 54
oilithreacht 13
O'Kearney 147
O'Keefe, Patrick 26, 136, 150, 171, 186, 188, 191, 192, 194, 196, 204, 205, 211, 221
Ó Laoide 223
Ó Maolfabhail 219, 221, 240
Ó Mathuna, Tadhg 22
Ó Muraíle 91
O'Neill, Con 190
O'Neill, Hugh 15, 40
O'Nolan 175
O'Rahilly, Michael Joseph 27
O'Reilly, Edward 26, 46
O'Reilly, Myles John 27, 185, 220
Ó Ríordáin, Seán 90
Ó Súilleabháin, Handbook of Irish Folklore 61
orality 237
ordenance 1
ordinare 1
ordnance, word derivation 1
Ordnance Survey 9, 17, 18, 24, 40, 46, 61, 62, 73, 135
   British 19, 49; and British view of ethnology 52; of Ireland 1, 19, 23, 49, 181
   headquarters of 1, 19, 22, 27, 171, 173, 186, 189; office of 15; protagonist of 57
Orient, the 56
Oriental Studies 78
Orientalism 57, 77, 83, 89
   Irish 56
Orientalist(s) 47, 78, 233
orthodoxy, religious 32
Orthographical Department, the 21, 22
orthographer(s) 80, 164, 235, 241
orthography 5, 26, 28, 81, 104, 171, 218, 220, 221, 222, 240, 241, 243, 244
osmosis 9
Ossian 56, 141, 142, 150, 238
Ossianic poems 55
Ossianic Society, the 46
other, the 13, 16, 18, 35, 95, 166, 229, 233
otherness 9, 84, 229, 250
Otherworld, the 13
outsider(s) 195
   British officers as 112
Ovid, *Metamorphoses* 237
Oxley, John 59

padhsán 104; see peasant
   see payson, pays, pysaum
paganus 103
paint, painting 16, 17, 106, 107
Papists 24
Paris 47, 52
Parliament 190–3, 210
   British 18, 19; Irish 83
pathology 127
Patrick's Day 143
patriotism 85, 157
   Irish 25
patronage 24
   imperial 22
paysan 103
pays 103
peasant, peasantry 4, 112, 142, 145, 154, 160, 163, 173, 203, 206, 236, 246, 247
   beliefs of 6, pagan 142; word derivation 103
Peel, Robert 20, 23, 25, 46, 94
Peelers 94
Pels 8, 52, 66, 73
Persia 56, 57
Petrie, George 24, 26, 27, 33, 55, 61, 62, 107, 146, 147, 157, 159, 160, 182, 221
   Description de l'Egypte 33
   Ancient Music of Ireland 22–3
   and art lessons 24
Petty, William 49, 102, 230
Philological Societies, Britain 44
philologist(s) 78, 227
philosophy 44, 223
photography 184
phrenology 42–3, 46
Physico-Historical Society, the 46, 51
physiognomy, physiognomic 42, 127, 128, 163, 202, 223
physiology 44
picturesque(, the) 129, 160, 163, 164, 184
   aesthetics of 107; conventions of 108
   see ithir, exile
Pictet, Adolphe 179
pilgrim(s), pilgrimage 13, 57, 66, 136, 199
   see ithir, exile
Pinkerton, John 158
Pit-Rivers, Augustus 59
Pitt, William 85
pittoresco 107
Plains of Tailteann, the 162
plantation 14
political economy 179
politics 17, 74, 86
   cultural 90
Poor Law 50, 85
Port Laoise 223
Portlee 119
portrait 16, 18–9, 20
Portlock, J. E., Colonel/ Captain 32, 46, 182
postcolonial 22, 90
power, powers 31, 34, 63, 74, 82, 85, 87, 95, 96, 97, 108, 223, 224, 226, 250
   colonial 81, 93; imperial 93; political 77; supernatural 134

Index

power relations 8, 30, 74, 79, 108, 225
place 102–29, 163, 194
  as property 231
(place) names 4, 5, 25–6, 27, 28, 61, 73, 81, 83, 137, 139, 144–5, 148, 152, 153, 157, 158, 180, 189, 195, 202, 204, 206, 208, 209, 210, 211, 214, 218, 222, 223, 225, 227, 229, 232, 234, 235, 236, 237, 238, 239, 240, 243, 245, 250
  assimilated 200; pronunciation of 171
Placenames Commission, the 223
Pococke, Richard, *Tour in Ireland* 57
potato, the
  as sign of Irishness 164; and Utilitarianism 179
practice, practices
  political 80; scientific 19
Pratt 64
Presbyterian(s) 81, 105, 128, 138, 210
print, medium of 16
Prichard, James Cowley 43, 53, 179, 213
  Researches into the Physical History of Man 53
  An Analysis of the Egyptian Mythology 53
  The Eastern Origins of the Celtic Nations 53
primitive(, the) 95, 161, 166, 173, 178, 234
  evolutionary category of 149
primitivism, cultural 63, 138
Príosún an Dubhaltaigh 201
pronunciation (Irish) 193, 194, 222, 236
progress 7, 30–1, 41, 42, 44, 76, 94, 95, 105, 113, 118, 121, 136, 158, 166, 209
prose 5, 20, 225, 234, 249
Protestant(s) 24, 46, 116, 120, 128, 176, 210, 235
  vs Catholic(s) 25, 105, 125; elite 54; unionist(s) 175
pseudo-science 43
Ptolemy, Claudius 13
Punch 86
Puritan 46
pysawn 104; *see* peasant

questionnaire 45, 51, 53, 55, 61
Queen Taillteann 220
Queen's College Cork 46; *see* University College Cork
Queen's County 186
Quetelet, Adolphe Lambert 63
  Essai de Physique Sociale 50
Quilkagh 162

race, racial(ist) 17, 18, 32, 43, 45, 53, 63, 84, 93, 102–29, 138, 142, 156, 163, 178, 203
  assumptions 106; difference 60, 88; instinct 59; jargon 145; origin 94
Raheja 205
Ramism 52
Ramus, Petrus 52
Ranke, Leopold von 6
rath 159, 220
Rath Aodha Mic Bruic 158
Rathlin Island, the 2

Ratoath 158
Raymoghy, Parish of 67
rebellion 17, 235
rectitude, moral
  and grammar/ pronunciation 105
reform 32, 113
  rhetoric of 94
Reform Act 85
reformers, imperial 126
refusal 165, 210; *see* resistance
Reid, Major 19
Der Reisende 51
relations
  cultural 17; Irish–British 212
relativism 82
  cultural 60
religion 20, 45, 88, 103, 113, 121, 125
Rennell, James 57
Report of the Commissioners appointed to inquire into the facts relating to the Ordnance Memoir of Ireland 25
reports 63, 102, 104
representation 6, 8, 10, 16, 17, 34, 56, 74, 107, 122, 123, 148, 223, 232
  canon of 79; mode of 225
repression, repressive 88, 224
rescue, rescuer(s) 95, 96, 123, 229–30
research 24, 40, 41, 54, 73, 75, 144, 175, 178, 202
  evolutionary English antiquarian and ethnographic 145; antiquarian/ ethnological/ ethnographic175, 181, 213
resistance 15, 89, 136, 137–8; *see* refusal
revolt 17
rhetoric, rhetorical 90–7, 136, 145, 164, 178, 208
  colonialist 17; imperial 140, 173
Rice, Thomas Spring 18
Richards 43, 57, 80, 136, 229
Ring, Co Waterford 119, 236
River Bann, the 143
River Murray, Australia 56
Robinson, Tim 4, 239
Rock of the Wilks, the 239
Roman Catholic(s) 24, 103, 118, 119, 120, 123, 124, 125, 140, 143–4
romantic movement 56–7
Roscommon 201; *see* County Roscommon
Ross Goill, Co Donegal 209
Rostrevor 109
Rotterdam 42
Roy, William, General 57
Royal Artillery, British 5, 18, 28
Royal Asiatic Society, the 44, 56
Royal College of Science, the 19
Royal Dublin Society, the 46, 51
Royal Engineers, British 5, 18, 20, 28, 61, 81, 103, 135, 136, 137, 190, 237, 247, 248
Royal Geographical Society, the 44, 93
Royal Irish Academy, the 22, 27, 46, 54, 182;
  see Irish Academy
Royal Society, the 57
'Ruadh' surname 119

Sadler, Michael Thomas 63
  Ireland: its Evils and their Remedies 50
Sahlins 7
Said, Edward 77, 83, 86
  Orientalism 94
Saldanha 28
Salemink 52
St Barry 201
St Enda 26
St John's Eve 59, 123, 143
S(ain)t Kevin's Shrine 164, 189
St Kieran 199
St Luroch's Well 192
St Molaise162, 186
St Patrick 144, 159, 190, 200, 246, 247
St Ronan 246
saints 13, 149, 192, 203
  Irish 158; names of 227
Sampson, George 102
Sandby, Thomas 15
savagery 123, 173
  anthropological ideas of 40
'savages' 41
scáird 196
Scárdán 196
Scarve Solus, battle of 208
scenery 102–14, 164
  romantic 107; and features of local economy and farming 108
Schama 15, 164
scholar(s), scholarship 46, 55
  Irish 5, 21, 26, 46, 54, 61, 151, 172, 175, 181, 210, 220, 232, 241; Gaelic Irish 45; German 52, 77; Scottish Gaelic 76
  vs romancers or shanachies 151
schools 63, 113, 114, 174–7
Scormore 190
science 6, 10, 18, 28, 44, 57, 77, 82, 92, 133, 134, 136, 156, 166
  colonial 29, 92; evolutionary 17, 20, 146, 190, 228, 246; Graeco-Roman 13; imperial 7, 17, 44, 58, 64, 133, 146, 165; popular 53; socio-cultural 40
  of the savages 2; of art/ art of 4, 9; and morality 127; *see* natural sciences, social sciences
Scot(tish/ ch) 24, 49, 106, 114–5, 116, 117, 125, 126, 129, 210
Scott, George 116, 124
Scotland 13, 15, 29, 45, 119
Scotticisms 106
Scor 196
scrofula 127
Sculboge 235
seanchas, seanchaí, Seanchie 47, 147, 156; *see* tradition
Sedwick, Adam 50
Seraphic Doctors, the 177
serfs, serfdom 85, 116
shanachie, shanachies 147, 151, 152, 154, 160, 190, 192, 237
Sharkey 208

Sheep Haven 209
Sheeptown 139
Sheridan 159
sí 143; *see* fairies
Síol Anmchadha 186
Sinclair, Sir John 29, 48, 56
  The Wealth of Nations 48
  Statistical Account of Scotland 49, 56
Sinn Féin 27
Sjoestedt 152
Skerry, Parish of, Co Antrim 114
sketch, sketches 17, 42, 182
Skurmore 196
Slanes, Parish of, Co Down 110
Slieve Baan 164
Slieve Russel 111
Slieve Snaght 162
Sligo 189
Sméaróid 150
Smith, Adam 42, 48, 63
  Smith, Charles 51
social sciences 75
'Société Ethnologique de Paris' 52
society, societies 179
  anarchic 139; British 85; colonial 104; colonized 93; industrial bourgeois 42; *see* bourgeoisie; learned 102; preindustrial European 104; rural 63, 104; scientific and scholarly 44, 46
South Asia 107
South Wales 59
Southey, Robert 84
Southampton 245
Spain 88
specificity, cultural 81
spelling 218, 221, 241
  Anglicized 26, 243; *see* Anglicization, orthography
Spurr 112, 238
Spivak 6
Spring Rice Commission, the 135
Spurzheim, J. C. 42
squandry, the 96; *see* gentry
Stage Irishman, the 3
Stagl 57, 102
Standard English 3
standardization 80, 240, 243
Stanhihurst 48
state formation 93
state nomadology 57
Stokes, J. 110, 113, 114, 123, 125, 128, 207
Stotherd, R., Lieutenant 121, 140
surveillance 94
  of state 34, 40; *see* control
Statistical Surveys, the 176
statistics , statistical 30, 31, 48–9, 55, 63, 67, 103, 107
  and travel 52
steam engine, invention of 126
stereotype, ethnic/ racial 66, 67, 128
Stevenson 33
Stewart, Dugald 53, 178, 179, 213

Index 279

Edinburgh lectures in moral philosophy 179
Stocking 41, 63
street nameplates, bilingual 223
Strokestown 200, 201
Stuart *see* Stewart
subjection 74–5, 84
subjectivity 35
sublimity 164, 184
superiority 43, 223
  cultural 42
superstition(s), superstitious 42, 47, 57, 66, 86, 95, 121, 133, 135, 136, 140, 142, 147, 148, 161, 165, 176, 203, 246, 247
surveiller 1
survey, 3, 4, 7, 9, 10, 11, 17, 18, 20, 21, 77, 93
  cartographic 57; ocular 172; topographical 18, 20
  accounts of, contemporary 170; archive of 5; authority of 212; completion of 22; Dublin agents of 20; equipment of 135; Irish agents of 59; lexicon of 144, 204; origin/ provenance of 9, 10; scientific aims of 59; scope of 25; technique of 6
  of India 126, 136; of Ireland 18, 45, 51 – British 53; of Scotland 29
  and classification 141; relations with the public/ London/ the Topographical Department 165–6; word derivation 1;
  see Ordnance Survey
surveyor(s) 2, 7, 9, 18, 20, 60, 78, 94, 96, 119–20, 133, 230
  civilian 59; nomadic 133–4; story about 133; *see* Boundary Surveyors
Sykes, W. H. 50
Synge, John Millington 26

Taghadoe 238
Tahiti 42
Tamlaght Finlagan, Parish of, Co Derry 111, 206
Tanderagee, Co Armagh 108
Tanderagee Castle 108
Tara 4
taste 17, 108, 110, 155
  English literary 106
taxation 18, 32, 34, 48, 88, 120, 162, 192, 229
  of Pope Nicholas 219
taxonomy 45, 65
Taylor, Edgar, *German Popular Stories* 143
Taylor, Sir Edward Burnet 60
Taylor, P., Lieutenant 140
technology 2, 4, 7, 20, 30, 66, 82, 117, 228
TeePetrie 27, 196
teleological 93
teleology, imperial 82
Telltown, Co Meath 220
Temple, Richard C. 59
Templecarn, Paarish of 66
Templemore, Co Derry 20
tenants 85
tensions 21, 22, 27, 108, 144, 146, 173, 236, 241

cultural 23; political 23; racial 23, 210; religious/ sectarian 23
tenure
  of land, system of 18
terra nullius 112
text 106, 248, 250
  and audience 245
thaumatrope, the 110
theodolites 2, 4, 20
theory 7, 10, 178
  cultural 7; ethnological 61
Thief's Stone, the 248
Thomas 92
Thomond 62
Thoms, William 51, 53
  Notes and Queries in Anthropology 57
thought, evolutionary 10; *see* Western
Tigh Tuaith 238
time 194
  historical 3
  new concept of 114; peasant view of 114
Times, the 86
Tithe affairs 211
Tobar Galáin Coilleáin 239
Tobar Gamhna 248, 249
Tobergollankillane 239
Tom Byrne 189
tomb of oblivion, the 156–60
Toneduff/ *tóin dubh* 244
Toomregan, Parish of, Co Fermanagh 111, 139, 140
topography, topographic 13–35, 42, 48, 54, 108, 151, 152, 157, 161, 173, 181, 223, 239
Topographical Department, the 21–2, 23–4, 103, 107, 136, 146, 166, 170, 220, 232, 238, 240
toponymy 223, 237
Tories 23, 24, 137
tourism, tourist(s) 17, 22, 107, 164, 184
Tower of London 1
townland, townlands 18, 28, 34, 58, 76, 137, 190, 196, 208, 218, 222, 229, 230, 231, 232, 240, 242, 245, 248
tradition 7, 8, 10, 47, 62, 63, 65, 67, 82, 86, 92, 96, 115, 133–66, 170, 172, 211
  classical 151; ethnological, ethnographical 10, 48; incoherent 95; national 5; oral 29, 54, 144–7, 166, vs written/literate 153, 159, 223, reliability of 147; prehistoric 157; vulgar 144–7
  rhetoric of 97, 178; as euphemism for primitiveness 166
traditional(ity) 173
  vs modern(ity) 6–7
'traditionary tales' 17
transactions and proceedings 102
Transactions of the Kilkenny Archaeological Society 51–2
translation 3, 6, 8, 10, 11, 17, 21, 29, 35, 47, 73–97, 106, 124, 174, 205, 212, 218–50
  five stages of 244
translator(s) 80, 171, 172, 234

transliteration 243
transparency 224–9, 250
travel, travelling 17, 32, 52, 55–8, 67, 73, 170, 202
  genre of 107; travel guides 17, travel reports 45
traveller(s) 57, 60, 102, 109, 160
  handbooks of 107; see explorer(s)
travelogue(s) 60
tree, the 111, 119
  as sign of civilization 110
Trevelyan, Charles 234
triangulation 19–20, 33, 34
Trinity College Dublin 22, 47
Troy 4
tuath 231; see countryside
Tullaghobegley, Parish of, Co Donegal 207
Tullow, Co Carlow 210
Tullyrusk, Parish of, Co Antrim 124
Tummock 136
tún 230
túnland 230; see townland
turas 199
Turnbull 34
Tynan, Parish of, Co Armagh 108
typology 127
  cultural 95, 96
Tyrone 196
Tyrrellspass, Co Westmeath 185

Ulster 5, 20, 91, 207
Union, the 85, 86
United Irishmen, the 136
universities 22
  Catholic 73, see Catholic University; Irish 46
University College Cork 2, 26, 46; see Queen's College Cork
unpredictability 96
unrest, political 144
  (and) agrarian 140, 165
Upper Cumber 133
Urney, Parsh of, Donegal 66
Usher, *Primordia* 219

Vallancey, Charles 26, 46, 57, 102, 146, 153, 173–4, 177
  Collectanea de Rebus Hibernicis 54
Vallancey School 174–7
valuation 26, 229
  of land 18, 34, 67
values 85, 112–13, 129
  and behaviour 85; imposition of 93, legitimization of 34

Varagnac, André 53, 170
Venuti 74, 80, 106, 228
vernacular, vernacularity 5, 7, 29, 48, 59, 114, 133, 134, 142, 157, 181, 204, 212, 226, 232; see culture
Viceroy see Lord–Lieutenant
Vidal 74
videre 1
Vienna 32
Virgil of Salzburg 13
vision
  normative 30; subjectivity of 30
Vivian, Sir Hussey 25
Völkerkunde 45
Völkskunde 45, 51
volkenkunde 52
voir 1
Volney, Constantin-François 55
voyage 13

Wakefield, Edward 90
Wakeman, William Frederick 26, 107, 182
Wales 17, 43
Walsh 235
war(s) 15
  Napoleonic 17, 49, 53
Ward, J. R. 110
watercolour 17
Waterford 51
Waterford city 27
Waters, Captain 219, 220, 240
Well of the Calf, the 249
Wellesley, Marquis 58
West, the 231
Western thought/notions 95, 232
Westminster 84
Wetherall, Robert 104
Wexford 235
Whately, Richard, Archbishop of Dublin 51
White, Gilbert 41, 47, 59
Wicklow 164
Williams, J. Butler 105, 125, 135
Williams, J. Hill 108, 109, 135
Windele, John 147
Wright, G. N.
  Guide to the Giant's Causeway 17
  Guide to Wicklow 17
  Guide to Killarney 17
writing 45, 102

Young 86, 87